Inorganic Experiments

Edited by J. Derek Woollins

Distribution:

VCH, P.O. Box 10 11 61, D-69451 Weinheim, Federal Republic of Germany

Switzerland: VCH, P.O. Box, CH-4020 Basel, Switzerland

United Kingdom and Ireland: VCH (UK) Ltd., 8 Wellington Court, Cambridge CB1 1Hz, United Kingdom

USA and Canada: VCH, 220 East 23rd Street, New York, NY 10010-4606, USA

Japan: VCH, Eikow Building, 10-9 Hongo 1-chome, Bunkyo-ku, Tokyo 113, Japan

ISBN 3-527-29235-7 (Hardcover)
ISBN 3-527-29253-5 (Softcover)

Inorganic Experiments

Edited by J. Derek Woollins

Weinheim · New York · Basel · Cambridge · Tokyo

Dr. J. Derek Woollins
Department of Chemistry
Loughborough University
of Technology
Loughborough, Leicestershire LE11 3TU
Great Britain

This book was carefully produced. Nevertheless, authors, editor and publisher do not warrant the information contained therein to be free of errors. Readers are advised to keep in mind that statements, data, illustrations, procedural details or other items may inadvertently be inaccurate. While authors, editor and publisher believe that the specifications and usage of chemicals, equipment and devices, as set forth in this book, are in accord with current recommendations and practice at the time of publication, they accept no legal responsibility for any errors, expressed or implied, with respect to material contained herein. Experiments are always performed at the readers' own risk.

Published jointly by
VCH Verlagsgesellschaft mbH, Weinheim (Federal Republic of Germany)
VCH Publisher, Inc., New York, NY (USA)

Editorial Director: Dr. Thomas Kellersohn
Production Manager: Dipl.-Wirt.-Ing. (FH) Bernd Riedel

Library of Congress Card No: applied for

A catalogue record for this book is available from the British Library.

Die Deutsche Bibliothek – CIP-Einheitsaufnahme

Inorganic experiments / ed. by J. Derek Woollins. – Weinheim;
New York; Basel; Cambridge; Tokyo: VCH, 1994
ISBN 3-527-29253-5 kart.
ISBN 3-527-29235-7 Gb.
NE: Woollins, J. Derek [Hrsg.]

Printed on acid-free and chlorine-free paper.

Composition: Filmsatz Unger & Sommer GmbH, D-69469 Weinheim
Printing: strauss offsetdruck gmbh, D-69509 Mörlenbach
Bookbinding: J. Schäffer GmbH & Co. KG., D-67269 Grünstadt

Printed in the Federal Republic of Germany

To Alex.

Foreword

I am delighted to write a foreword to this work, *Inorganic Experiments.* While many chemists may prefer to spend most of their time tapping on keyboards, I have always thought that making new compounds provides the real fun and enjoyment in chemistry.

While the experiments set out utilize known chemistry, any students who do even a modest selection of them should be given confidence to go into the laboratory not only to make starting materials but to do some original reactions with them as well.

Professor Woollins and the distinguished international group of contributors have provided a stimulating and instructive selection of experiments that should prove invaluable in teaching institutions.

July 1994

Professor Sir Geoffrey Wilkinson
Imperial College
London

Contents

List of Contributors

Milan Alberti
Department of Inorganic Chemistry, Masaryk University, Kotlarska 2, 611 37 Brno, Czech Republic

Jörg Albertsen
Institut für Anorganische und Analytische Chemie, Technische Universität Berlin, D-10623 Berlin, Germany

Stephen Anderson
Department of Chemistry, Imperial College, South Kensington, London SW7 2AY, UK

Michael A. Beckett
Department of Chemistry, University of Wales, Bangor, Gwynedd, LL57 2UW, UK

Neil M. Boag
Department of Chemistry and Applied Chemistry, University of Salford, Salford M5 4WT, UK

Manfred Bochmann
School of Chemical Sciences, University of East Anglia, Norwich NR4 7TJ, UK

Juan José Borrás-Almenar
Department of Inorganic Chemistry, Faculty of Chemistry, University of Valencia, Dr. Moliner 50, E-46100 Burjasot, Spain

Duncan W. Bruce
Centre for Molecular Materials and Department of Chemistry, The University, Sheffield S3 7HF, UK

Neil Burford
Department of Chemistry, Dalhousie University, Halifax, Nova Scotia B3H 4J3, Canada

Ian R. Butler
Department of Chemistry, University of Wales, Bangor, Gwynedd, LL57 2UW, UK

Roderick D. Cannon
School of Chemical Sciences, University of East Anglia, Norwich NR4 7TJ, UK

Tristram Chivers
Department of Chemistry, The University of Calgary, Calgary, Alberta T2N 1N4, Canada

Edwin C. Constable
Institut für Anorganische Chemie, Universität Basel, Spitalstraße 51, CH-4056 Basel, Switzerland

Darren Cook
Department of Chemistry, Imperial College, South Kensington, London SW7 2AY, UK

Eugenio Coronado
Department of Inorganic Chemistry, Faculty of Chemistry, University of Valencia, Dr. Moliner 50, E-46100 Burjasot, Spain

Andreas A. Danopoulos
Johnson Matthey Laboratories, Department of Chemistry, Imperial College, South Kensington, London SW7 2AY, UK

Kurt Dehnicke
Fachbereich Chemie der Universität Marburg, D-35032 Marburg, Germany

Daniel D. Doxsee
Department of Chemistry, The University of Calgary, Calgary, Alberta T2N 1N4, Canada

Simon R. Drake
Department of Chemistry, Imperial College, South Kensington, London SW7 2AY, UK

Dennis F. Evans
Department of Chemistry, Imperial College, South Kensington, London SW7 2AY, UK

Peter N. Gates
Department of Chemistry, Royal Holloway and Bedford New College, Egham, Surrey TW20 OEX, UK

John F. Gibson,
Department of Chemistry, Imperial College, South Kensington, London SW7 2AY, UK

Judith Gindl
Institut für Anorganische Chemie der Universität Göttingen, Tammannstraße 4, D-37077 Göttingen, Germany

Christopher Glidewell
School of Chemistry, University of St. Andrews, St. Andrews, Fife, KY16 9ST, UK

David M. L. Goodgame
Department of Chemistry, Imperial College, South Kensington, London SW7 2AY, UK

William P. Griffith
Department of Chemistry, Imperial College, South Kensington, London SW7 2AY, UK

Colin Greaves
School of Chemistry, The University of Birmingham, Edgbaston, Birmingham B15 2TT, UK

Ionel Haiduc
Chemistry Department, Babes-Bolyai University, RO-3400 Cluj-Napoca, Romania

Philip Hall
Department of Chemistry, Imperial College, South Kensington, London SW7 2AY, UK

Paul Hammerton
Department of Chemistry, Imperial College, South Kensington, London SW7 2AY, UK

Max Herberhold
Laboratorium für Anorganische Chemie, Universität Bayreuth, D-95440 Bayreuth, Germany

Anthony F. Hill
Department of Chemistry, Imperial College, South Kensington, London SW7 2AY, UK

Robert W. Hilts
Department of Chemistry, The University of Calgary, Calgary, Alberta T2N 1N4, Canada

Christopher J. Jones
School of Chemistry, The University of Birmingham, Edgbaston, Birmingham B15 2TT, UK

Thomas M. Klapötke
Institut für Anorganische und Analytische Chemie, Technische Universität Berlin, 10623 Berlin, Germany

Joseph W. Kolis
Department of Chemistry, Clemson University, Clemson, South Carolina, 29634-1905, USA

Simon J. Lancaster
School of Chemical Sciences, University of East Anglia, Norwich NR4 7TJ, UK

M. Elena Lasterra-Sanchez
Department of Chemistry, Imperial College, South Kensington, London SW7 2AY, UK

Dominique Lavabre
URA 470, Université Paul Sabatier, F-31062 Toulouse, France

Gaston Levy
URA 470, Université Paul Sabatier, F-31062 Toulouse, France

Paul D. Lickiss
Department of Chemistry, Imperial College, South Kensington, London SW7 2AY, UK

Alan G. Massey
Department of Chemistry, Loughborough University, Loughborough, Leicestershire LE11 3TU, UK

Sylvia Massey
Department of Chemistry, Loughborough University, Loughborough, Leicestershire LE11 3TU, UK

Jean-Claude Micheau
URA 470, Université Paul Sabatier, F-31062 Toulouse, France

D. Michael P. Mingos
Department of Chemistry, Imperial College, South Kensington, London SW7 2AY, UK

Kieran C. Molloy
School of Chemistry, University of Bath, Bath BA2 7AY, UK

Christopher P. Morley
Department of Chemistry, University College of Swansea, Singleton Park, Swansea SA2 8PP, UK

Josef Novosad
Department of Inorganic Chemistry, Masaryk University, Kotlarska 2, CZ-61137 Brno, Czech Republic

Anthony F. Orchard
Inorganic Chemical Laboratory, University of Oxford, Oxford OX1 2JD, UK

David J. Otway
Department of Chemistry, Imperial College, South Kensington, London SW7 2AY, UK

Timothy T. Paget
School of Chemistry, University of Bath, Bath BA2 7AY, UK

Ivan P. Parkin
Department of Chemistry, University College, Gower St, London WC1E 6BT, UK

Trenton M. Parks
Department of Chemistry, Dalhousie University, Halifax, Nova Scotia B3H 4J3, Canada

Véronique Pimienta
URA 470, Université Paul Sabatier, F-31062 Toulouse, France

Andrew W. G. Platt
Chemistry Division, Staffordshire University, College Road, Stoke-on-Trent, ST4 2DE, UK

Leslie Pratt
Department of Chemistry, Imperial College, South Kensington, London SW7 2AY, UK

David T. Richens
School of Chemistry, University of St. Andrews, St. Andrews, Fife, KY16 9ST, UK

Herbert W. Roesky
Institut für Anorganische Chemie der Universität Göttingen, Tammannstraße 4, D-37077 Göttingen, Germany

Antonín Růžička
Department of Inorganic Chemistry, Masaryk University, Kotlarska 2, CZ-61137 Brno, Czech Republic

Matthias Schrepfermann
Laboratorium für Anorganische Chemie, The Universität Bayreuth, D-95440 Bayreuth, Germany

Martin Schröder
Department of Chemistry, The University of Edinburgh, Edinburgh EH9 3JJ, UK

Ralf Steudel
Institut für Anorganische und Analytische Chemie, Technische Universität Berlin, D-10623 Berlin, Germany

Susan E. Thomas
Department of Chemistry, Imperial College, South Kensington, London SW7 2AY, UK

Inis C. Tornieporth-Oetting
Institut für Anorganische und Analytische Chemie, Technische Universität Berlin, D-10623 Berlin, Germany

David A. Widdowson
Department of Chemistry, Imperial College, South Kensington, London SW7 2AY, UK

Geoffrey Wilkinson
Johnson Matthey Laboratories, Department of Chemistry, Imperial College, South Kensington, London SW7 2AY, UK

Mark J. Winter
Department of Chemistry, University of Sheffield, Sheffield S3 7HF, UK

J. Derek Woollins
Department of Chemistry, Loughborough University, Loughborough, Leicestershire LE11 3TU, UK

Lesley J. Yellowlees
Department of Chemistry, The University of Edinburgh, Edinburgh EH9 3JJ, UK

G. Brent Young
Department of Chemistry, Imperial College, South Kensington, London SW7 2AY, UK

Zdirad Žák
Department of Inorganic Chemistry, Masaryk University, Kotlarska 2, CZ-61137 Brno, Czech Republic

1 Introduction

Chemistry remains a practical subject and for many of us the topics we recollect most sharply (and understand most thoroughly) are often derived from our own experiences in the laboratory. Meaningful experiments which develop laboratory skills, introduce interesting chemistry and are reliable are not always easy to find. This text seeks to address the problem for inorganic chemists. The following compilation of experiments in no way attempts to cover all of inorganic chemistry. However, I hope that there is sufficient range to demonstrate the majority of the important techniques in the context of some interesting and stimulating chemical examples. The experiments have generally come from laboratory courses where they have been tried and tested or they have been checked so we can optimistically assume that they 'work'. For convenience, the experiments have been classified (by me – not the authors) into 'introductory', 'intermediate' and 'advanced'. Clearly, laboratory course organisers must make their own assessment as to the level of difficulty of individual experiments in the context of their laboratory facilities, experience of the students, etc. In general, we have not described measurement methodology in great detail, again on the assumption that facilities differ from one laboratory to the next. Furthermore, some experimental arrangements differ depending on the origin of the submission. This is the case in research and in industry and I have made no effort to impose any house style, there is much to be learnt from the differences!

The experiments are usually prefaced by a section detailing any special safety precautions. Although this is an aid for the user, it should not be assumed that all aspects of safety have been dealt with – the laboratory course supervisor and the student performing the experiment **must** make their own assessment as to the hazards which the chemicals and procedures represent. Although we have made every reasonable effort to test experiments and to provide appropriate safety data and instructions, the authors and the editor do not assume any responsibilty or liability for any mishaps or accidents that may occur in the use of any part of this text as a laboratory manual.

2 Introductory Experiments

2.1 Preparation and Investigation of Some Coordination Compounds

Roderick D. Cannon and Manfred Bochmann

Special Safety Precautions

1. Ammonia solutions are irritating to skin and eyes, and the vapour must not be inhaled. Wear rubber gloves and safety spectacles when handling ammonia solutions and work in a fume cupboard. If ammonia is splashed on your skin, wash it off with plenty of water. Provided you do this immediately there is no cause for alarm.
2. Oxalic acid and oxalates are toxic and must not be ingested.
3. Hydrogen peroxide can cause burns, and skin contact must be avoided. Wear rubber gloves.

Note on the concentration of hydrogen peroxide. Bottles of hydrogen peroxide are traditionally labelled "20 vol", "100 vol", etc., meaning that one litre of the solution, when decomposed according to the equation $2H_2O_2 = 2H_2O + O_2$, yields 20 (etc.) litres of oxygen gas, at standard temperature and pressure. "20 volume" hydrogen peroxide contains *ca.* 6% H_2O_2 by weight.

If you are supplied with hydrogen peroxide labelled in this way, be sure to select the correct bottle. In your notebook, take care to record the concentration in the form specified on the label. In your report, however, be sure to mention the molar concentration as well as the "vols".

Experimental

a) Hexaamminenickel(II) tetrafluoroborate $[Ni(NH_3)_6](BF_4)_2$

Dissolve about 3 g of hydrated nickel chloride, $NiCl_2 \cdot 6H_2O$, in 5 ml of warm water, and filter if necessary to remove any insoluble matter. Then slowly add 5 to 6 ml of concentrated aqueous ammonia (specific gravity 0.88 g cm^{-3}), stirring until all the initial green precipitate of nickel hydroxide has dissolved. The clear deep blue liquid now contains a copious crystalline deposit of the violet nickel hexaammine chloride. If this does not appear, leave it to stand for 5 minutes in a beaker of ice. If the solution is not clear blue add another 1 ml of 0.88 ammonia.

Redissolve the crystals by careful addition of a minimum (less than 8 ml) of cold water, added in 0.5 ml portions with continuous stirring. Take care not to hydrolyse the complex with excess water. Finally, filter off any small insoluble residue.

Dissolve approximately 2.5 g of ammonium tetrafluoroborate in dilute aqueous ammonia (ca. 2 mol dm^{-3}) and add this to the solution of the nickel ammine chloride. The sparingly soluble hexaaminenickel tetrafluoroborate is immediately formed as a crystalline precipitate. Filter by suction and wash the precipitate with aqueous ammonia until the filtrate (solution) is colourless. Finally, wash with acetone. Allow the mauve crystals to dry in a desiccator. Record the yield in your notebook.

b) Potassium tris(oxalato)ferrate(III)

Dissolve 15 g of ammonium ferrous sulphate $(NH_4)_2Fe(SO_4)_2 \cdot 6H_2O$ in 50 ml of hot water acidified with dilute sulphuric acid. Add a hot solution of 7.5 g of oxalic acid in 50 ml of water. Cautiously heat the mixture to boiling and then allow the yellow precipitate of iron(II) oxalate, $FeC_2O_4 \cdot 2H_2O$, to settle. Cool the solution. Wash the precipitate with 30 ml water by decantation. Repeat the washing two to three times.
Now add a warm solution of 10 g of potassium oxalate monohydrate $K_2C_2O_4 \cdot H_2O$ in 30 ml of water. Add slowly, with a teat pipette, 25 ml of '20 volume' hydrogen peroxide (see Special Safety Precautions), stirring the solution continuously and keeping the temperature below 40°C. Any effervescence indicates that the solution is too hot. Then heat the mixture, which now contains some ferric hydroxide, to nearly boiling. The precipitate is now brown. Add a saturated solution of oxalic acid until the precipitate just redissolves. Filter the solution whilst still hot and finally add 30 ml of methylated spirits. Leave to crystallise in the dark.
When green crystals of the complex have appeared (probably the next day) filter the solution and wash the crystals in acetone. Allow them to dry at room temperature in the dark. Store the product in a glass specimen tube, wrapped in aluminium foil. **N.B.** – be sure to attach the label to the glass, and not to the foil!

Chemical Properties

The object of the following tests is to give you experience in observing and identifying the products from reactions of transition metal complexes. Make notes in your laboratory notebook of your observations, as you make them. In your report, summarise your observations and try to write equations and explanations for the reactions.
Thermal decomposition of hexaamminenickel(II) tetrafluoroborate. Heat 0.3 g of your product carefully in a large dry test tube and record your observations (colour changes, gas evolution, etc.). Test both the residues and the sublimates for both boron and fluorine. Suitable tests can be found in Vogel's *Qualitative Analysis.* (6th ed., revised by G. Svehla. Longman Scientific & Technical, 1987). Note that any test which might involve the evolution of HF must be carried out on a small scale and in a fume cupboard. Account for your results. Write an equation that satisfactorily describes the decomposition.

The following test is designed to compare the lability of the ammonia in the hexaamminenickel(II) and hexaamminecobalt(III) and cobalt(II) complexes. Obtain a sample of $[CoNH_3)_6]Cl_3$ from the chemical shelf. Add some sodium hydroxide solution to a little of the nickel(II) and cobalt(III) complexes in test tubes and cover each with a piece of moistened indicator paper. Observe what happens over several minutes and then reduce the cobalt(III) complex to cobalt(II) by adding a little aluminium powder to the alkaline solution. In which complex is the ammonia least labile?

Investigate the equilibria which are present in a solution of potassium ferrioxalate by testing separate portions with potassium thiocyanate for ferric ion and barium chloride for oxalate. What happens when an excess of a sodium oxalate or an ammonium fluoride solution is added to the ferrioxalate solution containing thiocyanate? Note your observations carefully and write equations to describe the equilibria you have found. Is ferrioxalate a labile complex?

Instructions for Course Organisers

1. Do **not** use "100 vol" hydrogen peroxide, 10 or 20 vol is sufficient.
2. Laboratory Technical staff need instructions to make saturated oxalic acid.
3. Supply references to the standard inorganic text book used in your course.

2.2 Gravimetric Analyisis of Hexaammine Nickel(II) Tetrafluoroborate, $[Ni(NH_3)_6](BF_4)_2$

Roderick D. Cannon and Manfred Bochmann

The quantitative determination of a material by precipitation followed by isolation and weighing a precipitate of known composition is called gravimetric analysis.

The success of the method depends upon several factors, especially the completeness of precipitation of the elements being analysed and the selectivity of the precipitation process. In addition, the precipitate should be readily collected and easily filtered. The conditions required for producing a pure precipitate of large granular particles which are easy to filter have been well investigated (see Van Weimann, *Chem. Rev.* **1925,** *2,* 207 and A. I. Vogel, *Quantitative Inorganic Analysis,* 3rd ed., Longman, **1966**).

It is essential that the precipitate obtained be pure, otherwise an accurate determination is not possible. If the precipitate is composed of very small crystals, the surface area of the precipitate may introduce appreciable errors. Furthermore, occlusion of 'foreign' ions during the process of crystal growth may occur, particularly if the precipitate is formed rapidly. The contamination of the precipitate by substances normally soluble in the mother liquor is termed *co-precipitation.* Because the size of the crystallites is, in general, larger the smaller the rate of precipitation, it is apparent that the effects of co-precipitation can be reduced by a slow rate of precipitation. As an added bonus, precipitates composed of larger crystallites are usually much more easily filtered than precipitates composed of fine particles. It is for these reasons that it is common practice to add a dilute solution of precipitating agent slowly, with stirring – the rate of precipitation depending on the amount by which the material to be precipitated exceeds the saturation value. The process of *digestion* (heating the precipitate and mother liquor on a steam bath for several hours) often decreases the effect of co-precipitation and gives more readily filterable precipitates, though it has little effect upon amorphous or gelatinous precipitates. An elegant method of maintaining only a small degree of supersaturation (giving large crystallite size) is *precipitation from homogeneous solution,* in which the precipitant is not added as such, but is slowly generated by a homogeneous chemical reaction within the solution. It is this technique which is to be used in the following experiments. The slow hydrolysis of urea that occurs in boiling aqueous solution will be used to slowly raise the pH of the solution by the generation of ammonia (Eq. 1).

$$CO(NH_2)_2 + H_2O \rightarrow CO_2 + 2NH_3 \qquad (1)$$

Nickel(II) will be precipitated quantitatively by the organic reagent dimethylglyoxime. The nickel ion is chelated by 2 molecules of demethylglyoxime, the overall reaction being given in Eqn. (2) Protons are released and the stability of the complex is such that at low pH, the equilibrium is displaced to the left while at pH above 5, it is displaced to the right. If the two reactants are mixed at pH 3, no precipitation takes place. Addition of urea and slow heating will raise the pH of the solution slowly, by virtue of Eq. (1), and the nickel dimethylglyoxime complex will slowly form. A beautifully crystalline precipitate can be obtained in this way.

Ni^{2+} + 2 HON=C(CH_3)–C(CH_3)=NOH ⟶ (2)

Bis(dimethylglyoximato)nickel

Experimental

Weighing bottle: prepare the weighing bottle by removing all dirt and grease, then wash it with tap water and distilled water, dry it in the oven (130 °C) and cool it for 10 minutes. Chemicals should be removed from a weighing bottle by careful tipping, the use of a spatula is undesirable as this is another source of error and possible contamination. Where several weighings are to be carried out, weighing by difference means that for n samples, only $n + 1$ weighings are required. It is useful to know the approximate weight of the empty bottle. NEVER STOPPER A HOT WEIGHING BOTTLE since after cooling, it will be found to be difficult to open without causing breakage.

Gravimetric Analysis: Use a sample of $[Ni(NH_3)_6](BF_4)_2$ prepared in Experiment 2.1. Be sure to write down in your notebook what you do, and all your weighings, at the time of the experiments.

Wash your weighing bottle with distilled water and dry in the oven for 10 minutes. Weigh accurately (to 4 decimal places) *two* samples of the nickel ammine salt (about 0.08 g) and dissolve each in 30 ml of distilled water. Adjust the pH to 2 or 3 by dropwise addition of 1 to 15 drops of conc. (12 M) HCl. Check the pH between drops with pH paper. Prepare 100 ml of a 1% solution of dimethylglyoxime solution in 1-propanol. Take 10 ml of this solution and heat it to 80 °C on a hotplate.

Add 4 g of urea to the Ni sample and place the solution on a hotplate at 80 to 85 °C. Check that the pH remains below 3. Slowly add the dimethylglyoxime solution to the nickel solution making sure that the pH remains below 3. If it rises, add 12 M HCl, dropwise. Cover the beaker with a watch-glass and heat for one hour at 80 to 85 °C. A red precipitate will form. Whilst waiting for this precipitation to occur, prepare the sinters for weighing (see below). After this period, check the pH to see whether it is above 7. If not, add a drop of ammonium hydroxide and check again. Having obtained a precipitate, remove the beaker from the hotplate and cool to room temperature by standing it in cold water. If a white precipitate appears here, it will be dimethylglyoxime. Dissolve it by adding 4 ml of 1-propanol and heating to 60 °C.

The sintered glass crucible: While the above reaction is in progress, start preparing the crucible as follows. Insert the sinter into the adaptor supplied, push the adapter through a rubber bung, and insert the rubber bung into the top of a 400 ml Büchner flask.

With the filter pump, suck a little concentrated nitric acid through the sinter. Turn off the pump and allow the acid to sit in the sinter. Suck plenty of distilled water through the system, followed by a little concentrated aqueous ammonia. (NB: Remove nitric acid from the Büchner flask first). A glass siphon with a small rubber bung should than be fitted into the top of the filter crucibles, as indicated in the diagram. Wash again with distilled water.

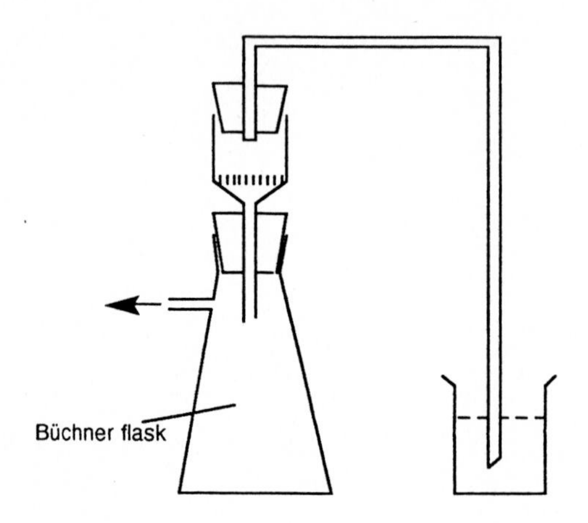

Remove the sinter with a pair of tongs and heat in an oven at 130 °C to constant weight. (Place the sinter in a large boiling tube, which has been labelled with your name, and place the tube in one of the test tube racks in the oven set at 130 °C). The hot crucible must be allowed to cool for 10 minutes before weighing is begun. Two consecutive weighings must agree within 0.2 mg.

Filter the solution on the prepared crucible using the siphon. Wash all the precipitate over with distilled water. If necessary, use a 'policeman' to dislodge particles adhering to the beaker. Remove the crucible with tongs, place in a boiling tube which carries your name on the test tube rack and dry for 1 hour at 130 °C. Allow to cool for 10 minutes and weigh. Repeat weighings until constant.

Calculation

Collect your results into a table as shown in Table **2.2-1** below. Calculate the percentage of nickel in nickel dimethylglyoxime, assuming the formula $Ni(C_4H_7O_2N_2)_2$, and thence calculate the percentage of nickel in your complex, following the procedure indicated in Table **2.2-2** below.

Table 2.2-1. Gravimetric determination.

	First determination	Second determination
Weight of clean dry weighing bottle		
Weight of bottle + complex		
Weight of complex		
Weight of sinter		
(i)		
(ii)		
(iii)		
(iv)		
Final weight		
Weight of sinter complex		
(i)		
(ii)		
(iii)		
(iv)		
Final weight		
Weight of complex		

Table 2.2-2. Calculations.

	First sample	Second sample
Weight of $Ni(dmgH)_2$ obtained		
Weight of Ni in the $Ni(dmgH)_2$		
Weight of ammine sample used		
Percentage of Ni in ammine sample		

Summarise your experimental procedure and make clear any difficulties. Set out your results and calculations concisely, *in tabular form.*

Estimate sources of error and give your final recommended value of the percentage of Ni in the ammine salt. Compare this with the theoretical percentage based on the formula given above.

2.3 Analysis of an Iron(III) Oxalate Complex

Roderick D. Cannon and Manfred Bochmann

Special Safety Precautions

1. Wear eye protection at all times. Concentrated hydrochloric acid is highly corrosive and causes severe burns to skin and eyes. Wear rubber gloves when handling it and work in a fume cupboard.
2. Potassium permanganate soluton is a powerful oxidant and the dilute solution is a disinfectant. Avoid skin contact, but if you do spill it on yourself, simply wash with cold water. Any remaining brown stain will soon disappear.
3. Mercury(II) chloride is highly toxic, as are mercury compounds in general. Mercury waste must be collected in specially labelled bottles as hazardous heavy metal residues.
4. Oxalic acid and oxalates are toxic and must not be ingested.

Iron(III) is nearly always determined by reduction to the divalent state followed by titration with permanganate or dichromate. Oxalate, however, would interfere and must be determined first. After titration of the oxalate with permanganate, two methods are available for the determination of the iron. The solution may be reduced with zinc amalgam in the presence of sulphuric acid, and the iron(II) titrated with permanganate; or the reduction may be effected by tin(II) chloride ('stannous chloride') and hydrochloric acid, and the iron(II) determined with dichromate. (Reduction of Fe^{III} by Sn^{II} is rapid only in hot solution in the presence of hydrochloric acid). Permanganate, a more powerful oxidising agent than dichromate, cannot be used, even in the cold, in such solutions owing to oxidation of the chloride to chlorine (a process which is catalysed by iron compounds). The stannous chloride method will be employed here.

Experimental

Determination of oxalate

The complex was prepared in Experiment 2.1. Weigh accurately about 0.35 g of the complex and dissolve it in dilute sulphuric acid. Heat the solution to about 60 °C and titrate when hot with the approximately 0.02 M solution of permanganate provided. Preserve the resulting liquid for the determination of iron.

The permanganate solution should be standardized against sodium oxalate in the following manner. Weigh accurately two portions of about 0.2 g of 'AnalaR' sodium oxalate into conical flasks. Dissolve each in water, acidify the solutions with suphuric acid and titrate with permanganate as above, at 60 °C.

Calculation

The half-reactions are

$M^{VII} + 5e^- = Mn^{II}$
$2CO_2 + 2e^- = [C_2O_4]^{2-}$

Hence the overall reaction may be written

$Mn^{VII} + 2.5\ [C_2O_4]^{2-} = Mn^{II} + 5CO_2$

Set out your calculations in tabular form as indicated below.

Standardisation			
Run	1	2	
Weight of sodium oxalate taken			g
Amount of sodium oxalate taken			mol
Amount of $[MnO_4]^-$ required			mol
Volume of $[MnO_4]^-$ solution used			ml
Concentration of $[MnO_4]^-$			mol dm^{-3}

Determination			
Run	1	2	
Volume of $[MnO_4]^-$ solution used			ml
Amount of $[MnO_4]^-$ consumed			mol
Amount of $[C_2O_4]^{2-}$ in sample			mol
Weight of $[C_2O_4]^{2-}$ in sample			g
Weight of sample			g
% of $[C_2O_4]^{2-}$			%

Determination of iron by potentiometric titration

To the solution obtained above, add 15 ml of concentrated hydrochloric acid. Heat the solution to the boiling point and add some stannous chloride solution *dropwise* until the yellow colour of the iron(III) complex is discharged; then add 2 to 3 drops of the reductant in excess. Cool the solution to room temperature and add quickly, in one portion, 10 ml of mercuric chloride solution. (The resulting precipitate of mercury(I) chloride should be pure white in colour). After use, all solutions containing Hg should be returned to the mercury waste bottle.

The potentiometric titration apparatus should already be set up in the laboratory. Wash the iron solution into a 400 ml beaker, put in a magnetic stirrer bar and place the beaker on the magnetic stirrer. Lower the standard calomel electrode and platinum disc electrode into the solution. (NB: Rinse the electrodes with distilled water after removing them form their storage solutions). Record the potential using the digital voltmeter. Now add 10 ml of 40% phosphoric acid while stirring and record the potential again. In your report, comment on any difference between the two potentials you have recorded.

Add 10 drops of sodium diphenylamine sulphonate indicator and titrate with the standard dichromate solution provided. Record both the potential *and* the colour of the solution in a table after the addition of each aliquot (0.5 to 1 ml) of dichromate. Plot a graph of potential against volume of titrant, *while the experiment is in progress.* Use the plot to decide the size of each successive aliquot to be added. As you approach the end point, you will get an early warning as the graph will begin to curve upwards. Take particular care in the region of the end point, where a large swing in potential will occur. Continue taking readings until the potential has levelled off appreciably beyond the equivalence point. Determine the equivalence point from the graph, and also from the colour change of the indicator.
NB: *It is essential to plot the graph, point by point, at the time you take the readings. Do not just list the numbers and plot them all up afterwards. If you do, you will almost certainly miss the end point.*

Calculation

The half reactions are

$Cr^{VI} + 3e^- = Cr^{III}$

$Fe^{III} + e^- = Fe^{II}$

Hence the overall reaction may be written

$Cr^{VI} + 3Fe^{II} = Cr^{III} + 3Fe^{III}$

Set our your calculations in tabular form as indicated below. Do not forget that one mole of dichromate contains two moles of Cr^{VI}!

	Using end point from graph	Using end point from colour change	
Volume of Cr^{VI} solution required			ml
Amount of Cr^{VI} used			mol
Amount of Fe reacted			mol
Weight of Fe			g
Weight of sample used in titration			g
Percentage of Fe in sample			%
Mean and error			

Determination of iron by colorimetry

Colorimetry may be used to directly estimate the amount of ferric ion in the complex. The salt is first photoreduced to form ferrous ions; then a reagent (1,10-phenanthroline, another chelating ligand) is added which will bind to the metal, forming an intense red colour. The intensity is compared with those of solutions of known concentrations of iron.

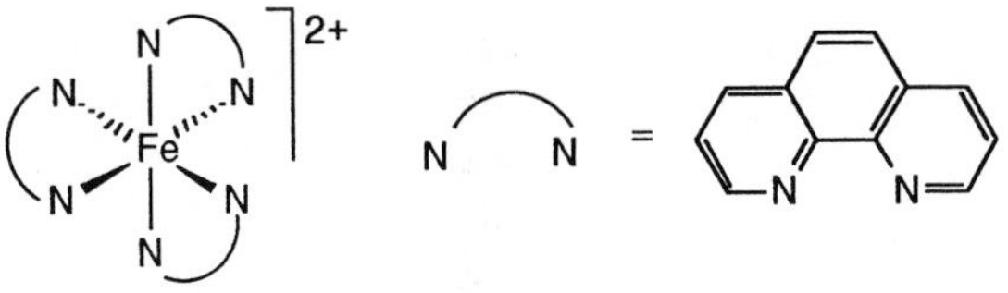

iron(II) tris(phenanthroline) complex

Accurately weigh a portion of about 0.120 g of potassium ferrioxalate. Dissolve it in distilled water and make up to 250 ml. Label this solution 'A'. Take 2 ml of solution A and add 5 ml of 0.5% phenanthroline solution, plus 5 ml of pH 4 buffer. Make this up to 20 ml with distilled water and label this solution 'B'. Mix solution B well and transfer a portion to the colorimetry tube provided. Photoreduce the iron(III) by exposing to normal light for at least 1 hour. The equipment for this is set up in the laboratory. Prepare the five ferrous standards as follows. Take 2 ml of 0.0005 M $FeSO_4$, add 5 ml of the phenanthroline solution and 5 ml of the buffer solution. Make up to 20 ml with distilled water. Repeat this using 2, 3, 4, 5 and 6 ml of 0.0005 M $FeSO_4$. Label these solutions 'C2', 'C3', 'C4', 'C5', and 'C6' respectively.

Compare the colour intensities of all the solutions using the colorimeter. For each solution, first set the absorbance reading to zero with the tube out of the light beam, then read the value with the tube in the light beam. Move the tube out of the beam and reset the absorbance to zero if necessary. In this way, measure three readings of absorbance for each sample. Tabulate the results in your notebook as you go along and calculate the average absorbance of each solution. Plot a graph of absorbance, A, against iron concentration, [Fe], for the five solutions C2 to C6. *Be sure to plot this graph while the experiment is in progress.* It is expected to be a straight line, or at any rate a smooth curve. If some points deviate significantly, you may consider it advisable to either repeat the measurement or make up fresh solutions.

Calculation

From your graph, read off the concentration of solution B. Hence, calculate the concentration of solution A, the amount of iron in solution A, the weight of iron and the percentage of iron in the complex.

Summarise your procedures and indicate any difficulties, set out your results neatly in tables and discuss the sources of error. Compare your final values for percentage of Fe with each other. Calculate the mole ratio of oxalate to iron in the complex. Calculate the percentage by weight of potassium in the complex by a) assuming the mole ratio K/Fe = 3 and b) assuming $K/C_2O_4 = 1$. Do the totals of Fe, C_2O_4 and K add up to 100%?

Notes for Course Organisers

The potentiometric titration apparatus consists of a standard calomel electrode (commercial) and a bright platinum electrode made from a small (1 cm^2) piece of Pt foil joined to a short length of Pt wire which, in turn, is fused into a glass tube. The potentiometer is a digital voltmeter capable of reading to ± 0.1 mV. The "equipment" for photolysis is an ordinary 'Anglepoise' lamp!

2.4 Characterisation of a Copper Oxalate Complex $K_aCu_b(C_2O_4)_c \cdot dH_2O$

Andrew W. G. Platt

The copper(II) ion is mildly oxidising and can be used as the starting material for the synthesis of complexes of copper(I) as well as copper(II). Since the chemistry of the copper ion is modified by the presence of the surrounding ligands, in this experiment, you will prepare and characterise the complex and investigate the effect of the oxalate ligands on the chemistry of the copper ion.

Experimental

Dissolve 6 g of copper(II) sulphate pentahydrate in 15 cm^3 of hot water and add to this a solution of 17.6 g of potassium oxalate hydrate in 100 cm^3 hot water. Allow to cool to room temperature and finally to about 10 °C in ice. Filter the product by suction and dry thoroughly in air. Record the weight obtained.

Determination of the oxalate content: Weigh accurately about 0.25 g of your product into a conical flask and add 50 cm^3 of dilute sulphuric acid. Heat to boiling and tritrate with 0.02 mol dm^{-3} potassium permanganate solution. Note that during the initial stages of the titration the solution is cloudy due to undissolved complex. This clears during the titration. Ensure that the solution is still hot as the endpoint is approached. Heat to boiling if the reaction appears to be slow. Take care not to overshoot the endpoint as the presence of a large excess of permanganate makes the copper determination difficult.

Determination of the copper content: Ensure that the solution obtained above contains no excess permanganate by boiling until the purple colour is discharged. To the cooled solution, add 2 g of potassium iodide and titrate the liberated iodine with 0.05 mol dm^{-3} sodium thiosulphate. When the colour of the resulting suspension is pale yellow, add a few drops of starch and continue titrating until the blue colour begins to fade. Add 0.5 g of potassium thiocyanate and swirl the contents. The blue colour should intensify as iodine, adhering to the solid CuI, is liberated. Continue the titration until a white suspension is obtained. Note that the addition of the potassium thiocyanate should be carried out very near to the endpoint.

Repeat both determinations until consistent titres per gram are obtained and hence calculate the copper to oxalate ratio in the complex. From these results deduce the likely values of *a, b, c* and *d*. You will have to deduce the oxidation state of the copper to do this. You have seen several indications of this already, but the qualitative tests below, and a determination of whether the complex is paramagnetic or diamagnetic will help to confirm this.
Use the equations below to help in the calculations.

$$2Cu^{2+} + 2I^- + 2e^- \rightarrow 2CuI$$

$$2[S_2O_3]^{2-} + \quad 2e^- \rightarrow [S_4O_6]^{2-}$$

$$2CO_2 + 2e^- \rightarrow [C_2O_4]^{2-}$$

$$I_2 + 2e^- \rightarrow 2I^-$$

$$[MnO_4]^- + 8H^+ + 5e^- \rightarrow Mn^{2+} + 4H_2O$$

Qualitative work
Prepare a solution of the product in distilled water and add a few drops of potassium iodide solution. Test for the presence of iodine by adding a little chloroform and shaking the tube. Acidify with dilute sulphuric acid and shake the tube.
Compare the above with the reaction of a dilute solution of copper sulphate. Account for the differences in terms of the structure of the complex ion.

2.5 Vanadium Alum: An Experiment in Electrosynthesis

David T. Richens

Vanadium alum, $NH_4V(SO_4)_2 \cdot 12H_2O$, is a member of an extensive series of isomorphous complexes. Unlike most compounds of V(III), it is fairly resistant towards oxidation in the crystalline state.

The preparation involves two separate reductions: from V(V) to V(IV) using SO_2 and from V(IV) to V(III) in an electrochemical reduction apparatus. These steps are accompanied by characteristic colour changes.

Experimental

Required materials NH_4VO_3, conc. H_2SO_4, SO_2 cylinder, magnetic stirrer, electrochemical reduction apparatus, 12V d.c. battery charger, mercury (designated fume cupboard should be available).

Cautiously mix 8 cm^3 of concentrated sulphuric acid with water contained in a 250 cm^3 conical flask and dilute to about 60 cm^3. In the **fume cupboard**, clamp the flask and insert a broad tube delivering SO_2 from the cylinder so that the end of the tube is close to the surface of the liquid, *i.e.* just dipping below it. Consult a demonstrator about the operation of the SO_2 cylinder. Warm the solution to about 50 °C then add 12 g of ammonium metavanadate, NH_4VO_3, *in small portions* while constantly stirring using a magnetic stirrer. Allow each portion of the initally formed brick-red divanadium pentoxide to dissolve by reaction with SO_2 before further addition. When all the solid has been added, **boil** the deep blue solution of vanadyl sulphate, $[VO(H_2O)_5]SO_4$, to expel excess SO_2.

To set up the electrochemical reduction, begin by carefully pouring mercury into the base of the apparatus (with the tap shut!) to form the cathode (Fig. **2.5-1**). The mercury pool should cover the bottom. Transfer the vanadyl sulphate solution to the reduction vessel and introduce the glass cooling coil within which is placed the anode. The anode is a piece of lead sheet dipping into dilute sulphuric acid in a porous pot. Run tap water through the cooling coil during the reduction and stir the solution occasionally. The electrolysis is carried out at a steady current density of about 0.1 amp cm^{-2} Hg until the solution is **deep green** with no remaining tint of blue. Maintain the bath temperature at 30 °C throughout. (A major cause of failure is incomplete electrochemical reduction – monitor by running a visible spectrum* every 20 min on a sample of the reaction solution between 850 and 300 nm or until the absorption band of V(IV) at 750 nm has **disappeared**).

The mercury is then drawn off through the tap and put aside to be rinsed and returned to the prep room. The vanadium(III) solution is passed through a funnel containing a plug of

* Take out 1 cm^3 of solution and make up to 10 cm^3 with water (volumetric flask) before recording UV-visible spectra.

glass wool and collected in a flask that can be stoppered. Making sure that no mercury droplets contaminate the solution, add 10 cm^3 of a **saturated** solution of ammonium sulphate and place the labelled flask in the refrigerator for a few hours (see a demonstrator) or preferably overnight. Seeding the crystallisation with a friend's sample sometimes helps.

The blue-violet crystals which separate are filtered at the water-pump and collected. The crystals are best dried on a piece of filter paper since desiccation over silica gel for more than 10 min can cause loss of lattice water and resulting effluorescence. Note the characteristic rhombic geometry of the alum crystals. The sample is to be submitted with your report when the magnetic and spectroscopic studies are completed. Record the yield and store in a well stoppered vial, but **not** a desiccator.

Magnetic susceptibility. Determine the magnetic moment of your sample as a fine powder.

Visible reflectance spectrum. Record the visible *'d-d'* spectrum of vanadium alum using a finely powdered sample in the form of a Nujol mull. Correlate the blue-violet colour of the alum with the reflectance spectrum you have recorded.

Report the yield and value of μ_{eff} and interpret the UV-visible and magnetic data in relation to the oxidation state and stereochemistry of the vanadium (III) ion.

Questions

1. Why is SO_2 a particularly convenient reducing agent in the present synthesis?
2. What is the coordination environment of V^{3+} in the alum? Compare with the environment existing for V^{3+} in crystalline $VCl_3 \cdot 6H_2O$.
3. Why is a solid state UV-visible spectrum recorded using the crystalline alum, rather than the usual absorption spectrum measured in solution?

References

General vanadium chemistry: F. A. Cotton, G. Wilkinson, *Advanced Inorganic Chemistry,* 5th ed., **1988** p. 665–681.

Structure of alums: A. G. Wells, *Structural Inorganic Chemistry.*

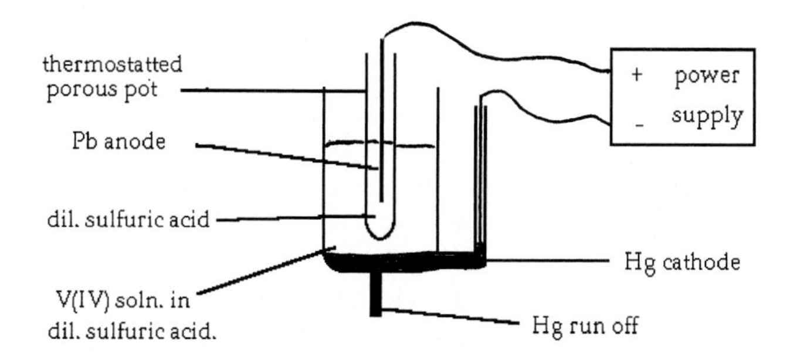

Figure 2.5-1. Electrochemical Reduction.

2.6 Ammonium Dichromate, Chromium(III) Oxide, Potassium Chromate and Potassium Tetraperoxochromate (V)

Antonín Růžička and Zdirad Žák

Special Safety Precautions

1. Chromates and dichromates are highly toxic and possibly mutagenic. Do not inhale the dust from these compounds, avoid the contamination of skin with their solutions.
2. Ammonium dichromate decomposes on heating.
3. Concentrated hydrogen peroxide is corrosive and can cause unpleasant burns. Always use rubber gloves.

Conversion (double displacement) reactions are among the most common of the simple chemical reactions and bear a special importance in the preparation of inorganic salts. The method depends on the fact that not only do the solubilities of the salts change with temperature, they also change at different rates for different salts. In a solution, inorganic salts are always present in the form of simple ions. When the conditions of a solution containing two or more salts are changed (*e. g.* lowering the temperature, evaporating the solvent), the salt with the cation/anion combination which is the least soluble at the given conditions can be crystallised. It is therefore possible to find such conditions of temperature, concentration and solvent that allow, from a solution of two salts, the respective salts with mutually exchanged

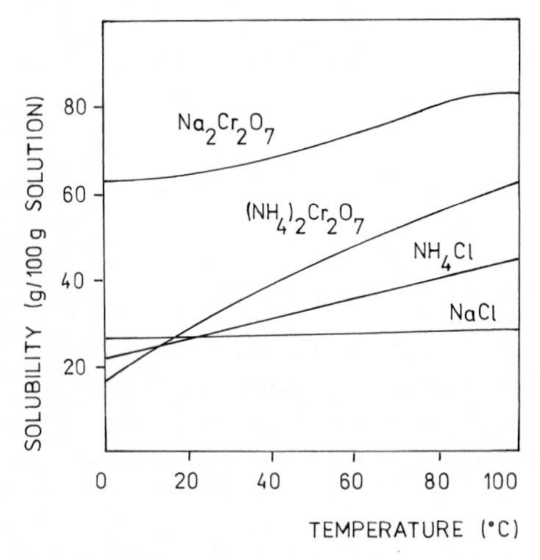

Figure 2.6-1. Solubility curves.

ions to be crystallised. The most simple examples of double displacement reactions are precipitations that lead to substances which are practically insoluble in a given solvent.

In the following experiment, we shall prepare ammonium dichromate by this conversion procedure. Ammonium dichromate decomposes to N_2 and Cr_2O_3 at temperatures above 185 °C and thus small quantities of pure chromium(III) oxide can be prepared. We shall demonstrate the formation of substances in an oxidising flux by the preparation of potassium chromate from Cr_2O_3. From K_2CrO_4, we shall further prepare a relatively stable compound of Cr(V): red-brown potassium tetraperoxochromate (V), $K_3[Cr(O_2)_4]$.

Experimental

a) $(NH_4)_2Cr_2O_7$

The conversion process can be expressed by Eqn. (1).

$$Na_2Cr_2O_7 + 2NH_4Cl \rightarrow (NH_4)_2Cr_2O_7 + 2NaCl \quad (1)$$

From the solubility curves of sodium dichromate and ammonium chloride in Figure **2.6-1**, it follows that on cooling a suitably concentrated solution of $Na_2Cr_2O_7$ and NH_4Cl (molar ratio 1:2) to 20 °C, only $(NH_4)_2Cr_2O_7$ can crystallise from the solution.

Dissolve the appropriate amounts of $Na_2Cr_2O_7 \cdot 2H_2O$ and NH_4Cl required for the preparation of 20 g of $(NH_4)_2Cr_2O_7$ as determined from the solubility curves + 15 ml of water at 60 °C (see Note 1). Filter the solution through a heated filtration funnel (see Note 2) and cool the filtrate in an ice bath 0–2 °C. Filter the crystalline mass of $(NH_4)_2Cr_2O_7$ using a sintered glass funnel and a vacuum filter flask. Wash the product with 10 ml of ice water and suck the mother liquid well off. Dry in an oven at 100 °C.

Notes

1. From the two salts, $Na_2Cr_2O_7 \cdot 2H_2O$ and NH_4Cl, the latter is less soluble. Therefore, the necessary amount of water is that as read from the solubility curve for NH_4Cl at 60 °C.
2. Make the heated funnel ready before you prepare the solution of $Na_2Cr_2O_7 \cdot 2H_2O$ and NH_4Cl. Be sure that both salts are completely dissolved.
3. Due to the toxicity of chromates, do not pour the mother liquid in a water sink. Dispose of as instructed by the supervisor.

Test for Cl^- *ions:* transfer a spatula tip of dry $(NH_4)_2Cr_2O_7$ to a test tube and dissolve it in 2 ml of water. Add a few drops of 5% solution of $AgNO_3$. By adding a few ml of diluted HNO_3 (conc. $HNO_3 : H_2O = 1:3$), dissolve the brown precipitate of $Ag_2Cr_2O_7$. If Cl^- ions are present in the sample, you will see a white precipitate of AgCl in a yellow solution.

b) Cr_2O_3

$$(NH_4)_2Cr_2O_7 \rightarrow Cr_2O_3 + N_2 + 4H_2O \quad (2)$$

Transfer 5 g of dry ammonium dichromate to a 500 ml Erlenmeyer flask. Clamp the flask securely about the neck in a horizontal position so that all the dichromate is at the lowest part of the flask. This prevents finely dispersed Cr_2O_3 from being blown out of the flask by the stream of nitrogen and water vapours. Gently heat the part of the flask where the dichromate is collected with a gas burner. Keep the burner in your hand and remove it when the decomposition starts. After the flask has cooled down, add 300 ml of water, stir and allow to settle. Decant the supernatant, transfer the suspension of Cr_2O_3 to a Büchner funnel, wash on the filter with water and remove the liquid by suction. Dry in an oven at 100 °C. Although it takes up to several hours to dry Cr_2O_3 completely (why?), a still slightly damp oxide can be used for the preparation of K_2CrO_4.

c) K_2CrO_4

One of the methods for the preparation of alkaline chromates is an oxidation of chromium(III) oxide in a melt of alkaline nitrate or chlorate(V) and hydroxide or carbonate according to the Eqns. (3) and (4). In the laboratory, we can prepare a small amount of K_2CrO_4 by an oxidation of finely dispersed Cr_2O_3, prepared by the decomposition of $(NH_4)_2Cr_2O_7$ by air in a melt of KOH (Eqn. 5).

$$Cr_2O_3 + 4KOH + 3KNO_3 \rightarrow 2K_2CrO_4 + 3KNO_2 + 2H_2O \quad (3)$$

$$Cr_2O_3 + 2K_2CO_3 + KClO_3 \rightarrow 2K_2CrO_4 + KCl + 2CO_2 \quad (4)$$

$$2Cr_2O_3 + 8KOH + 3O_2 \rightarrow 4K_2CrO_4 + 4H_2O \quad (5)$$

Transfer all of the prepared Cr_2O_3 to an iron or nickel crucible, add a stoichiometric amount of KOH (on the assumption that the decomposition of ammonium dichromate gives oxide with a 100% yield), place the crucible on a clay triangle and heat intensively with a gas burner. Stir the melt frequently with an iron or nickel wire and heat for a further 30 minutes. The bottom of the crucible should be heated to a dull red colour during this period. Allow to cool, dissolve the product in 25–30 ml of water and filter into a beaker filled with 100–150 ml of ethanol. Filter precipitated K_2CrO_4 using a sintered glass funnel *in vacuo,* wash with ethanol (15 ml) and dry at 100 °C.

Iodometric titration: use standard procedure and calculate the percentage of the salt.

d) $K_3[Cr(O_2)_4]$

Potassium tetraperoxochromate(V) is formed by the action of hydrogen peroxide on potassium chromate(VI) in a strongly alkaline solution (Eqn. 6).

$$2K_2CrO_4 + 9H_2O_2 + 2KOH \rightarrow 2K_3[Cr(O_2)_4] + O_2 + 10H_2O \quad (6)$$

Dissolve 2.0 g of K_2CrO_4 and 2.0 g of KOH in 25 ml of water in a 100 ml Erlenmeyer flask. Immerse the flask in a ice/salt cooling mixture (or dry-ice/alcohol mixture) and wait until the contents solidify into a slush. Without disrupting the cooling, add 13 ml of a 30% solution of H_2O_2 dropwise under intensive stirring. Allow to stand in the cooling bath for a further hour, stirring from time to time. After this time, remove the flask from the bath and allow to warm up until the contents melt. Using a sintered glass funnel, vacuum filter the precipitated red-brown salt, wash with ethanol (15 ml) and dry in air.

Qualitative determination of O_2^{2-} *ions:* hydrogen peroxide forms with chromates(VI) in an acid solution giving a blue colouration ("chromium blue") which can be extracted in ether. Perform the following experiment.

Transfer 5 small crystals of K_2CrO_4 to a test tube, dissolve them in 2 ml of water, add a few drops of diluted H_2O_2 (3%), 2 ml of diethyl ether and a few drops of diluted sulphuric acid (3 ml water + 1 ml conc. acid). Shake the test tube and watch the blue colouration of the ether layer.

Perform the same reaction with the prepared $K_3[Cr(O_2)_4]$, but without hydrogen peroxide.

References

H. Hecht, Präparative Anorganische Chemie; Springer Verlag Berlin 1951 I. Gerassimov, Z. Anorg. Allg. Chem. 1930, *187,* 321

2.7 Linkage Isomerism: An Infra-Red Study

David T. Richens and Christopher Glidewell

A number of ligands are **ambidentate**, that is, they can coordinate to a central ion in more than one way. Some examples are NCS^-, which can coordinate via N or S; NO_2^-, which can coordinate via N or O; and acac$^-$(($CH_3CO)_2CH^-$), which can coordinate via C or O. In this experiment, the ligand NO_2^- is investigated and a pair of linkage isomers having this ligand bonded to cobalt via both N and O is prepared and distinguished.

Experimental

Required materials: Cobalt(II) carbonate, ammonia solution, ammonium carbonate, ammonium chloride, HCl, sodium nitrite, diethyl ether, ethanol, pH meter plus electrode, dry KBr.

a) $[Co(NH_3)_5Cl]Cl_2$

Dissolve 5 g of cobalt(II) carbonate in 8 cm^3 of **hot** concentrated hydrochloric acid, filter, cool and add to a mixture of 60 cm^3 of 10% ammonia and 12 g of ammonium carbonate in 60 cm^3 of cold water. Draw a rapid stream of air* through the solution for three hours, then add 38 g of ammonium chloride and evaporate the mixture on the rotary evaporator until a fairly thick sludge is obtained. Acidify the product with dilute hydrochloric acid, using a pH meter and a glass electrode, stirring constantly until carbon dioxide evolution has ceased. Neutralise with ammonia and add an excess (2 cm^3) of conc. ammonia solution. Warm the solution on the water bath for about half an hour, dilute to 100 cm^3 and add 75 cm^3 of concentrated hydrochloric acid. Heat the resultant mixture on the water bath for 30–45 minutes and cool when the product separates. Filter off the product and wash with dilute hydrochloric acid and then with alcohol. Dry over silica gel in a desiccator and record the yield. Check the UV-visible spectrum in water against that of an authentic sample provided by the demonstrator.

b) The two linkage isomers of $[Co(NH_3)_5NO_2]Cl_2$ (in fume cupboard)

Isomer **I**: dissolve 1.5 g of $[Co(NH_3)_5Cl]Cl_2$ in 15 cm^3 of H_2O to which 5 cm^3 of 6 mol dm^{-3} ammonia has been added. Warm on water bath until salt dissolves**, filter, cool and acidify with dilute HCl. Add 2 g of $NaNO_2$ and heat gently until the red precipitate first formed has dissolved. Cool and add 20 cm^3 of conc. HCl carefully since there is considerable effervescence. Cool the solution in ice, filter off the brown-orange crystals, wash them with alcohol and finally with ether. Dry in a desiccator over silica gel.

* NB The solution **must** be basic before bubbling air through; add more ammonia if necessary.

** NB It may be necessary to boil to ensure complete dissolution of the starting material.

Isomer **II:** dissolve 1.5 g of $[Co(NH_3)_5Cl]Cl_2$ in 25 cm^3 of water to which 5 cm^3 of concentrated ammonia has been added, warming gently if necessary.** Filter and add 6 mol dm^{-3} HCl dropwise until the solution is just neutral to universal indicator paper† (about 15 cm^3 is needed). Add 1.5 g of $NaNO_2$ to the cold solution plus 1.5 cm^3 of 6 mol dm^{-3} HCl and allow to stand for 1-2 hours in ice. Cool in ice, filter off the salmon-pink product as rapidly as possible, wash with ice water and alcohol and finally with ether, and then dry in a desiccator.

Record the IR spectra of both isomers immediately after preparation, and also of $[Co(NH_3)_5Cl]Cl_2$ using KBr discs. Do not grind the sample for too long a time – grind the KBr alone first, then add the sample and grind just long enough to mix thoroughly. Then expose small samples of each to (i) heat (oven), and (ii) bright daylight (sunlight if possible or use a UV lamp) for a few days. Record the spectrum of each after treatment.
Report the yields of all of your products and assign the IR bands as far as possible to vibrations involving the NH_3 and NO_2^- ligands. Identify the two isomers, determine which is the more stable thermodynamically, and explain the changes occurring. Propose a mechanism for the isomerisation.

References

W. G. Jackson, G. A. Lawrencer, P. A. Lay and A. M. Sargesen, *J. Chem. Soc., Chem. Commun.* **1982**, 70.
This experiment is loosely based on W. H. Hohmann, *J. Chem. Educ.* **1974**, *51*, 553. and W. M. Philips, S. Choi, J. A. Larrabee, *J. Chem. Educ.* **1990**, *67*, 267.

** NB It may be necessary to boil to ensure complete dissolution of the starting material.
† It is important that this solution is not alkaline.

2.8 Preparation and Complexation of Tris(3,5-dimethylpyrazoyl)hydroborate

Manfred Bochmann

Special Safety Precautions

In this experiment hydrogen is liberated and carbon monoxide is used. All parts to this experiment should be performed in a fume cupboard.

Binary boron hydrides are an extremely interesting class of compounds and form a multitude of structures (see, for example, N. N. Greenwood, A. Earnshaw, *Chemistry of the Elements,* Pergamon Press, **1984,** 171). However, they are very difficult to handle. The simplest, BH_3, behaves as a Lewis acid and forms, for example, an adduct with THF. It also adds a hydride anion to give BH_4^-. In this anion the hydrogen has **hydridic** character, *i. e.* it reacts with a proton source to liberate H_2. This reaction principle is utilised in part a).

Experimental

a) Potassium tris(3,5-dimethylpyrazolyl)hydroborate

The tetrahydroborate anion reacts with the acidic hydrogen of pyrazole and its derivatives with liberation of H_2 and formation of B–N bonds. Bis-, tris- and tetra-pyrazolyl borates can be made which act as versatile anionic chelate ligands towards transition-metal ions. With 3,5-dimethylpyrazole, which we use here, steric crowding only allows substitution of a maximum of three hydrides (Eqn. 1).

$$K^+BH_4^- + 3\,C_5H_8N_2 \rightarrow K^+[HB(C_5H_7N_2)_3]^- + 3\,H_2 \qquad (1)$$

($C_5H_8N_2$ = 3,5-dimethylpyrazole).

The reaction is monitored by measuring the amount of H_2 evolved. Assemble the apparatus as in Figure **2.8-1**, consisting of a 100 ml flask with magnetic stirring bar, connected **via** a wash bottle (as a suck-back trap) to a water-filled 2 litre measuring cylinder in a half-filled bowl for the determination of the volume of hydrogen gas evolved. Fill the bowl to between 1/3 and 1/2 full and the cylinder completely with water, cover the cylinder top with tight plastic or cling-wrap and quickly stand it upside-down in the bowl. Have suitable clamps ready beforehand. Now remove the plastic without letting air in. Mark the level of any air in the cylinder so that it can be subtracted later.

Caution: Hydrogen can be explosive – make sure to naked flame or sparks are in the neighbourhood. Work in an efficient fume cupboard. Is the wash bottle connected the right way around?

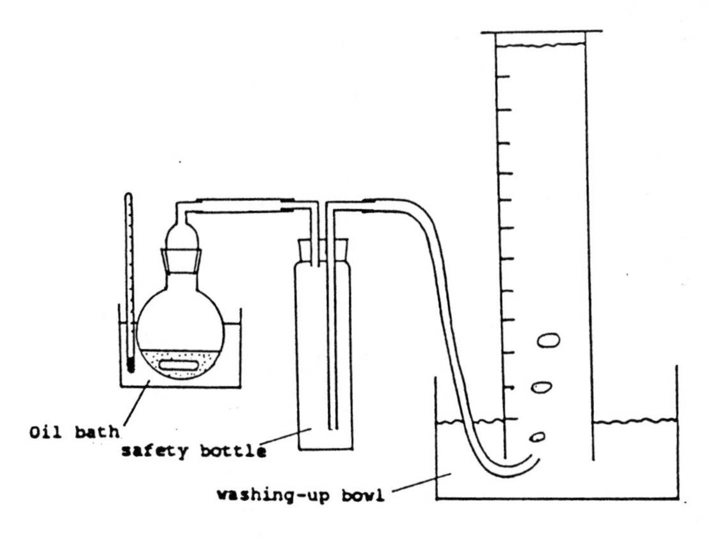

Figure 2.8-1.

Quantities:
Potassium tetrahydroborate (M_r =) 18.5 mmol = g.
3,5-Dimethylpyrazole (M_r =); 73 mmol = g.
ml H_2 expected: = mmol H_2
ml H_2 found: = mmol H_2

Place powdered potassium tetrahydroborate (18.5 mmol) and 3,5-dimethylpyrazole (73 mmol) in the flask, connect to the wash bottle and heat on a silicone oil bath at 230 °C. The whole reaction should be completed within 1 to 1 1/2 h. The stirred mixture melts at 130–140 °C oil bath temperature, and hydrogen evolves. Determine how many equivalents of H_2 you expect and how much you collect. Towards the end of the reaction, the product will solifidify. Allow to cool to *ca.* 100 °C and add 50 ml toluene. Filter off while hot and wash the white crystalline residue with more toluene (2 × 50 ml) and finally with petrol (b. p.: 40–60 °C) to remove excess pyrazole. Dry *in vacuo* for 10–20 min. (rotary oil pump).

Record the ^{1}H NMR spectrum (in D_2O) and the IR spectrum (Nujol mull). Assign the NMR spectrum and the significant bands of the IR spectrum.

Pyrazolylborates act as excellent chelating, mono-anionic ligands towards transition metals, and many complexes are known. In some cases, unusual complexes can be stabilised in this way. The complex prepared in part b is such an example. Generally, complexes of Cu(I) with carbon monoxide are very unstable.

b) A copper carbonyl complex of tris(3,5-dimethylpyrazolyl) hydroborate

Caution: CO is highly poisonous. Work in an efficient fume cupboard.
Stir 2 mmol of finely powdered copper(I) iodide in 50 ml acetone on an ice bath. Stopper the flask with a "suba-seal" and bubble CO through the inlet and outlet needles at a moderate rate for at least 5 min. It is important to keep the flask under CO and exclude air as much

as possible during the reaction. Then quickly add 2 mmol of the K[HB(Me_2pyr)$_3$] and continue CO treatment. The suspension soon becomes clear. Stop stirring, continue to bubble CO through until the mixture turns cloudy and crystals begin to form. Stop the CO stream, but leave the flask under a CO atmosphere on the ice bath for *ca.* 1 h. Filter off the colourless crystals, wash with a little acetone and dry *in vacuo.*

Measure the ^1H NMR spectrum (in $CDCl_3$) and the IR spectrum. Assign the spectra and compare with the data obtained for K[HB(3,5-Me_2pyr)$_3$].

References

S. Trofimenko, *Chem. Rev.* **1972,** *72,* 497.
M. I. Bruce, A. P. P. Ostazewski, *J. Chem. Soc., Dalton Trans.* **1973,** 2433.

2.9 Tetraiodotin(IV) and its Triphenylphosphine Complex

Leslie Pratt und T. Derek Woolins

Special Safety Precautions

Tin compounds, iodine, phosphines and chloroform are all extremely harmful, and acetic acid and its anhydride are very corrosive. Do not inhale them or allow them to contact the skin. Perform all operations in a fume cupboard.

Most of the heavier main group (p-block) elements have the capacity to expand their coordination number beyond that of their normal covalency by forming complexes with neutral ligands. The triorganophosphines (R_3P) are an important class of coordinating ligands, especially in the chemistry of the more electron-rich ('softer') metals. It is often found that complexation by a phosphine will decrease the reactivity of the metal species towards, for example, hydrolytic decomposition, as is the case here. The preparation features a convenient non-aqueous solvent procedure.

Experimental

a) Tetraiodotin(IV), SnI_4 (stannic iodide)

Glacial acetic acid (25 cm^3) and acetic anhydride (25 cm^3) are placed in an oven-dried 100 cm^3 round-bottomed flask. Sheet tin (0.5 g, 4.2 mg-atom), cut into small pieces, and iodine (2 g, 7.88 mmol) are added. A reflux condenser and $CaCl_2$ drying tube are fitted and the contents gently boiled on a sand bath until a vigorous reaction begins. When this subsides, boil the liquid until all (or most) of the tin metal has been consumed. The heating is then discontinued, the condenser removed and the hot supernatant liquid *carefully* (gloves – *highly* corrosive) decanted into a hot, oven-dried conical flask. The mixture is allowed to cool, and the orange crystals filtered rapidly under suction. The sample product is recrystallised by redissolving in the minimum volume of warm chloroform and then cooling to ice temperature. The recrystallised product is filtered off under suction and allowed to dry in a vacuum dessicator. The product is placed in a tube with a polythene cap and sealed with paraffin.

b) Tetraiodobis(triphenylphosphine)tin(IV), $(Ph_3P)_2SnI_4$

SnI_4 (1.0 g, 1.6 mmol) and Ph_3P (0.9 g, 3.4 mmol) are each dissolved in separate portions of dry chloroform (10 cm^3). The two solutions are then mixed and allowed to stand in a stoppered flask for about 25 minutes. The dark orange crystalline product is filtered off and dried in a vacuum desiccator.

Exercises

1. Write equations for the formation of SnI_4 and $(Ph_3P)_2SnI_4$. Calculate your yields.
2. Why are stannic halides rendered more inert by complexation?
3. Draw the isomeric forms of $(Ph_3P)_2SnI_4$. Suggest how you might differentiate between the different isomers.

2.10 Silicon Oxygen Compounds and Siloxane Polymers

Paul D. Lickiss

Special Safety Precautions

1. Chlorosilanes R_nSiCl_{4-n} (R = alkyl, aryl, *etc.*; $n = 1$–4) hydrolyse readily in the atmosphere to give HCl which is highly corrosive. Hydrolytic reactions should, therefore, be carried out in a fume cupboard. This also applies to the opening of commercial bottles of chlorosilanes, which often have a simple screwcap lid. Spillages of chlorosilanes should be treated with sodium carbonate.
2. Dimethyldichlorosilane, diphenyldichlorosilane and the organic solvents diethyl ether, ethanol, hexane, toluene, ethyl acetate and *t*-butanol are all volatile and highly flammable and should be treated with appropriate care.
3. An oil bath at 200 °C is potentially dangerous. Check that the oil in the bath is suitable for using at such high temperatures for several hours without smoking. The smoke from such a bath may be toxic and should not be inhaled. A silicone oil bath is best for such heating, which is best carried out in a fume cupboard.
4. Concentrated hydrochloric acid and sodium hydroxide solutions are both very corrosive and contact with skin, eyes, *etc.* should be avoided.

Since the 1940's, the interest in non-silicate silicon oxygen compounds (siloxanes and silanols) has grown enormously and they have been transformed from being laboratory curiosities to the products of a billion dollar a year industry. The interest in this type of compound stems from their useful combination of properties. The polysiloxanes, usually known as silicones (as their structure was initially thought to be $R_2Si{=}O$, *i.e.* analogous to ketones), may be prepared on a large scale and they are generally chemically and biologically inert, thermally stable, and have useful surface and electrical properties. These properties also vary by a relatively small amount with temperature, which allows them to be used over a wide temperature range. Although silanols are not bulk industrial products, the study of them is important as they are the intermediates from which polysiloxanes are made. The preparations below show how steric effects are important in the stabilization of silanols towards condensation and how rings and polymers containing Si–O bonds may be made.

Experimental

a) Hydrolysis of Me_2SiCl_2

Prepare (in a fume cupboard) a solution of dimethyldichlorosilane (10 cm^3) in diethyl ether (20 cm^3) in a 100 cm^3 conical flask. Cool the solution in an ice/water bath and add water

(20 cm^3) dropwise using a pipette or burette taking care to swirl the solution as the addition is carried out. Do not allow the solution to get warm. After the addition is complete, separate the two layers using a separatory funnel and then put the aqueous layer back into the separatory funnel and extract it with hexane or 60–80 petroleum ether (15 cm^3). Combine the two organic layers and neutralise the HCl in them with sodium bicarbonate solution. Then separate off the organic layer and dry it over $MgSO_4$. The drying agent should then be removed by filtration and the organic solution placed in a clean, dry and pre-weighed round bottomed flask for use on the rotary evaporator. Remove the volatile materials under reduced pressure and reweigh the flask to obtain the yield of the oily siloxane product.

Calculate the yield of oil and record its IR spectrum. Although the Si – Cl bond is strong, silyl chlorides are much more readily hydrolysed than alkyl chlorides. Why is this?

b) A high molecular weight siloxane

Transfer the oil obtained in the experiment above to a 50 or a 100 cm^3 round-bottomed flask with a B24 neck (if it is not already in one) and add about 7% by weight of boric oxide and mix thoroughly for 2–3 minutes. Fit an air condenser or a water condenser with no water flowing through it to the flask and place the flask in a preheated oil bath at 200 °C for 3 hours. Allow the product to cool and then pour it out onto a preweighed watchglass, you may need to scrape the material out with a spatula.

Calculate the yield of the product. Roll the product into a ball and record its percentage bounce. Leave the material to stand (for several days preferably) and see what effect this has on its appearance and on the percentage bounce. What gives the polysiloxane its unusual properties?

c) Diphenylsilanediol

Prepare a mixture of toluene (2 cm^3), *t*-butanol (4 cm^3) and water (16 cm^3) in a 50 cm^3 round-bottomed flask and place it in a water bath at room temperature. Stir the mixture for 5 min, allowing it come to room temperature. Add a solution of toluene (2 cm^3) and diphenyldichlorosilane (4 cm^3) to the stirred aqueous mixture with a pipette at such a rate as to keep the reaction mixture below 25 °C. When the addition is complete, stir the reaction mixture for a further 5 min. Remove the white solid formed by filtration and wash it with water (3 cm^3) and diethyl ether (3 cm^3). Air dry the product.

Record the yield and melting point of the silanediol and record its IR spectrum for comparison with the product obtained from the hydrolysis of dimethyldichlorosilane. Diphenylsilanediol is an example of an organometallic compound that has some beneficial biological effects. It has been found to have anticonvulsant properties and it is an efficient antiepileptic agent. It does, however, like many drugs, have toxic side effects.

d) Octaphenylcyclotetrasiloxane

Add two drops of a 4 M aqueous sodium hydroxide solution to a solution of diphenylsilanediol (1 g) in absolute ethanol (10 cm^3). Boil the resulting mixture under reflux for 20 min during which time a white precipitate should form. Cool the flask to room temperature and remove the solid product by filtration. Crystallise the product using a minimum of warm ethyl acetate. Cooling the ethyl acetate solution with an ice-water/salt bath after removal of the first crop of crystals should afford a second crop of product.

Record the yield and melting point of siloxane and record its IR spectrum for comparison with the silanediol starting material.

e) Hexaphenylcyclotrisiloxane

Prepare a mixture of diphenylsilanediol (1 g), diethyl ether (15 cm^3) and concentrated hydrochloric acid (0.5 cm^3) and boil under reflux for three hours. Cool the reaction mixture to room temperature and remove the ether layer carefully with a Pasteur pipette and dry it over $MgSO_4$. Remove the drying agent by filtration and then remove the volatile materials from the remaining diethyl ether solution using a rotary evaporator to leave an oily white solid. Dissolve the product in a minimum of ethyl acetate and cool the product in an ice-water/salt bath to obtain crystals of the trisiloxane product.

Record the yield, melting point and IR spectrum of the product for comparison with the other materials made.

References

For a review of the preparations, properties and uses of silicones see F. O. Stark, J. R. Fallender, A. P. Wright in *Comprehensive Organometallic Chemistry, Vol. 2,* (Eds.: G. Wilkinson, F. G. A. Stone, E. W. Abel), Pergamon, Oxford, **1982,** p. 305.

For IR analysis of organosilicon compounds, see D. R. Anderson in *Analysis of Silicones,* (Ed.: A. Lee Smith), Wiley-Interscience, New York, **1974**, Chpt. 10; L. J. Bellamy *The Infra-red Spectra of Complex Molecules,* 3rd ed., Chapman and Hall, London, **1975,** Chpt. 20.

For the X-ray crystal structure of $Ph_2Si(OH)_2$ see, J. K. Fawcett, N. Camerman, A. Camerman *Can. J. Chem.,* **1977,** *55,* 3631.

For a discussion of the biological effects of organosilicon compounds including $Ph_2Si(OH)_2$ see, R. Tacke, H. Linoh in *The Chemistry of Organic Silicon Compounds, Part 2* (Eds.: S. Patai, Z. Rappoport), Wiley-Interscience, Chichester, **1989,** Chpt. 18.

For a discussion of the structure of $(Ph_2SiO)_4$ see, J. F. Hyde, L. K. Frevel, H. S. Nutting, P. S. Petrie, M. A. Purcell *J. Amer. Chem. Soc.,* **1947,** *69,* 488.

For the X-ray crystal structure of $(Ph_2SiO)_3$ see, N. G. Bokii, G. N. Zakharova, Yu. T. Struchkov *J. Struct. Chem.,* **1972,** *13,* 267 (English translation of *Zhur. Strukt. Khim.,* **1972,** *13,* 291).

2.11 Preparation and NMR Identification of Two Phosphorus Esters

Christopher Glidewell

Special Safety Precautions

1. Phosphorus trichloride (PCl_3) is very moisture sensitive and is toxic by inhalation. Handle it **only** in a fume cupboard.
2. Carbon tetrachloride is toxic by inhalation or contact.

Phosphorus forms many oxo-acids, most of which form many esters. In this experiment, two esters are prepared and identified.

Phosporus(III) chloride reacts with alcohols in the presence of base to form trialkyl phosphites (Eqn. 1). In the absence of base, the reaction between ROH and PCl_3 takes a different course. You are asked to identify the product formed (compound **A**) under these conditions when $R = CH_3$, and also to identify the reaction product of **A** with benzylamine and carbon tetrachloride (compound **B**).

$$PCl_3 + 3\,ROH + 3\,B \rightarrow P(OR)_3 + 3\,B \cdot HCl \qquad (1)$$

Experimental

a) Preparation of A

Set up, in a fume cupboard, a 250 cm^3 3-necked flask equipped with a magnetic stirrer, ice bath, dropping funnel and condenser. Attach a tube from the top of the condenser to the water pump and fit a $CaCl_2$ tube to the dropping funnel.

Put 100 cm^3 CCl_4 and 9.6 g MeOH into the flask, and 13.7 g PCl_3 into the dropping funnel. Turn on the water pump with the release tap open. Add the PCl_3 dropwise with stirring. When addition is complete, continue stirring with the water pump still on for a further hour. Remove the solvent on the rotary evaporator. Distill under oil-pump vaccuum to obtain **A**. Record the distillation conditions and the yield.

b) Preparation of B

Into a conical flask containing 100 cm^3 CCl_4, place 1.1 g of **A**, then 2.2 g of $PhCH_2NH_2$. Mix well and leave to stand overnight. Filter off the white solid (this is hygroscopic, so do not pump for too long), dry a small sample in a vacuum desiccator and record its melting point and infra-red spectrum (Nujol). Attempt to identify it. Meanwhile, evaporate the filtrate to obtain **B** and record the yield. Check the purity of both **A** and **B** by g.l.c.

c) Spectroscopy of ***A*** *and* ***B***

For each of your products **A** and **B**, as well as for an authentic sample of $(MeO)_3P$, record an infra-red spectrum, a ^1H NMR spectrum in the range $-2 \rightarrow +13$ ppm, and ^{31}P NMR spectra (with and without proton decoupling) in the range $0 \rightarrow +200$ ppm. Assign all the NMR absorptions and identify **A** and **B**. Suggest a mechanism for the formation of **A** and **B** (the identity of the white solid discarded during the preparation of **B** may be helpful here).

Predict mechanistically the products of the reaction between MeBr and $P(OMe)_3$. There are two phosphorus containing products formed in the reaction between $PhCOCH_2X$ (X = halogen) and $P(OMe)_3$: predict mechanistically what these might be.

2.12 The Preparation of Potassium Peroxodisulphate, $K_2S_2O_8$

Zdirad Žák and Antonín Růžička

Special Safety Precautions

1. Concentrated sulphuric acid is a highly corrosive substance. Avoid contacts with skin, it can cause severe burns. Protective gloves and safety goggles should be worn all time.
2. Potassium peroxodisulphate is a strongly oxidising agent. Its contact with inflammable substances should be avoided.

One of the possible methods of the preparation of peroxocompounds is an anodic oxidation. In this experiment, we shall prepare potassium peroxodisulphate, $K_2S_2O_8$, by the anodic oxidation of a saturated $KHSO_4$ solution in diluted H_2SO_4 at 0–3 °C (Eqn. 1).

$$2\,KHSO_4 \;-\; 2e^- \;\rightarrow\; K_2S_2O_8 + 2\,H^+ \tag{1}$$

Since peroxodisulphate is poorly soluble in this medium, it collects on the bottom of the electrolyser.

Experimental

a) $K_2S_2O_8$

The solution of $KHSO_4$ will be prepared by dissolving K_2SO_4 in 38% H_2SO_4. Prepare 140 ml of 38% acid (*Warning:* pour acid in the water and not *vice versa!*) and dissolve 22 g of K_2SO_4 in the still hot solution. Cool the solution first with tap water and then in an ice bath to 3 °C. Excess $KHSO_4$ crystallises from the solution on cooling. Meanwhile, set up the electrolyser according to the drawing in Figure **2.12-1**. Place a 1000 ml beaker on a magnetic stirrer, insert the electrolyser (with magnetic stirring bar) in the beaker and fill the space between the electrolyser and the beaker with an ice-salt mixture. Decant the cool saturated solution of $KHSO_4$ into the electrolyser (save the solid for further work) and switch on the electric source. Adjust the current through the electrolyser to approx. 1 Amp. (the current density on the Pt anode should be *ca.* 10 Amp cm^{-2}). Under continuous stirring, allow to electrolyse for 90 minutes. Be sure that the temperature of the solution does not rise above 3 °C, supply fresh ice and salt if necessary. Disconnect the electric leads, take the apparatus apart and transfer the precipitated $K_2S_2O_8$ onto a sintered glass funnel and filter. Wash with ethanol (20 ml) and diethyl ether (20 ml) and dry in desiccator over anhydrous $CaCl_2$ *in vacuo.* Weigh the dried product and determine the contents of peroxodisulphate in the product by iodometric titration.

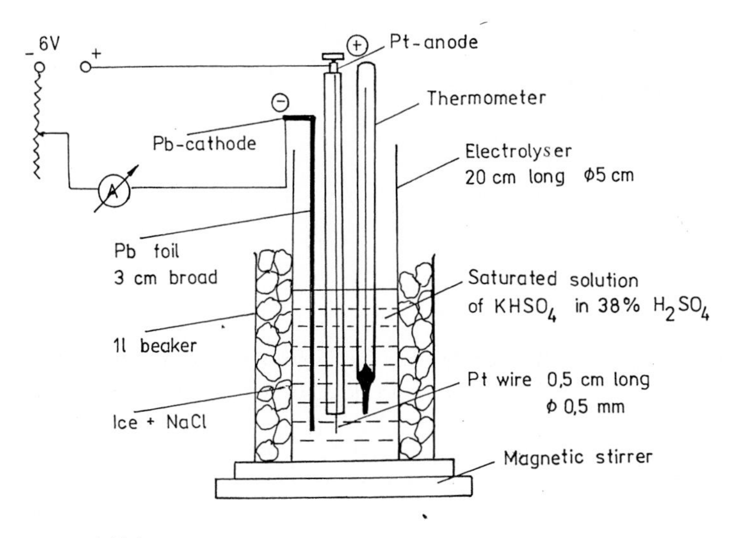

Figure 2.12-1.

b) Determination of the $K_2S_2O_8$ Purity

Potassium peroxodisulphate is a strong oxidant. Water solutions of peroxodisulphates oxidise iodides (in the presence of NH_4Cl catalyst) to iodine which is determined by the titration with thiosulphate. The reactions can be described by Eqns. (2) and (3).

$$S_2O_8^{2-} + 2I^- \rightarrow 2SO_4^{2-} + I_2 \tag{2}$$

$$I_2 + 2S_2O_3^{2-} \rightarrow S_4O_6^{2-} + 2I^- \tag{3}$$

Insert 3–4 g KI and 3–4 g NH_4Cl in a 300 ml Erlenmeyer flask with ground stopper. Add 150 ml of water, dissolve and heat the solution to 30–40 °C. Quantitatively wash a weighed sample (see Note) into the lukewarm solution, close the flask with the glass stopper and allow to stand for 15 minutes. The solution will turn brown with liberated iodine. Open the flask, rinse off the stopper into the flask and titrate with a 0.1 M solution of $Na_2S_2O_3$ until faintly yellow. Add 5 ml of a starch solution and continue the titration until the solution is completely decoloured. Calculate the percentage of $K_2S_2O_8$ in the prepared product.
Note: The amount of weighed sample should be equivalent to 15–20 ml of 0.1 M $Na_2S_2O_3$.

c) Recovery of K_2SO_4

Transfer the solid phase which remained after the decantation of the satured $KHSO_4$ solution in the electrolyser to a sintered glass funnel and vacuum filter. Wash with ethanol (20–30 ml) and ether (20–30 ml) and dry in a desiccator *in vacuo.* The isolated crystalline substance is not usually pure $KHSO_4$. Depending on the concentrations of K_2SO_4 and H_2SO_4 in the solution, it has a general composition $aK_2SO_4 \cdot bKHSO_4$. Detailed knowledge about the equilibrium between the solid phase and its saturated solution is obtained from the study of the $K_2SO_4/H_2SO_4/H_2O$ phase diagram. Weigh the dried salt and determine its exact composition by alkalimetric titration.

Weigh about 1.8 g of the salt to the nearest 0.1 mg and transfer it quantitatively into a 100 ml volumetric flask. Add 50 ml of water to dissolve the salt and then fill the flask to the mark. Stopper the flask and mix it by inverting at least three times. Pipette 20 ml aliquots (3 ×) and titrate each with 0.1 M NaOH. From the average consumption of NaOH, calculate the content of $KHSO_4$ in the salt. Calculate the amount of KOH necessary to neutralise all $KHSO_4$ to K_2SO_4 in the isolated salt. Dissolve the salt in a minimum of water in a 600 ml beaker and neutralise with a 20% solution of KOH. Cool with tap water and precipitate K_2SO_4 from the solution by addition of approx. 200 ml of ethanol. Filter and allow to dry in air.

References

H. Hecht, Präparative Anorganische Chemie; Springer Verlag Berlin 1951.

2.13 $K_2ON(SO_3)_2$ (Fremy's Salt) and $[Ph_3PCH_2C_6H_4CH_2PPh_3]ON(SO_3)_2$

David M. L. Goodgame

Special Safety Precautions

0.880 Ammonia *is* caustic. Glacial acetic acid is highly corrosive and flammable. Acetone is inflammable and an irritant. Ethanol is toxic and is inflammable. Since dry Fremy's salt may *explode* (mildly), store it over a small quantity of ammonium carbonate in an open tube with a cotton wool stopper.

The violet $ON(SO_3)_2^{2-}$ ion in solution is paramagnetic, with one unpaired electron. It is a useful one-electron oxidising agent in organic chemistry. In solid Fremy's salt, dimerisation of the anion occurs. The monomeric anion can be stabilised in the solid state by using a large dipositive cation.

Experimental

a) Potassium nitrosodisulphonate, $K_2ON(SO_3)_2$

Well cool a 15 cm^3 aqueous solution of sodium nitrite, $NaNO_2$ (5.8 g, 0.08 mole) in an ice bath, and add crushed ice (35 g) with continuous stirring. A solution of sodium metabisulphite, $Na_2S_2O_5$, (7.3 g, 0.04 mole) in water (15 cm^3) is added next, followed by glacial acetic acid (3.5 cm^3). *After 4–5 minutes* remove *ca.* 0.5 cm^3 of the liquid. Dilute this and add stock starch solution. A 0.1 M solution of iodine is now added dropwise. The *first* drop should produc a blue colour, showing that all bisulphite has reacted and the preparation can continue. The mixture is rendered alkaline by adding 0.88 ammonia (2.5 cm^3) and then stirred continuously, in the ice bath, during addition of an ice cold solution of $KMnO_4$ (2.1 g, 0.013 mole) in water (65 cm^3). The precipitate of MnO_2 is filtered off under suction using Celite and a sintered glass funnel, keeping the solution as cold as possible. To the violet filtrate is added a saturated solution (85 cm^3) of potassium chloride (33 g/100 cm^3 of water). The orange-yellow product is filtered under suction, washed several times with 5% KOH solution, then twice with ethanol containing *ca.* 5% v/v 0.88 ammonia and finally with acetone. Air should not be drawn through the solid, but instead spread the solid on a watch glass and allow the acetone to evaporate over 10–15 minutes. The product is dried in a desiccator over ammonium carbonate and calcium oxide. Write equations describing the formation of Fremy's salt, and record the yield obtained.

b) *p*-Xylylene-bis(triphenylphosphonium)-nitrosodisulphonate,
$Ph_3PCH_2C_6H_4CH_2PPh_3)ON(SO_3)_2$ [$= cat^{2+}ON(SO_3)_2^{2-}$]

$Cat^{2+}Br_2$ (0.15 g, 0.00019 mole) is heated to boiling in water (6 cm^3) and then allowed to cool somewhat without precipitation of the cat^{2+} salt. Fremy's salt (0.05 g, 0.00019 mole) is now quickly added and the mixture stirred. The violet product is filtered off under suction, washed with small amounts of water and acetone and air dried on the filter for a few minutes.

Add a little $K_2ON(SO_3)_2$ to aqueous KI slightly acidified with acetic acid. Explain your observations. Explain the difference in colour between Fremy's salt and the cat^{2+} salt.

References

F. A. Cotton, G. Wilkinson, *Advanced Inorganic Chemistry,* wiley interscience, 4th ed. p. 431.
D. M. L. Goodgame, *J. Chem. Educ.* **1969,** *46,* 724.

2.14 Polyiodides, Me_4NI_x

Andrew W. G. Platt

There are many examples of polyhalide ions, anions containing halogen atoms only. These can be considered as being derived from a simple ionic halide, *e. g.* KI and a molecular halogen such as I_2 to give the triiodide ion I_3^-, pentaiodide I_5^-, etc. ions. Many other examples are known such as tribromide and trichloride ions and those which contain more than one type of halogen atom such as ICl_2^-, etc.

By far the most important polyhalide ion is the triiodide ion. The formation of I_3^- is responsible for the increase in solubility of iodine in water in the presence of iodide ions. The intense blue colour observed in iodine thiosulphate titrations is due to a complex formed between the starch and the triiodide ion.

Although I_2 forms the I_3^- ion in solution with any source of I^-, stable complexes can only be isolated in the presence of large cations. One of the reasons for this is that the lattice energies of solids which have a large anion (*i. e.* the polyhalide) and a small cation such as Na^+ or K^+ tend to be low compared with those of NaI or KI. Thus, attempted crystallisation of NaI_3 from solutions of iodine and sodium iodide tends to lead to isolation of either very unstable polyhalides, which decompose at room temperature, or of sodium iodide.

In this experiment, you will attempt to prepare polyiodides using the tetramethyl ammonium ion, $(CH_3)_4N^+$, as the large cation.

Experimental

a) Tetramethylammonium triiodide

Finely powder 1 g of tetramethylammonium iodide and place in a beaker with 25 cm^3 of alcohol an add 1.3 g of iodine. Heat the mixture on a steam bath until all the white solid has dissolved. Allow to cool slowly to room temperature and finally cool in ice. Filter under suction, wash with a little diethyl ether and dry in air. Record the weight obtained and calculate the percentage yield.

b) Analysis

One of the accepted ways for chemists to prove that they have made the compounds they claim is by obtaining satisfactory elemental analyses. This usually means determining the percentage by weight of one or more of the elements present in the compound. In the case of the compound you have synthesised, iodine is the easiest element to determine.

Weigh accurately about 0.2 g of the product into a flask and add about 20 cm^3 of ethanol. Swirl the contents of the flask and add 50 cm^3 of a 0.05 M silver nitrate solution in 10 cm^3 portions. After the addition, the mixture consists of a fine yellow precipitate of silver iodide which must be coagulated before filtration. This is done by heating the mixture on the steam bath until all the precipitate has settled and the supernatant is clear. Filter the silver iodide into a dry preweighed sintered glass crucible, ensuring that all the precipitate is collected. Make a note of the number etched onto the glass for identification. *Do not stick labels or write on the crucibles.* Wash the precipitate with a little water followed by a little ethanol and dry to constant weight.

From the weight of the dry silver iodide obtained, calculate the weight of iodine present (hence the amount of iodine in the original weight of compound) and calculate the percentage of iodine in your sample of Me_4NI_3. Compare your value with the theoretical percentage and comment on the result.

Using a similar procedure to that above, prepare and analyse a sample of tetramethylammonium pentaiodide.

Questions

1. What are the shapes of the I_3^- and I_5^- ions?
2. What would be the products of the thermal decomposition of $KICl_2$ and KBrICl? Explain your answer.

2.15 Interhalogen Compounds

Ivan P. Parkin
Adapted from the Open University CHEM777 Summer School as prepared by the S343 course team

Special Safety Precautions

1. Chlorine gas is toxic. The whole cylinder should be securely clamped inside a fume cupboard. If in any doubt whatsoever about safe operation of the cylinder, consult a demonstrator.
2. Iodine is toxic, grind the crystals in a fume cupboard.
3. Wear rubber gloves (marigold) at all times. Iodine monochloride (ICl) is corrosive. If any is split, douse it immediately with thiosulphate solution (preferably) or water. If any ICl(l) gets onto your hands, wash them immediately in dilute thiosulphate solution first and then with water.
4. Chloroform is toxic by inhalation and contact.
5. All apparatus should be dry (oven dried) prior to use otherwise the ICl(l) will decompose.
6. All reactions must be carried out in a fume cupboard.

The halogens form many compounds amongst themselves in binary combinations that may be neutral or ionic, for example, BrCl, IF_5, Br_3^+, I_3^-. Tertiary combinations are possible but occur only in polyhalide ions such as, for example, $IBrCl^-$.

Neutral interhalogen compounds are of the type XY_n where n is always odd and Y is always the lighter halogen. The compounds are all diamagnetic with the valence electrons present either as bonding pairs or unshared pairs. In these complexes, the X atom can be considered to have a net oxidation state of n^+.

In this experiment, you will prepare the interhalogen ICl and react this with caesium bromide to form a polyhalide. You will determine the formula of the polyhalide by gravimetric, titrimetric and spectroscopic methods.

Experimental

a) Iodine monochloride, ICl

In this experiment, as described below, you will prepare iodine monochloride, ICl, by passing dry chlorine through a known mass of iodine until it increases in mass by an appropriate amount.

$I_2(s) + Cl_2(g) = 2ICl(l)$

Grind up some iodine using a dry pestle and mortar. Then transfer 8 g into your internal-seal trap Fig. **2.15-1**. The easiest method of transfer is to lay the trap on its side and attach a small glass funnel (with a very short piece of vinyl tubing) to the side-arm. Then stand the trap containing the iodine upright in a beaker as shown in Figure **2.15-1**.

Weigh the supported trap on the rough balance in the fume cupboard, and hence work out the mass of the trap when all the iodine has been converted to iodine monochloride.

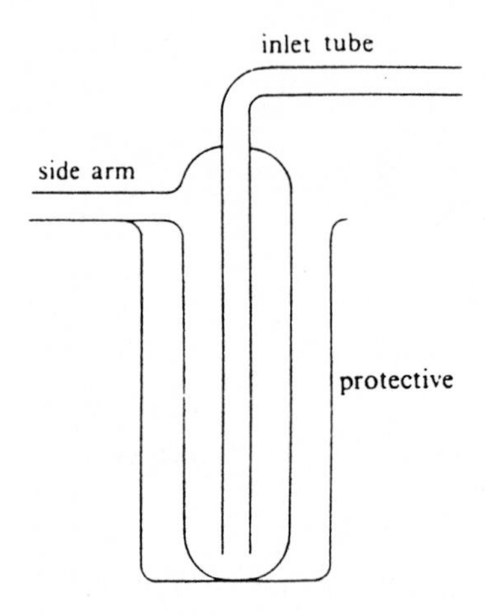

Figure 2.15-1. The internal-seal trap.

Pass dry chlorine into the trap slowly (so as not to blow out your product) and shake or swirl the trap gently. After a couple of minutes, weigh the flask on the balance in the fume cupboard to see how near you are to the required mass. Continue passing in chlorine until this mass is reached, *but no longer.* (With excess chlorine, the reaction will continue to give iodine trichloride, ICl_3, so you must be careful not to overshoot.)

Using great care, pour your product through the side-arm of the trap into a **dry** 25 cm^3 conical flask with a glass stopper. **Stopper it** and label it with your name and the mass. *This must be kept and used in a fume cupboard.*

Run the electronic spectrum of your product in a solution of dry $CHCl_3$. Using a dropping pipette, take only one drop (that is, a minimal amount) of ICl in 10 ml of solvent. If the solution is too concentrated, dilute it as necessary. Don't forget to keep the solutions, and the iodine monochloride, stoppered. Use a 1 cm silica cell, and obtain the spectrum from 700–325 nm. Compare your spectrum with the electronic spectra of the halogens (Fig. **2.15-2**). Which halogen spectrum does the spectrum of your product resemble? Why?

b) A polyhalide

To investigate this reaction, take 2 g of finely powdered CsBr (use a pestle and mortar) and pour it into your iodine monochloride in the conical flask in the fume cupboard. Stir the mixture very thoroughly with a glass rod for two minutes, ensuring that the CsBr has dissolved

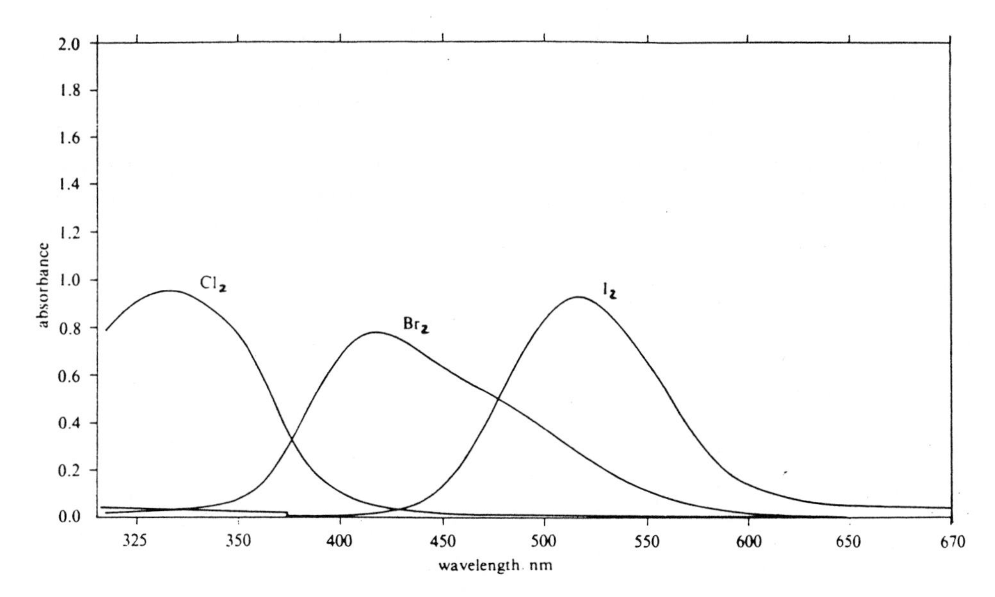

Figure 2.15-2. Electronic spectra of chlorine (0.013 3 M* in CCl_4), bromine (0.004 4 M in CCl_4) and iodine (0.000 9 M in CCl_4). For all three spectra the path length was 10 mm.

completely. Then add 20 cm^3 of chloroform to the mixture, stir well, stopper and leave to stand for ten minutes with occasional shaking. Stir thoroughly again to try to disperse any brown lumps in the product.

To purify your product, transfer it to a small beaker and wash it thoroughly with several small (~5 ml) aliquots of chlorofom (in the fume cupboard), decanting off and discarding the washings, until the final chloroform wash is almost colourless. Filter using a clean dry Büchner funnel. When the residual chloroform has evaporated, transfer your product to a dry sample tube. Stopper it and reweigh. Keep the tube stoppered to prevent damp air from entering.

c) Thermal decomposition of the polyhalide

Before you go on to identify the halogen(s) in your product from (b), first heat with a Bunsen burner a *small* portion in a test-tube in the fume cupboard, gently, then strongly, watching to see what happens. Try to devise a method of collecting and identifying the gas given off. Heat until there is no further change, and then identify the residue by heating a small amount on nichrome wire (what colour is produced) and by dissolving the solid in water (2 ml) and adding dilute $AgNO_3$ (aq). Discuss your results.

d) Quantitative analysis of the polyhalides

Caesium analysis

It is possible to determine the percentage of caesium in the polyhalide complex quantitatively using *gravimetric analysis.* The method is to take a known amount of the complex in solution, react it with a suitable reagent to form an insoluble caesium compound, filter this off quantitatively, dry it and weigh it. Caesium is determined gravimetrically as caesium tetraphenylborate, $CsB(C_6H_5)_4$, using a solution of sodium tetraphenylborate as precipitant.

Take two no. 4 sintered glass crucibles, label them clearly A and B and with your initials using a soft pencil. Then put them in the oven to dry at 100–120 °C. Weigh accurately about 0.5 g *(but not more than 0.6 g)* of the polyhalide and dissolve it in 40 cm^3 of 1 M HCl. Carefully transfer the solution to a 100 ml volumetric flask and make up to the mark with distilled water. Shake well and pipette 25 ml of the solution into each of two clean 250 ml beakers. To *each beaker* add ≈0.1 g of NaI (the amount on the round end of a standard spatula), stir until it dissolves and then add 20 ml of 0.05 M $Na_2S_2O_3$ (sodium thiosulphate or 'hypo'). If the solution does not clear completely, add more thiosulphate solution until it does.

Add 35 ml of distilled water and then place the beaker in an ice-water bath for about 10 minutes. *Slowly* add 40 ml of sodium tetraphenylboron solution with stirring. Stand for 10–15 minutes (but not more than an hour) in the ice-water bath. While the beakers are cooling, take your two crucibles from the oven and put them in a dessicator to cool, then weigh them on an accurate balance. Filter each precipitate through a weighed sintered glass crucible, making sure that *all* the precipitate is transferred. Wash the precipitate several times with small amounts of distilled water. Dry the precipitate at 100–120 °C for 1 hour. Cool in a dessicator and reweigh.

Calculation: Caesium tetraphenylboron contains 29.39 per cent of caesium.

$$\text{percentage Cs in complex} = \frac{\text{mass of precipitate (g)} \times 29.39}{\text{mass of sample (g)}}$$

Iodine analysis

As a confirmation of the formula of the polyhalide, it is possible to determine quantitatively the amount of iodine in a weighed sample using a volumetric method.

Dissolve an accurately weighed sample of the polyhalide (≈0.5 g) and acidify with a few ml of dilute HCl. (This releases the iodine from the complex.) Titrate the released iodine *immediately* with standard 0.05 M sodium thiosulphate using starch indicator.

The procedure for this titration is to run the sodium thiosulphate into the brown solution until it becomes straw coloured. At this stage, add a few drops of starch indicator. Continue titrating until the blue colouration disappears and the solution becomes colourless.

Only freshly prepared or properly preserved starch indicator should be used, and the *same volume* of starch solution should be added to each titration.

The starch must not be added until just before the end-point is reached, because at high iodine concentration some iodine may remain adsorbed even at the end-point.

The equations for the above reactions are:

$$I^+(aq) + I^-(aq) = I_2(s)$$

$$2S_2O_3^{2-}(aq) + I_2(s) = S_4O_6^{2-}(aq) + 2I^-(aq)$$

Calculate the percentage of iodine in your polyhalide.

Exercises

1. From your observations in part c, what elements are present in the polyhalide?
2. From your calculations in part d, together with your observations in part c, suggest a molecular formula for the polyhalide.
3. Based on the VSEPR rules, predict the shape of the polyhalide anion.

2.16 Simulation of Atomic and Molecular Orbitals

Alan G. Massey and Sylvia Massey

This experiment describes the use of magnets to simultate *s, p* and *d* atomic orbitals from which a wide variety of molecular orbitals can be derived. The technique gives students an idea of the molecular orbitals' shape and stresses the importance of the symmetry labels.

It is well known that iron powder or iron filings can be used to trace out the magnetic fields of magnets placed directly underneath a flat surface. A short note (see References) has described the use of such a system to simulate σ and π orbitals of diatomic molecules. However, this technique suffers from the major drawback that to simulate a bonding σ (l*s*) orbital, for example, two bar magnets have to be placed vertically so two *unlike* poles are in contact with the underside of the surface, *ie* $\boxed{\mathrm{N}}\boxed{\mathrm{S}}$. This is, of course, different from the bonding interaction of two atomic l*s* orbitals where the symmetries of their wave functions have to be identical.

However, if the magnets are placed about 2 cm away from the underside of a flat aluminium disc, a different situation occurs: the iron powder, on tapping the disc gently, is found to move away from the regions where one would expect the highest magnetic field to be.

Presumably, the magnets cause the small particles of iron to become magnetised and hence align themselves in the direction of the field but, because they are all aligned in the same direction, the particles repel each other. If the permanent magnets are the correct distance from the aluminum disc, the 'magnet-iron powder' interaction will be weaker than the 'iron powder-iron powder' repulsion and hence on tapping the disc the magnetised iron particles move away from each other.

The ideal nature of this system arises from the fact that when constructing, say, a σ (l*s*) bonding orbital the 'symmetries' of the two magnetic fields under the disc have to be identical – now exactly analogous to the overlapping of two atomic l*s* orbitals. This situation arises because of the strong repulsion between all the identically magnetised iron particles. Thus, the σ (l*s*) bonding orbital can be simulated by placing two vertical bar magnets under the disc with their poles either $\boxed{\mathrm{N}}\boxed{\mathrm{N}}$ or $\boxed{\mathrm{S}}\boxed{\mathrm{S}}$. When the magnets are placed $\boxed{\mathrm{N}}\boxed{\mathrm{S}}$, a σ^* (l*s*) antibonding orbital results.

Thus, the technique not only gives the student an idea of the molecular orbital's shape but also he is forced to discover that those mysterious + and – signs on atomic orbitals – their symmetry labels – are very important in the bonding context. Since l*s*, 2*p* and 3*d* atomic orbital representations are available, it is possible to simulate virtually any molecular orbital, the only constraint being that the molecular orbital has to be represented in only two dimensions.

Experimental

Materials required

The ideal magnet length is about 1–1.5 cm and it is important that they are closely matched in strength otherwise the orbital patterns become distorted. The six-inch diameter aluminum disc used to view the magnet field patterns can be supported on a cork ring of the type used for holding round-bottomed flasks. The magnets are glued to 'Lego' pegs so that they can easily be moved about on a 'Lego' peg-board placed underneath the cork ring and yet are held firmly enough by friction so that they do not move each other by their magnetic interaction. The *d*-orbital array of magnets can be built up each time it is required but it is much more convenient to have the four (identical) magnets permanently glued together on one peg.

When constructing aromatic ring systems such as benzene or naphthalene, the peg-board has to be cut into small pieces and glued back into the required shape because the normal peg pattern does not allow hexagonal formations to be made (less satisfactorily, the magnets can be held in pieces of plasticine stuck to the bench top). The distance between the magnets in the aromatic rings, or indeed in any of the formations, depends largely on the strength of the magnets available and thus has to be determined by a little trial and error. Magnets used as carriage couplings in 'Lego' train sets are ideal.

The best way to apply the iron powder to the perfectly horizontal disc is to sprinkle it evenly from a sugar sifter. When the magnets are in position, a typist's rubber pencil is placed point down in the centre of the disc and tapped gently with a hard object until the iron powder takes up the correct shape; any disturbance of the powder distribution caused by the pencil can be rectified by careful addition of more powder from a non-magnetic spatula. However careful one is during experimentation it is soon found that the magnets collect clusters of iron filings which tend to distort the magnetic fields; these filings can be removed by 'dabbing' the magnets with a piece of plasticine.

a) The 1*s atomic orbital*

Place a magnet on the peg-board with its north pole facing upwards. Put a cork ring on the peg-board so that the magnet is approximately central and cover with the aluminum disc. Sprinkle iron filings onto the disc until the central area is thinly covered. Finally, vibrate the disc by placing a typist's rubber pencil vertically at the centre and tapping the pencil top lightly with a small, hard object. The filings move away from the area of high field to leave a circular space of symmetry N (for north) which represents a 1*s* orbital.

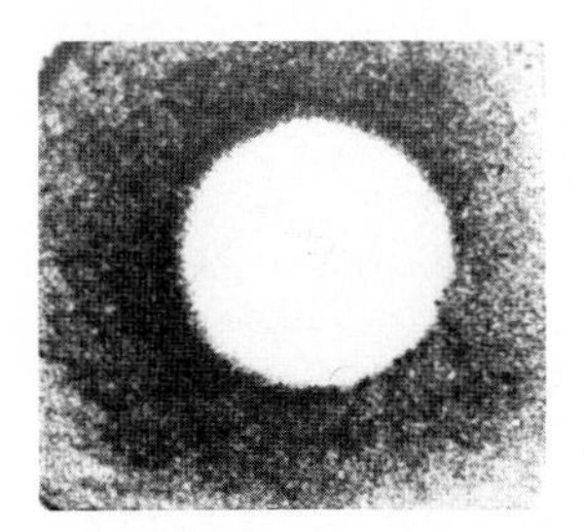

b) The 2p atomic orbital

Place two magnets upright on the peg-board so that their exposed poles are north for one magnet and south for the other. Put the cork ring and aluminum disc in place and, after sprinkling iron filings on the disc surface, tap the surface as described previously. Two clear areas develop in the iron filings, one of which has N symmetry and the other S.

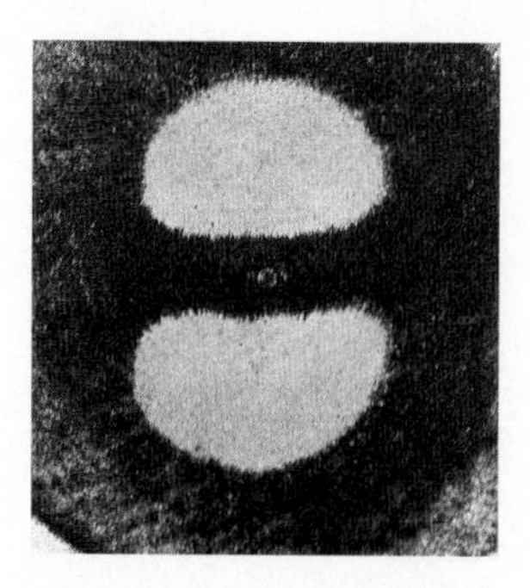

c) The 3d atomic orbital

This orbital simulation requires a group of four magnets arranged to stand upright in the form of a square so that their exposed poles alternate round the square. The developed pattern shows four equal areas of clear space among the filings which alternate in N or S symmetry and thus may be used to demonstrate a 3*d* atomic orbital.

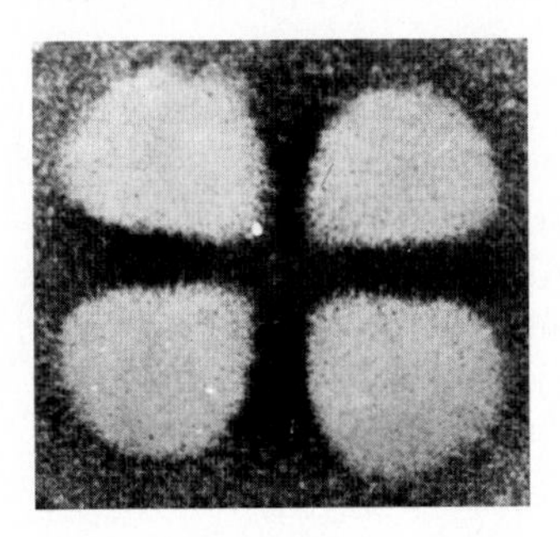

d) Nodes

It will be noticed that the 2*p* and 3*d* orbital patterns have the orbital lobes separated by bands of iron filings. These bands represent nodes in the orbitals. The clear spaces in the iron filings change their N or S symmetry from one side of a node to the other in the same way that actual orbitals change the symmetry labels (+ or −) of their lobes as a nodal line or nodal plane is crossed.

The above basic magnet arrangements can now be used to investigate a variety of molecular orbital shapes in both organic and inorganic species.

e) σ (1s) orbital in H_2

Place two vertical magnets about 2.5–3.0 cm apart with their north poles uppermost and then generate the resulting magnetic pattern on the disc as described above. Repeat the experiment with one of the magnets inverted so that its south pole is uppermost; which molecular orbital does this latter pattern simulate?

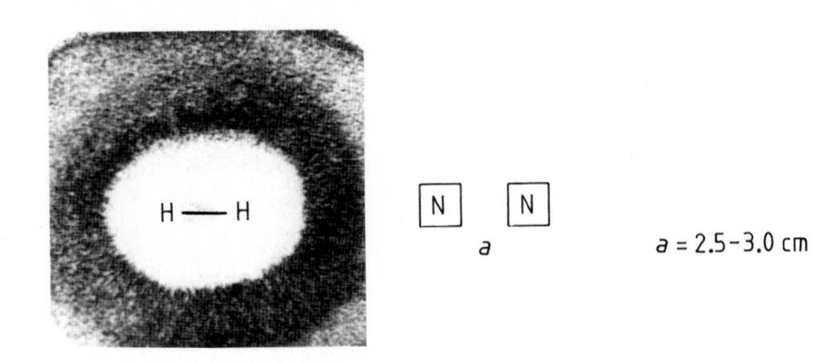

f) π and π orbitals in ethene*

A π-bonding orbital in ethene, $CH_2=CH_2$, arises from the sideways overlap of two 2*p* orbitals, one on each carbon atom. Remembering that the symmetry of the interacting lobes in the 2*p* orbitals have to match, decide on a suitable arrangement of vertical magnets which will mimic a bonding π-orbital. Develop the magnetic field on the aluminum disc and verify that it resembles the two lobes expected of a π-bonding orbital.

Demonstrate a π* antibonding orbital and convince yourself that it has the same lobal symmetry as a 3*d* atomic orbital.

g) P–O π-bonding in $OPCl_3$

The P–O bond in phosphoryl chloride, $OPCl_3$, can be considered to involve a component which arises from an interaction between an empty phosphorus *d* orbital and a filled oxygen 2*p* orbital. Demonstrate the resulting molecular orbital arising from the following magnet arrangement.

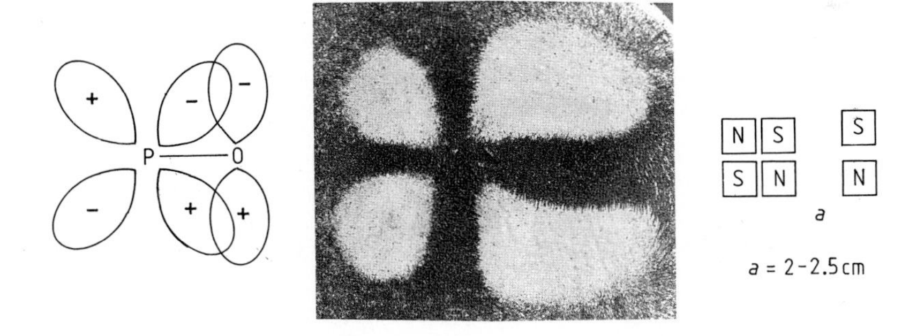

h) Back-bonding in transition-metal carbonyls

It is thought that metal carbonyls such as $Cr(CO)_6$ achieve a considerable increase in stability from an interaction between filled *d* orbitals on the metal and empty π* antibonding orbitals on the CO ligands. The shape of the molecular orbitals resulting from this back-bonding can be illustrated by developing the magnetic field patterns using the following magnet arrangement.

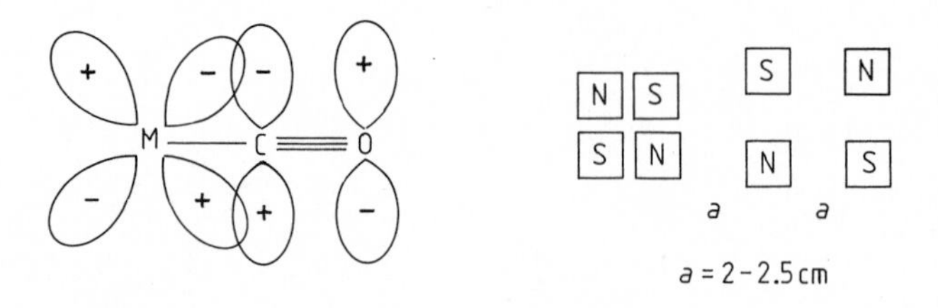

i) Back-bonding in transition-metal ethene complexes

This supplementary bonding in metal ethene complexes arises from an interaction between a filled metal *d* orbital and the empty π^* antibonding orbital of ethene in the following manner.

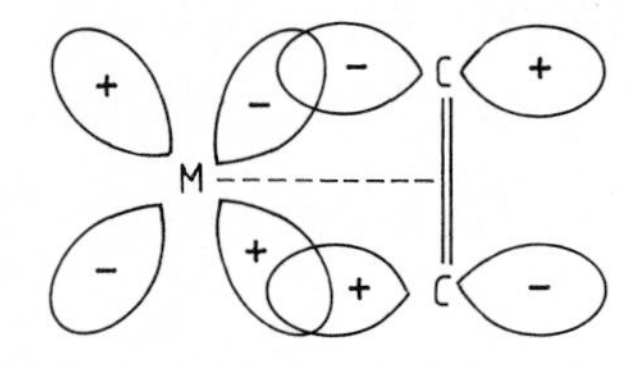

Deduce the correct magnet arrangement and develop the resulting magnetic field pattern.

j) The π molecular orbitals in benzene

The six carbon atoms in a benzene ring each have a $2p$ orbital lying vertical to the plane of the ring. Theory tells us that when n atomic orbitals interact in a molecule they generate n molecular orbitals, several of which are of high energy and are not normally occupied by electrons. Demonstrate, using six magnets placed at the corners of a regular hexagon, the shapes of the six π orbitals which occur in a benzene molecule. (Note that there is a nodal plane in the benzene π orbitals which is defined by the plane of the ring. This simulation gives only a planar, top view of the π orbitals; decide in each case what shape and symmetry each orbital lobe would have below the C_6 plane).

k) π bonding in transition-metal complexes

The metal t_{2g} orbitals and certain of the ligand p orbitals in octahedral complexes have identical symmetry and can thus interact to give supplementary metal–ligand π bonding. This can be simulated for the four ligands lying in the equatorial plane of a complex by using the magnet arrangement shown below. Note how this interaction has the effect of increasing the effective size of the original metal d orbital. This illustrates the nephelauxetic, or electron-cloud expanding, concept encountered during the interpretation of transition-metal d-d spectra.

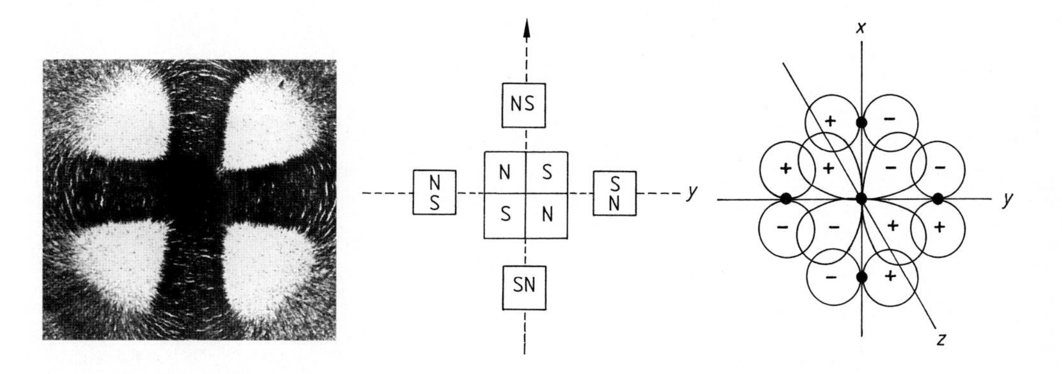

ACKNOWLEDGEMENT
We thank the Royal Society of Chemistry for granting permission to reproduce parts of an article which appeared in *Education in Chemistry* **1976,** p. 111.

References

L. F. Druding *J. Chem. Educ.,* **1972,** *49,* 617.

3 Intermediate Experiments

3.1 Ferrocene, (η^5-C_5H_5)$_2$Fe, and Its Derivatives

Leslie Pratt and J. Derek Woollins

Special Safety Precautions

Dicyclopentadiene, cyclopentadiene and dimethylsulphoxide are poisonous. Avoid breathing vapour and skin contact. The KOH/ether mixture prepared in this experiment is very corrosive. Acetic anhydride is an irritant and should be used in a fume cupboard

The discovery in 1951 of bis(pentahaptocyclopentadienyl)iron, commonly known as ferrocene, with its unusual "sandwich" type structure and remarkable thermal inertness resulted in a vast amount of research into the synthesis and reactions of related organometallic compounds. A variety of organic reactions can be carried out on the cyclopentadienyl rings, which have aromatic character. In this experiment, ferrocene is prepared, acetylated and then the acetyl group is reduced. The various stages are studied by ^{1}H NMR spectroscopy.

Experimental

a) Ferrocene

Cyclopentadiene dimerises readily at room temperature to dicyclopentadiene by a Diels-Alder reaction. Accordingly, the commercial sample must be first 'cracked' on the day you plan to use it. The apparatus for doing this is set up in a fume cupboard and it may be convenient for you to carry out the cracking process in collaboration with other people doing this experiment. Check that the flask contains at least 60 ml of dicyclopentadiene (100 ml if more than one sample of cyclopentadiene is needed). Collect the distillate that condenses in the range 42–44 °C and keep it cooled with an ice bath around the receiving flask. (You will need about 8.5 ml).

While you are cracking the dicyclopentadiene, fit a three-necked 500 ml flask with a mechanical stirrer in the central neck, a 100 ml dropping funnel fitted to one side neck by means of a side-arm adaptor (connecting the side arm to a nitrogen supply), and a Liebig condenser in the other side neck, with the exit end of the condenser connected to a bubbler to prevent air entry and to monitor the nitrogen flow rate. Charge the flask with diethyl ether (100 ml) and *flake* potassium hydroxide (40 g), stir well and flush with nitrogen. (If this preparation is not carried out in a fume cupboard, a safety screen should be used). Meanwhile, dissolve finely powdered iron(II) chloride tetrahydrate (10 g) in dimethylsulphoxide, degassed by bubbling nitrogen through it (40 ml, avoid skin contact; vigorous stirring for about an hour is usually required). Keep the iron(II) chloride solution in a sealed container to prevent oxidation.

With vigorous stirring and under a slow stream of nitrogen, add 8.5 ml cyclopentadiene to the KOH/ether mixture. After *ca.* 15 min, discontinue the nitrogen flow and add the iron(II)

chloride solution dropwise. The reaction is exothermic and the ether may boil. When this subsides, restore a slow nitrogen flow. Replace any ether lost by evaporation. Continue stirring for a further 30 min. (**Care:** The KOH/solvent mixture is extremely corrosive). Decant the ether layer and wash the dark residue in the flask with 50 ml of ether. Combine the ethers and wash it with 2 M HCl (2 × 40 ml), to neutralise any hydroxide, and then with water (2 × 40 ml). Carefully evaporate off the ether to deposit orange crystals of ferrocene. Purify a small sample by sublimation using a Petri dish and lid on a warm hotplate *in a fume cupboard.* (Careful: too rapid heating or cooling of the dish may crack it). Remove the golden crystals of ferrocene periodically from the Petri dish lid. Record the m. p. and write equations for the reactions involved in the preparation.

Examine the solubility of ferrocene in a) water, b) dichloromethane, c) toluene and account for your observations in terms of the structure and bonding of the molecule. Add ferrocene (0.1 g) to water (5 ml) followed by concentrated nitric acid (5 ml) – *extreme* caution. Shake the tube gently for 2 min and record your observations.

b) Acetylation

Add crude ferrocene (3 g) to acetic anhydride (10 ml) in a 50 ml round-bottomed flask provided with a calcium chloride or silica gel guard tube. Carefully add orthophosphoric acid (2 ml) dropwise with shaking. Heat the mixture on a steam bath for 20 min, then pour the hot mixture onto crushed ice (80 g) with stirring. Wash your flask out with some additional ice and combine the aqueous materials. When all the ice has melted, neutralise the solution with solid sodium bicarbonate, cool the mixture in ice for 20 min and then filter off the brownish yellow solid. Dry it in a vacuum desiccator.

Recrystallise your product from cyclohexane or petroleum ether and check its purity by TLC (repeat the recrystallisation and TLC examination if necessary, and draw a representation TLC development in your report, including R_f values). Record the m. p.

c) Reduction of acethylferrocene

Dissolve acethylferrocene (1 g) in ethanol (15 ml) and add water (5 ml). Dissolve sodium tetrahydroborate (0.8 g) in water (4 ml) and add this solution slowly to the stirred solution of acethylferrocene. After the solution has stood for 15 min, it may be pale yellow, but often the colour does *not* become paler and stays brown. If necessary, add further portions of sodium tetrahydroborate to complete the reduction. Add water (100 ml) and extract the aqueous mixture with diethyl ether (2 × 20 ml). Dry the ether extracts over magnesium sulphate or calcium chloride, filter and then evaporate off the solvent on the rotary evaporator. Recrystallise the product, which is often an oil, from 40/60 petroleum ether and dry it in a vacuum desiccator. Record the m. p.

Study the ^{1}H-NMR spectra of acetylferrocene and its reduction product and interpret them. What is a plausible mechanism for the acetylation, given that ferrocene reacts some 3×10^6 times faster than benzene?

3.2 The Use of Organolithium Reagents in the Preparation of Ferrocene Derivatives

Ian R. Butler

Special Safety Precautions

Before starting this experiment you must read the manufacturers safety data sheets on the chemicals you will be using. Rubber gloves and proper eye protection must be worn throughout the lab work.

1. *n*-Butyllithium Solutions in Hexane. This experiment deals with the use of *n*-butyllithium solutions which are extremely flammable. This reagent poses no danger when handled properly, but it is important that the correct handling procedure is adopted and that it is treated with respect. This is without doubt the most commonly used lithium based reagent in the organic and organometallic research laboratories. A typical reaction of *n*-butyllithium is shown below:

   $$R-H + n\text{-BuLi} \rightarrow RLi + n\text{-BuH}$$

   where H is a relatively acidic proton.
   The reaction of *n*-butyllithium with water is a particularly violent one, therefore it is important that all apparatus that you will use is absolutely dry. In addition, all solvents must be predried. For details on how to work with air sensitive compounds see Aldrich Technical Information Bulletin Number AL-134, which is supplied with the reagent.
2. Other Reagents. *N,N,N',N'*-Tetramethylethylenediamine (TMED), chlorodiphenylphosphine, chloro-tri-*n*-butyltin and dimethylformamide (DMF) are all highly toxic and all transfers should be carried out in a properly vented fume cupboard. Nickel chloride hydrate and ferrocene are toxic and should also be weighed out and handled in a fume cupboard.
3. Solvents. Hexane(s), toluene, diethyl ether and methylated spirits are all highly flammable and must be kept away from sources of ignition. Dichloromethane and deuterated chloroform are toxic by inhalation or contact and should be handled in a fume cupboard.
4. Alumina. Alumina (Al_2O_3) powder can be harmful if ingested – always use in a well vented fume cupboard.
5. Residues. All residues should be disposed of in an appropriate waste receptacle.

The objective of this exercise is to gain an insight into some practical inorganic chemistry as you may find it in a research environment. The time spent waiting in the initial stages of the synthesis can be profitably used in the library. In this experiment, you will prepare some synthetically useful organometallic compounds and a metal complex of one of these. Your are required to work in pairs, although each person in a pair will prepare a different product. You will be able to practice working under an inert atomosphere (N_2), which will be extremely useful for your future research careers.

Although the general area of organometallic chemistry is over 100 years old, it was only in the early 1950's that major developments were made which turned it into one of the most intensely studied fields today. One of the milestones was the discovery of ferrocene, independently by Pauson and Keally and Miller, and the later realisation of its unusual bonding features by Wilkinson and Fischer. It was the first example of a sandwich compound in which the iron metal is bound to the 10 equivalent carbon atoms (in organometallic notation, $2 \times \eta^5$) in two aromatic cyclopentadienyl rings. The chemistry of ferrocene is dominated by its electrophilic aromatic substitution reactions. In essence, ferrocene can be treated as behaving like an aromatic organic compound with respect to electrophiles, the only difference being that ferrocene is typically about 100,000 times more reactive than benzene for example. Since the individual cyclopentadienyl rings in ferrocene are held apart at a constant distance, the ferrocene nucleus is a useful backbone for the design of ligands capable of chelating to a metal centre. For example, disubstitution of ferrocene as shown below with two electron donating groups D will result in a compound which has the potential of binding to a metal centre

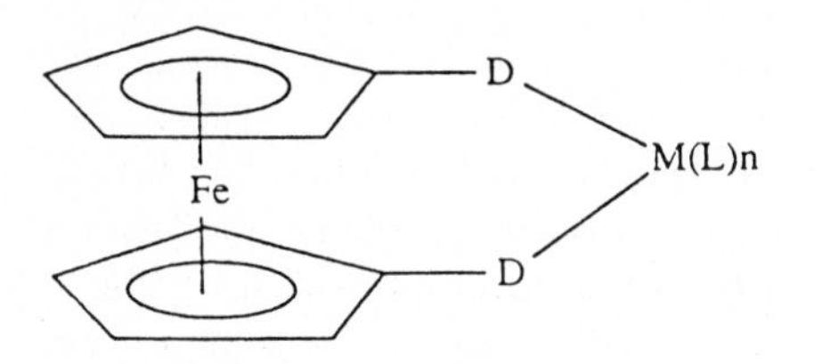

D = electron donor atom or group

M = metal (usually bound to some other ligands. L)

The experimental details are intentionally non-exhaustive in that you may have to make slight modifications to solvent volumes, etc. as might happen in a research environment. In this way, you will learn to make necessary adjustments according to prevalent conditions. In this experiment, you will functionalise the ferrocene by generation of dilithioferrocene in the form of its TMED adduct. You will then prepare derivatives from this intermediate.

Experimental

Lithiation of Ferrocene

This part of the procedure must be carried out under a dry nitrogen atmosphere (work in pairs with each person carrying out one experiment).

Ferrocene (1.86 g, 10 mmol) is placed in a 250 ml Schlenk tube. The tube is then evacuated and refilled with nitrogen (or flushed with nitrogen for 2 minutes), then dry hexane (or hexanes) (30 ml) is added. A solution of *n*-butyllithium in hexane (1.6 M, 13 ml) is then added using a syringe. The correct method of using the syringe and the experimental set up is shown here:

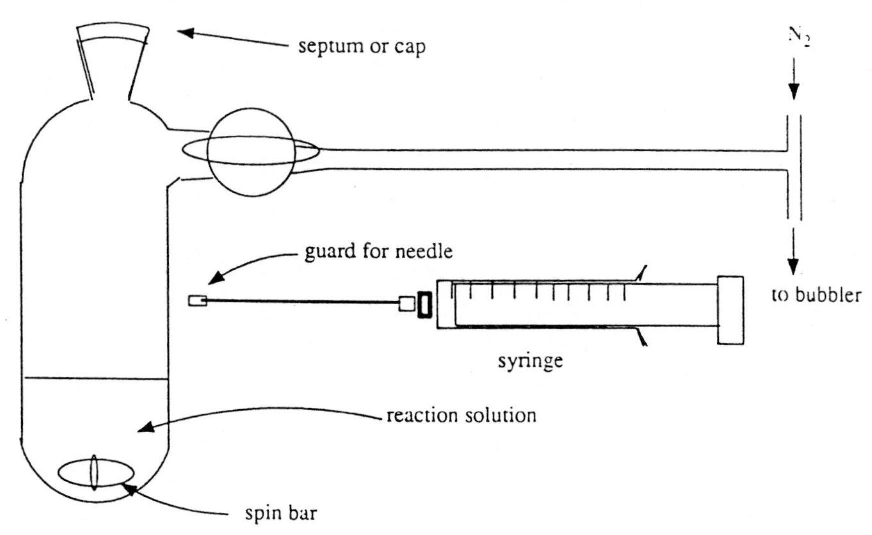

Figure 3.2-1.

When using a syringe, always guard against accidentally stabbing yourself by making sure the tip of the syringe has a cap on it except when actual transfers of reagent are being performed. Firstly, fill the syringe with nitrogen by placing the tip of the needle into the neck of your Schlenk tube and slowly pulling back the plunger to the desired volume. Next, withdraw the syringe and expel the nitrogen and repeat the procedure, this time filling the syringe with a slightly larger volume of nitrogen than the volume of the solution you will require (approx. 16 ml) – do not expel the nitrogen this time. One person then holds the bottle of *n*-butyllithium steady, while the other carefully inserts the needle through the septum on the top of the bottle. Keeping the tip of the needle above the headspace, inject the nitrogen in the syringe into the bottle, then allow the tip of the syringe to go below the surface of the liquid and withdraw 13 ml of the *n*-butyllithium solution. Slowly extract the needle and immediately transfer the *n*-butyllithium into the Schlenk tube. The person holding the bottle should recap it as soon as the syringe has been removed. The syringe should now be washed by removing the plunger and rinsing both parts under running water – a slight hiss may be heard due to the reaction of water with traces of residual *n*-butyllithium.

Now add *dry* tetramethylethylenediamine (1.2 g, approx. 30 mmol) to the well stirred solution (using a Pasteur pipette). Cap the Schlenk tube and allow the reaction mixture to stir overnight, making sure the butane produced in the reaction is vented through the bubbler.

** Repeat the experiment reversing roles with your partner**

After the solution has been stirred overnight, you will observe the formation of a pale orange precipitate of 1,1-dilithioferrocene · TMEDA. This is a pyrophoric solid, therefore you will continue directly without its isolation. Outlined below are three different experimental procedures. Each pair of workers will carry out reaction a) and either b) or c). Cool the Schlenk tube in an ice bath, then using a Pasteur pipette add either
a) chlorodiphenylphosphine (4.6 g, 21 mmol) or
b) chloro-tri-*n*-butyltin (6.67 g, 21 mmol) or

c) dry diethyl ether (60 ml) then dimethylformamide (1.6 g, excess) dropwise. After the addition, allow the mixture to warm up to room temperature and stir for a further 2–3 h then hydrolyse carefully by the slow addition of water (50 ml).

a) The chlorophosphine procedure: bis-diphenylphosphinoferrocene

An orange-brown precipitate will form. Decant the supernatant hexane layer or remove the supernatant hexane layer by syringe and wash the solid with methylated spirtits (2 × 10 ml) and hexane (3 × 30 ml). Finally redissolve the residue in approx 20 ml hot toluene (*ca.* 80 °C). Add hot hexane until the solution becomes turbid (use steam bath to warm solvents). The solution is then cooled to room temperature to give fine orange crystals of the product, which should be isolated by filtration. Record the melting point and submit your sample for ^1H NMR analysis using $CDCl_3$ as the solvent (a few mgs of your product are required). Interpret the spectrum as far as possible. Record the weight and % yield of your product.

Preparation of a Nickel Complex

A solution of the bis(diphenylphosphino)ferrocene (dppf) (0.55 g, 1×10^{-3} mol) in methylated spirits (20–30 cm^3) is placed in a round-bottomed 2-necked flask equipped with a reflux condenser, a nitrogen inlet (placed on top of the condenser), and a magnetic stirrer. The solution is then brought to a gentle reflux and $NiCl_2 \cdot 6H_2O$ (0.20 g, 8.5×10^{-4} mol), dissolved in hot methylated spirits (15 ml), is added. A fine green precipitate will form immediately. This can be separated by filtration while the solution is still warm and washed with cold methylated spirits, then air dried. Record the weight and % yield of the product.

b) The stannyl reaction: bis(tri-n-butylstannyl)ferrocene

Add 30 ml dichloromethane, then separate the organic layer and discard the aqueous fraction. Dry the organic fraction over anhydrous $MgSO_4$ (~5 g) for 20 min, or overnight if necessary. Filter off the $MgSO_4$ and wash with 10 ml dichloromethane to ensure the product is fully removed from it (you should observe the leaching of the yellow colour). Concentrate the organic fraction to a few ml using the rotary evaporator. Record the crude yield of product which is an oil.

Make up a column packed with neutral alumina (approx. 100 g) by making a slurry of the alumina in hexane or hexanes in a beaker and fill the column slowly with this. (Some workers may try a dry packed column – ask the demonstrator which you should use.)

Allow the solvent volume to run down to just above the top of the alumina, then apply a little (1.0 g) of the crude product in a small quantity of hexane. (If you use a dry packed column, the reaction mixture (in a little hexane) can be applied directly to the top of the column.) Let this run into the column, then add a few ml more of hexane. Again allow this to run into the column, then fill the column with the eluting solvent (hexane). Collect the product bis(tri-*n*-butylstannyl)ferrocene as the first pale-yellow/orange fraction to elute from the column. Obtain the ^1H NMR spectrum of this and any subsequent products and access their purity. Store the product sample(s) in a dark bottle(s) under nitrogen. If your product is not pure, you will observe peaks in the spectrum attributable to ferrocene and mono(tri-*n*-butylstannyl)ferrocene. Assign all the peaks in the ferrocenyl region of the spectrum (3.5–5.0 ppm) and estimate the product purity.

c) Reaction with dimethylformamide: ferrocene-bis-carboxaldehyde

Add a further 30 cm^3 of water to the dichloromethane reaction mixture (60 cm^3). The dark red organic layer should then be separated (separatory funnel) and flash chromatographed through a plug of alumina in a Büchner funnel, the alumina being washed with more dichloromethane to ensure your product is fully extracted. Dry the organic layer over anhydrous magnesium sulphate (30 min minimum time), filter and reduce the volume until a precipitate is observed. [If an oil is observed add a small quantity of diethyl ether to the oil and place the solution in a freezer in a sealed container to crystallise.] Cool the solution by placing the product in a stoppered flask in the refrigerator or freezer. This should give dark red-brown crystals which can be filtered off, washed with ether and dried under vacuum. Obtain a ^1H NMR spectrum of your product (in $CDCl_3$) and interpret the data. Record your product yield.

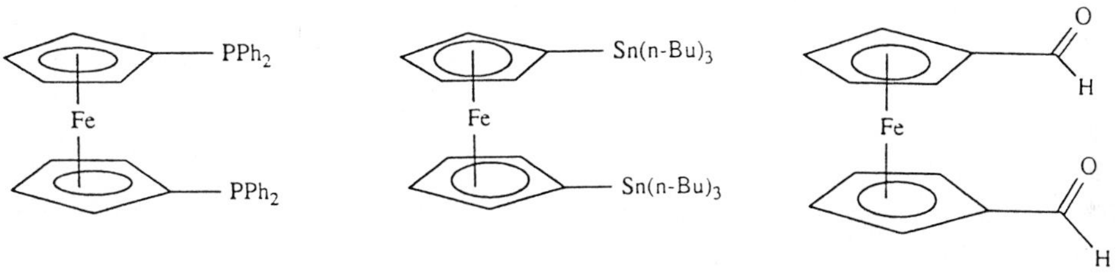

In addition to the ^1H NMR spectra of the products, decide which other spectroscopic or physical techniques would be useful to carry out on your product(s), *e.g.* IR for the ferrocene-bis-carboxaldehyde, magnetic susceptibility for the nickel complex of the bis(diphenylphosphino)ferrocene, mass spectroscopy, etc. If the appropriate resources are available, carry out the respective analyses or obtain the data if the technique you choose is not "hands on".

Exercises

1. What is the easiest method of assessing the purity of your samples by ^1H NMR?
2. In the experiment, you have used *n*-butyllithium solutions. It would have been possible to use a *t*-butyllithium solution equally well from a chemical standpoint. Explain why this particular reagent would not be a good choice from a safety point of view and give reasons to support your argument.
3. In each of the three preparations, both of the cyclopentadienyl rings have been substituted. Why does disubstitution of one ring not occur?
4. What are the possible coordination geometries of the nickel complex that you have prepared? Of these, which is most likely? How could you determine the correct geometry?
5. From a commercial point of view, the compounds which you have prepared are a very saleable commodity. If you had to sell your product, how much would be a reasonable charge for the compound you have made? (Base your answer on a discussion with several others in the class taking into account the cost of reagents, solvents, labour, etc.).
6. Can you draw a reaction mechanism to account for the formation of the bis-aldehyde product?

References

D. F. Shriver, *The Manipulation of Air-Sensitive Compounds,* McGraw-Hill Book Company, New York, **1969.**

Lithiation

M. D. Rausch, D. J. Clappenelli, *J. Organomet. Chem.* **1967,** *10,* 127.

M. D. Rausch, G. A. Moser, C. F. Meade, *J. Organomet. Chem.* **1973,** *51,* 1.

I. R. Butler, W. R. Cullen, J. Ni, S. Rettig, *Organometallics* **1984,** *4,* 2196.

Phosphine complexes

A. W. Rudie, D. W. Lichtenberg, M. L. Katcher, A. Davison, *Inorg. Chem.* **1978,** *17,* 2859

I. R. Butler, W. R. Cullen, T.-J. Kim, S. J. Retting, J. Trotter, *Organometallics* **1985,** *4,* 972.

Tin compounds

M. E. Wright, *Organometallics* **1990,** *9,* 853.

I. R. Butler, S. B. Wilkes, S. J. McDonald, L. J. Hobson, A. Taralp, C. P. Wilde, *Polyhedron* **1993,** *12,* 129.

Bis aldehyde

U. T. Mueller-Westerhoff, Z. Yang, G. Ingram, *J. Organomet. Chem.* **1993,** *463,* 163.

General

I. R. Butler, W. R. Cullen, *Organometallics* **1986,** *5,* 2537.

3.3 Complexes of π-Bonding Arene Ligands

Andrew W. G. Platt

Special Safety Precautions

Compounds such as $FeCl_3$ and $AlCl_3$ are extremely corrosive. Avoid all contact with the skin and wash any spillages immediately with cold water. Both mesitylene and cyclohexane are flammable materials, do not carry out the reflux using a naked flame.

The first complexes of neutral aromatic hydrocarbons were synthesised in 1919 by the reaction of phenyl magnesium bromide with anhydrous chromium chloride. At the time, it was assumed that the product was a polyphenyl chromium compound, and it was only in 1954 that the true nature of the complex as bis-benzene chromium(0), **1**, was deduced.

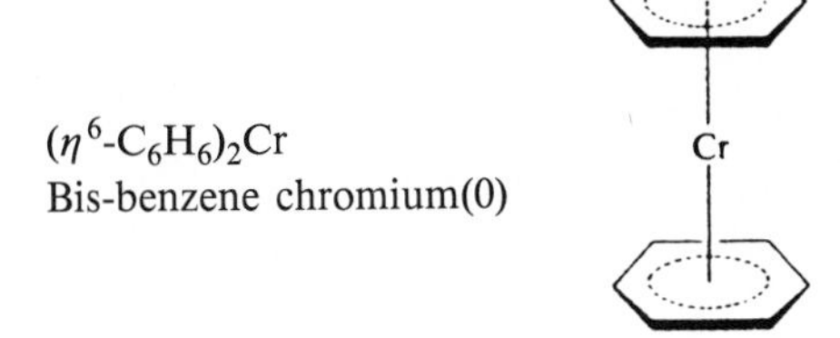

$(\eta^6\text{-}C_6H_6)_2Cr$
Bis-benzene chromium(0)

A more general synthesis for this type of compound was then devised as outlined in Eqn. (1).

$$3\,CrCl_3 + Al + AlCl_3 + 6\,C_6H_6 \rightarrow [Cr(\eta^6\text{-}C_6H_6)_2]^+\ AlCl_4^- \quad (1)$$

The reduction of the cation gives the neutral chromium(0) complex.

This general method can be extended to other metals and to other aromatic hydrocarbons. For example, $(\eta^6\text{-arene})_2M$ complexes are known for M=Ti, V, Nb, Cr, Mo and W. Cationic derivatives such as $Fe(arene)_2^{2+}$ and $Mn(arene)_2^+$ are common and isoelectronic with the zero valent chromium systems.

Experimental

Bis-mesitylene iron(II) hexafluorophosphate, $[(\eta^6\text{-}C_6H_3(CH_3)_3)_2Fe^{2+}][PF_6^-]_2$

The reaction of mesitylene (1,3,5-trimethylbenzene) with anhydrous iron(II) chloride and aluminium chloride produces the bis-mesitylene cation as the $AlCl_4^-$ salt which is then readily converted into the hexafluorophosphate. However, it is more convenient to prepare the complex according to Eqn. (2). This is then converted into the PF_6^- salt which is easier to handle.

In this preparation, several of the reagents are sensitive to moisture. The reaction can be left at any of the points indicated with a * until the next laboratory session as long as the flask is stoppered to exclude atmospheric moisture.

$$3\,FeCl_3 + Al + 5\,AlCl_3 + 6\,mesitylene \rightarrow 3\,[Fe(mesitylene)_2][AlCl_4]_2 \quad (2)$$

Note that as both $FeCl_3$ and $AlCl_3$ are hygroscopic, the time that these reagents are exposed to the air is best kept to a minimum. Plan your work accordingly and make sure that the lids are replaced on the reagent bottles immediately. The aluminum chloride should be yellow in colour. If the sample contains grey or white material, it is likely that some hydrolysis has occured.

Weigh out about 2.5 g of anhydrous $FeCl_3$ (the exact amount is not important so long as the weight taken is known). Place in an "quickfit" round-bottomed flask equipped with a magnetic stirrer bar, and cover with about 30 cm^3 of cyclohexane. Add the required amount of $AlCl_3$ and Al powder and a slight excess of mesitylene, use Eqn. (2) to calculate the amounts. (Note that it is best to finely powder the $AlCl_3$ in a dry mortar and pestle before addition to the reaction flask). *

Stir the mixture vigorously whilst heating to reflux on an oil bath for about 2 h. Cool the mixture to room temperature * and carefully decant the solvent from the solid product. To the remaining solid add a slurry of KPF_6 in water. Stir the mixture vigorously, making sure that all the dark coloured material is dislodged from the sides of the reaction flask. (You may need to scrape off any obstinate pieces with a spatula). Filter the mixture and wash with water and then a little ethanol and dry at the pump to leave a pale orange-tan coloured solid. Record the weight obtained and calculate the percentage yield. Record the infrared spectrum of the product (as a KBr disc), and of mesitylene for comparison.

Exercises

1. Briefly describe the function of the reagents used in the preparation.
2. Discuss the bonding of mesitylene to the iron and show how the infrared spectrum of the product supports this theory.
3. In what way would the bonding in this compound be likely to differ from that in the isoelectronic chromium(0) compounds.
4. Bearing the answer to 3 in mind, arrange the following in the order of expected metal to ligand bond strengths for iron(II): benzene, 1,3,5-trimethoxybenzene and hexacyano benzene (think of the electronic properties of the substituents).

3.4 Organotin Chemistry

Kieran C. Molloy and Timothy T. Paget

Special Safety Precautions

1. Diethyl ether is highly flammable. Keep away from flames.
2. Carbon tetrachloride is toxic by inhalation and contact.
3. Bromine is toxic by inhalation and causes severe burns to eyes and skin. Use only in a fume cupboard. Wear rubber gloves at all times.
4. All mercury compounds are potentially poisonous, particularly Hg(II). Solid residues should be collected and stored in a designated waste container. Avoid inhalation of powder.
5. Vacuum distillations should be carried out in a fume cupboard and behind a safety screen. The distillation flask should be allowed to cool before air is admitted to the apparatus.

Organotin compounds have been widely used as PVC stabilisers, agrochemicals, wood preservatives, anti-fouling paints and precursors for the chemical vapour deposition (CVD) of electrically conducting films of SnO_2. Recent work has shown that certain di-organotin compounds show anti-tumour activities in excess of that of *cis*-platinum. Our understanding of organotin compounds is aided by the wide variety of physical methods which can be applied to their study. ^{119}Sn (8.6% abundance) has $I = 1/2$, and has an NMR chemical shift range of *ca.* 1000 ppm. In addition, couplings to both this nucleus and the less abundant ^{117}Sn (7.6%, $I = 1/2$) can be seen in the corresponding ^{1}H and ^{13}C NMR spectra, and provide valuable information about the coordination sphere about the metal. ^{119}Sn is Mössbauer active, and is the most widely studied Mössbauer nucleus after ^{57}Fe. Finally, with ten stable isotopes, the largest number for any element in the Periodic Table, mass spectral fragments show characteristic isotopic distribution patterns making assignments relatively facile.

In this experiment, you will first prepare a tetra-organotin, [Ph_3MeSn], using a Grignard reagent and then convert this compound to an organotin halide (either $Ph(Me)SnCl_2$ or $Ph(Me)SnBr_2$), the usual starting materials for the synthesis of functionalised organotins, by the selective electrophilic cleavage of Sn–C (phenyl) bonds, and finally use this latter derivative as a precursor for the synthesis of an organotin heterocycle, Ph(Me)SnS.

$$Ph_3SnCl + MeMgBr \rightarrow Ph_3MeSn + MgBrCl$$

$$Ph_3MeSn + 2Br_2 \rightarrow Ph(Me)SnBr_2 + 2PhBr$$

$$Ph_3MeSn + 2HgCl_2 \rightarrow Ph(Me)SnCl_2 + 2PhHgCl$$

$$Ph(Me)SnX_2 + Na_2S \rightarrow Ph(Me)SnS + 2NaX$$

$$X = Cl, Br$$

a) Methyltriphenylstannane, $CH_3(C_6H_5)_3Sn$

A 3-necked 250 ml oven-dried round-bottomed flask, fitted with reflux condenser and $CaCl_2$ guard tube, and a 150 ml pressure-equalising dropping funnel, is charged with 0.51 g (21 mmol) of magnesium. The metal is just covered with dry diethyl ether and then a few drops of methyl iodide are added. The formation of the Grignard reagent can be initiated by the addition of a crystal of iodine if necessary. When Grignard formation has commenced, a further 25 ml of solvent is added to the reaction mixture, and a solution of methyl iodide (3.00 g, 21 mmol) added dropwise from the funnel at a rate which maintains solvent reflux. When Grignard formation is complete (*ca.* 1 hour), a solution of chlorotriphenylstannane (5.79 g, 15 mmol) in 100 ml ether is added dropwise from the funnel. When the addition is complete, the mixture is refluxed for 1 h. After allowing the solution to cool, water is added slowly to decompose the remaining Grigard reagent.

The reaction mixture is then transferred to a separatory funnel, the organic layer isolated and dried over anhydrous magnesium sulphate. After filtering the inorganic salts, the ether is removed on a rotary evaporator, and the residue recrystallised from ethanol. Record the yield, melting point and ^1H NMR spectrum of your product. Interpret the Mössbauer spectrum of the product, shown in Figure 3.4-1.

b) Methylphenyldibromostannane, $CH_3(C_6H_5)SnBr_2$

To a stirred solution of Ph_3MeSn (2.00 g, 5.5 mmol) in 25 ml carbon tetrachloride (100 ml round-bottomed flask) add dropwise a solution of bromine (1.75 g, 11 mmol) in the same solvent (25 ml) over a 30 min period. Continue stirring for at least 1 h (preferrably overnight), during which time the colouration due to the halogen will have largely disappeared. The solvent and bromobenzene are removed on a rotary evaporator, leaving the product as a yellow oil which can be purified by vacuum distillation. Record the yield and ^1H NMR spectrum of the product. Interpret the mass spectrum of the product shown in Figure 3.4-2, using the isotope distribution patterns shown in Figure 3.4-5. Account for the appearance of the Mössbauer spectrum of $Me(Ph)SnCl_2$ shown in Figure 3.4-4.

c) Methylphenyldichlorostannane, $CH_3(C_6H_5)SnCl_2$

To a solution of Ph_3MeSn (3.1 g, 8.4 mmol) in acetone (30 ml), cooled in an ice bath, add 4.62 g (17.0 mmol) of mercury(II) chloride in small portions. After addition of the solid, remove the ice bath and allow the suspension to continue stirring for at least 1 h (preferably overnight). The solvent is then distilled on a rotary evaporator, and the residue stirred for 10 min with 40 ml petroleum ether (60/80). Phenylmercuric chloride is separated by filtration. Solvent removal from the filtrate as before yields the product as a light yellow oil. A white solid is obtained after purification of this material by vacuum distillation. Record the yield, melting point and ^1H NMR spectrum of the product. Interpret the mass spectrum of the product shown in Figure 3.4-3 using the isotope distribution patterns shown in Figure 3.4-5. Account for the Mössbauer spectrum of the product shown in Figure 3.4-4.

d) Methylphenyltin sulphide, $CH_3(C_6H_5)SnS$

A solution of **either** $Me(Ph)SnBr_2$ (0.93 g, 2.5 mmol) **or** $Me(Ph)SnCl_2$ (0.71 g, 2.5 mmol) in ethanol (15 ml) is added dropwise to a suspension of $Na_2S \cdot 9H_2O$ (0.90 g, 3.7 mmol) in the same solvent (15 ml). After refluxing the mixture for 2 h, the solution should be cooled and evaporated to dryness on a rotary evaporator. Ether (30 ml) and water (30 ml) are added to the solids, and the mixture shaken in a separatory funnel. The ether layer is separated, dried over anhydrous $MgSO_4$, filtered and the solvent evaporated *in vacuo* to yield the product as a pale-yellow oil. Record the ^1H NMR of the product and identify the products formed by analysis of the Sn – Me signals.

References

A. G. Davies, P. J. Smith in *Comprehensive Organometallic Chemistry* (Eds.: G. Wilkinson, F. G. A. Stone, E. W. Abel), Pergamon Press, Oxford, **1982**, p. 519.

M. Gielen, P. Lelieveld, D. de Vos, R. Wittem in *Metal Based Anti Tumor Drugs* (Ed.: M. Gielen), Freund Publishing House, Tel Aviv, **1992**, p. 29.

B. Wrackmeyer, *Chem. Brit.,* **1990** *26,* 48.

P. G. Harrison, *Chemistry of Tin,* Blackie, Glasgow, **1989.**

D. B. Chambers, F. Glocking, M. Weston, *J. Chem. Soc. (A)* **1967,** 1759.

K. C. Molloy, *Adv. Organomet. Chem.* **1991,** *33,* 196.

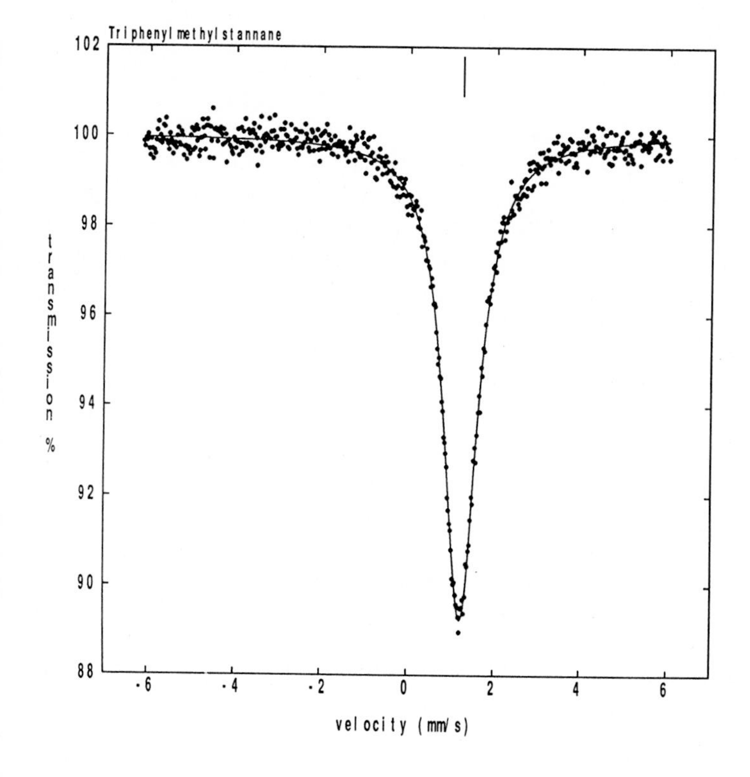

Figure 3.4-1. Mössbauer spectrum of $Ph_3(Me)Sn$.

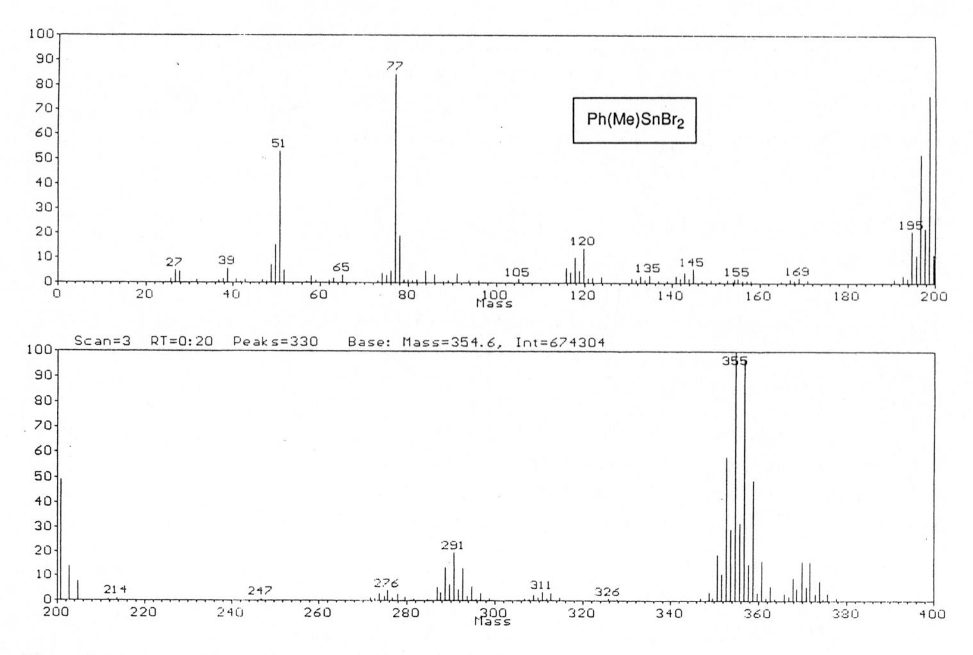

Figure 3.4-2. Mass spectrum of Ph(Me)$SnBr_2$.

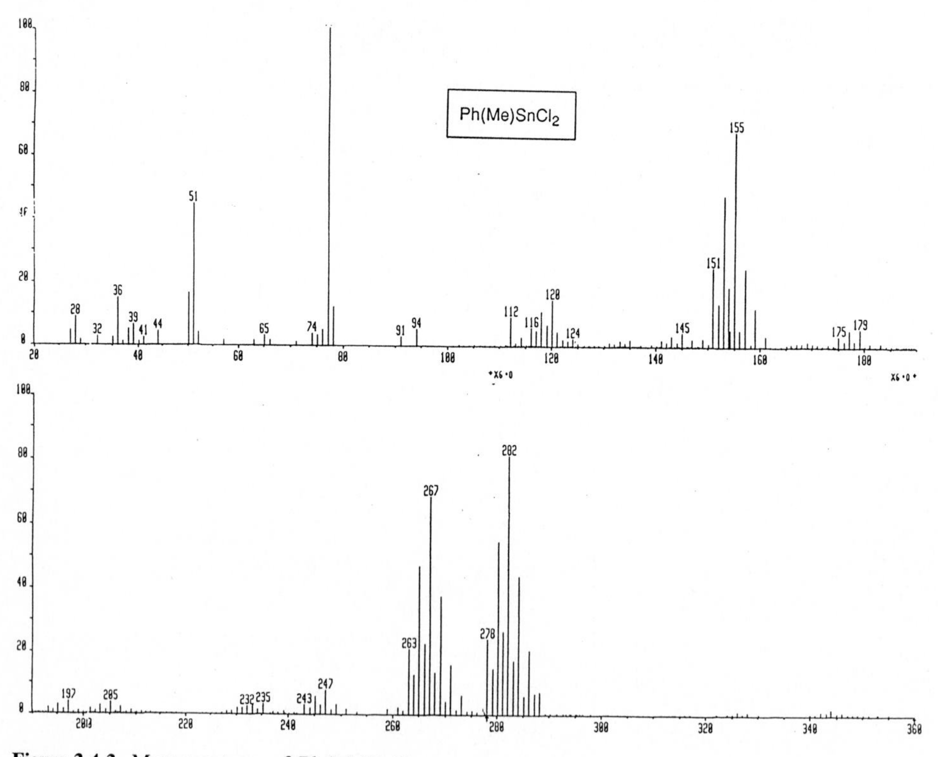

Figure 3.4-3. Mass spectrum of Ph(Me)$SnCl_2$.

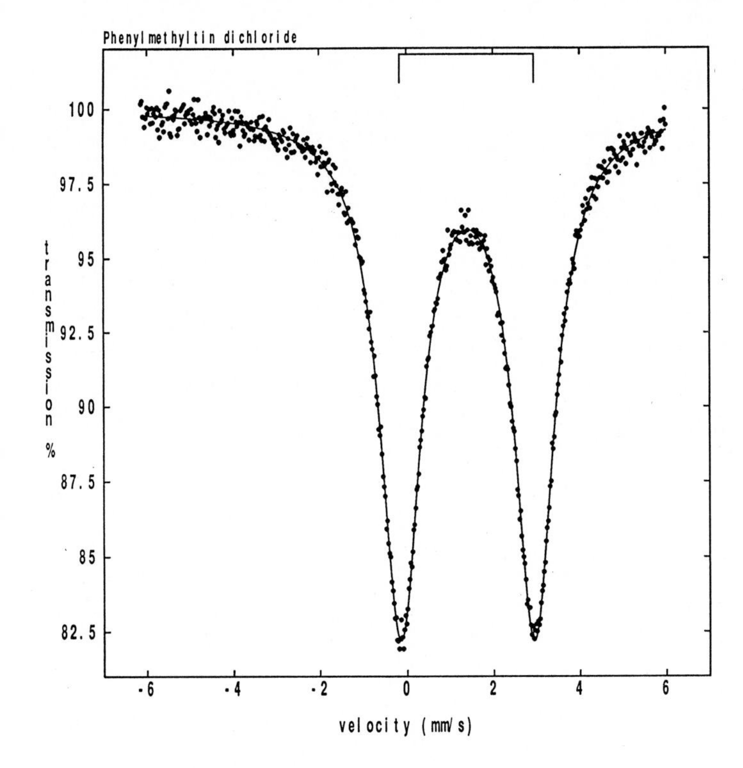

Figure 3.4-4. Mössbauer spectrum of $Ph(Me)SnCl_2$.

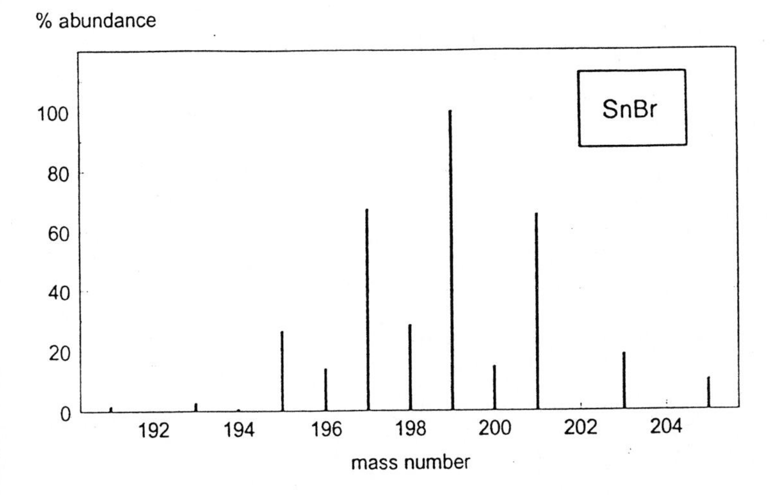

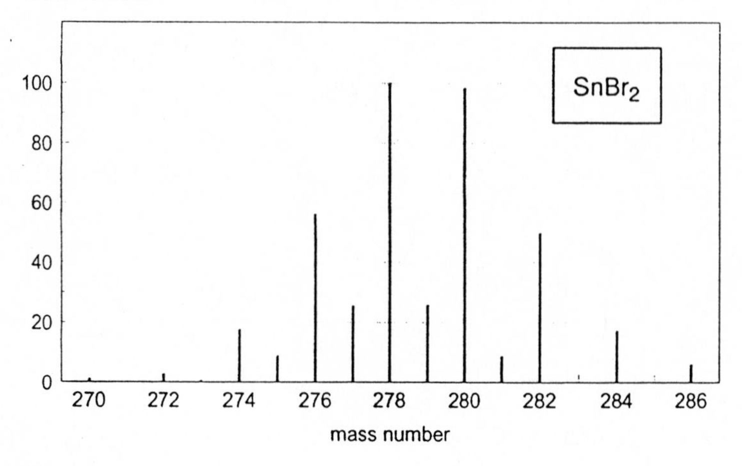
% abundance
100
80
60
40
20
0
270
272
274
276
278
280
282
284
286
mass number
SnBr2

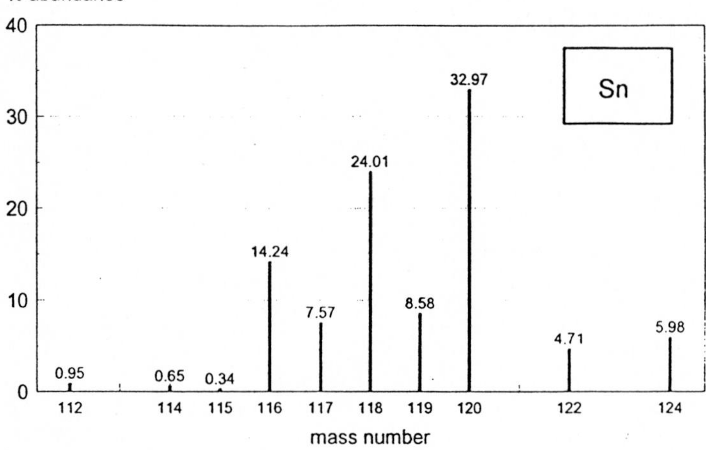
% abundance
40
30
20
10
0
0.95
0.65
0.34
14.24
7.57
24.01
8.58
32.97
4.71
5.98
112
114
115
116
117
118
119
120
122
124
mass number
Sn

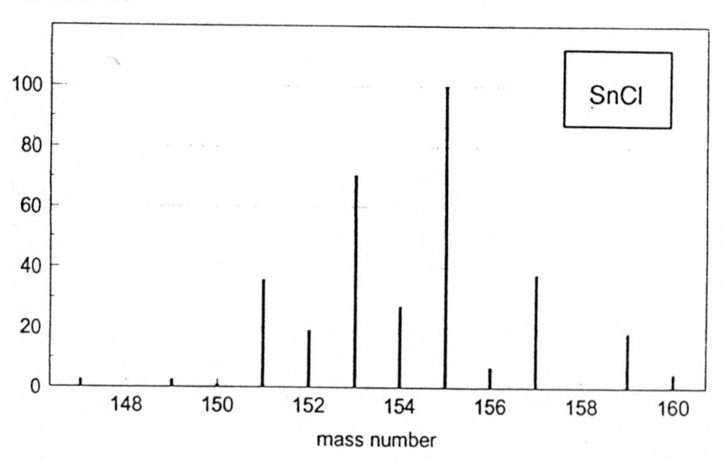
% abundance
100
80
60
40
20
0
148
150
152
154
156
158
160
mass number
SnCl

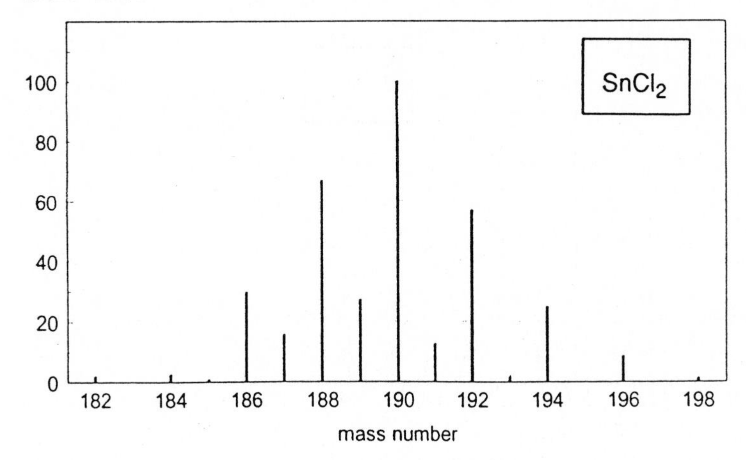

Figure 3.4-5. Isotope distribution patterns for Sn, SnX and SnX_2 (X = Br, Cl).

3.5 Nickel Catalysed Cross-Coupling of Alkylmagnesium with Haloarene

G. Brent Young

Special Safety Precautions

All halocarbons, phosphines and nickel compounds should be regarded as irritant and toxic. Wear gloves and avoid inhalation. Diethyl ether is highly flammable. Organomagnesium solutions are caustic and may be flammable and toxic. Avoid ingestion.

The metal mediated formation of carbon–carbon bonds is an important aspect of organic synthesis. In this experiment, selective coupling of two precursors can be achieved by the formation and rearrangement of labile organonickel species participating in a catalytic cycle. Organonickel complexes, though, may have a variety of rearrangement options which may depend, for example, on the nature of the ancillary ligands (phosphines, in this case) and which may lead to loss of regioselectivity. You will make and study two dichlorobis(phosphine)nickel(II) complexes and examine their suitability as catalysts for coupling of 2-propylmagnesium chloride with chlorobenzene. Most of the manipulations must be carried out with strict exclusion of oxygen, and this experiment is intended as a preparation for sophisticated anaerobic handling techniques. *It is especially important* to read the experimental procedure throughly before embarking on this experiment. Careful planning is vital to smooth operation.

Experimental

a) The dichloro[1,2-bis(diphenylphosphino)ethane]nickel(II) catalyst

Equip a 100 cm^3 round-bottomed flask with a reflux condenser and magnetic stirrer bar. Add $NiCl_2 \cdot 6H_2O$ (0.71 g) to ethanol (10 cm^3) in the flask and warm the mixture in an oil bath, with stirring, until all the nickel complex has dissolved. Remove the oil bath and carefully add 1,2-bis(diphenylphosphino)ethane (dppe, 1.20 g) to the warm *but not boiling* solution. Precipitation of product should occur immediately. Replace the condenser and restore the oil bath and continue stirring at reflux for two hours. Allow the mixture to cool, filter off the product and wash with ethanol and diethyl ether before drying in a vacuum desiccator. Record the yield.

b) The dichlorobis(triphenylphosphine)nickel(II) catalyst

$NiCl_2 \cdot 6H_2O$ (1.5 g) is dissolved in water (1.5 cm^3) and added to glacial acetic acid (30 cm^3). To this solution is added a suspension/solution of triphenylphosphine (3.25 g) in glacial acetic acid (15 cm^3) and the mixture stirred until all the phosphine has dissolved (*ca.* 1 hour). Keep the precipitate in contact with the solution for *ca.* 24 hours, filter and wash thoroughly with diethyl ether. Dry at 70 °C under vacuum for 2–3 hours.

c) 2-(Chloromagnesio)propane

All glassware should be oven dried (but do not place plastic stopcock keys in ovens). Equip a 250 cm^3 2- (or 3-) necked flask with reflux condenser, pressure-equalising addition funnel and magnetic stirrer bar. A glass T-piece adaptor should be fitted to the top of the reflux condenser, one arm connected to the nitrogen supply and the other to a mineral oil (Nujol) bubbler (a Dreschel bottle is best). Any remaining openings can be closed by glass stoppers. The whole apparatus can be flushed with nitrogen by removing the stopper(s) for *ca.* 2 minutes while a brisk current of nitrogen is flowing.

Magnesium turnings (8.3 g) are placed in the flask and dry diethyl ether (45 cm^3) is added. (Nitrogen is kept flowing during addition). A solution of 2-chloropropane (9.5 cm^3) in dry ether (40 cm^3) is prepared and thoroughly mixed in the addition funnel, and a small quantity (*ca.* 5 cm^3) is added to the magnesium. Reaction is initiated by introduction of a few drops of 1,2-dibromoethane. Once effervesence is observed, the chloropropane solution is added at a rate sufficient to maintain a gentle reflux. During reflux, the nitrogen supply should be reduced to a slow steady flow. Continue stirring for 15–30 minutes after addition is complete. Increase the nitrogen flow, remove the additon funnel and replace it by a rubber ("subaseal") septum.

Immediately before employing any organomagnesium reagent it should be standard practice to estimate its concentration. Using a 5 cm^3 syringe with a narrow gauge 18 cm needle, withdraw exactly 2 cm^3 of the solution and inject into distilled water (20 cm^3) in a 100 cm^3 conical flask. Add a few drops of a suitable indicator (phenolphthalein or bromothymol blue) and titrate against 0.1 M aqueous HCl. Hence, estimate the concentration of RMgCl, assuming that OH^- arises only from Mg–C hydrolysis. Dismantle and clean the syringe (see next section).

d) Catalysed cross-couplings

This apparatus should be set up during preparation of the organomagnesium reagent. All glassware should be oven dried. Equip each of two 100 cm^3 3-necked round-bottomed flasks with a pressure-equalising addition funnel, a reflux condenser and a magnetic stirrer bar. A plug of glass wool should be wedged (not *too* tightly) in the outlet tube from the reservoir of each funnel, above the stopcock. Each condenser should be fitted with a glass T-piece adaptor, one arm of which is connected to the nitrogen supply and the other to a mineral oil bubbler vent (the two assemblies can be joined "in-series").

Fit "subaseal" septa to both additon funnels and close any other openings with glass stoppers. Make sure the whole apparatus is thoroughly flushed with briskly flowing nitrogen (the "subaseals" should each be removed for *ca.* 1 minute to allow flushing of the funnel).

To each flask, add chlorobenzene (4.15 cm^3) and diethyl ether (10 cm^3). Into one flask, place (dppe)$NiCl_2$ (0.36 g) and into the other $(Ph_3P)_2NiCl_2$ (0.44 g). Stir both mixtures and cool in ice.

After determination of the concentration of *i*PrMgCl in ether, introduce equal portions of this solution (40×10^{-3} moles) to each addition funnel, using a syringe with a wide bore 18 cm needle. Try not to draw up large solid particles in the syringe. *Care* should also be taken at this stage *not* to *i)* inject yourself or anyone else with the solution and *ii)* spill the solution, especially on your skin. Spillage is best avoided by measuring the correct amount into the syringe by ejecting gas bubbles back into the source flask with the syringe pointing upwards, and then drawing enough nitrogen into the syringe to fill the needle and introduce a little gas above the solution. The needle is now removed from the source flask and inserted into the addition funnel, and then, with the barrel still pointing up, the gas is expelled first, followed by the liquid. (Ensure that the stopcock on the funnel is already closed). Draw some gas back into the syringe and remove the needle from the septum. The procedure (which is standard practice for anaerobic manipulation) is repeated for the other funnel. If in *any* doubt, *ask a demonstrator.* As soon as transfer is complete, take the syringe to a fume cupboard and *remove the plunger.* Leave the assembly for 5–10 minutes, then wash both components with water, followed by IMS (or methanol).

Add the organomagnesium solution – cautiously at first in case of an exothermic reaction – to the cooled, stirred solution of PhCl. The glass wool will act as a filter for any suspended solids. When addition is complete, remove the ice bath. Dry the exteriors of the flasks and substitute oil baths. Heat the mixtures at gentle reflux with stirring for 4 hours. During reflux, reduce the nitrogen flow to a slow steady trickle (*ca.* one bubble per second).

Unused organomagnesium solution must be quenched by cooling and slow (at first) addition of water (the hydrolysis is exothermic). Unreacted magnesium metal can be destroyed by introduction of dilute aqueous HCl. (fume cupboard).

e) Analysis of cross-coupling reactions

After cooling the reactions in ice baths (be careful to turn up the nitrogen supply during cooling so as to avoid sucking the bubbler oil back into the flasks), water (20 cm^3) is introduced first into the addition funnels *via* syringe, then into the reaction mixtures, slowly at first, to quench any unreacted *i*PrMgCl. More ether (20–30 cm^3) can now be added to aid in separation. Anaerobic handling is no longer necessary. Each reaction flask is treated identically. The contents are poured into a separatory funnel (filtration may be necessary to remove excessive amounts of solid) and the organic layer separated. The aqueous phase is shaken with fresh diethyl ether (25 cm^3) and the ether fractions combined and washed successively with saturated aqueous sodium thiosulphate (2×20 cm^3) and distilled water (2×20 cm^3). The organic phase is then dried over anhydrous $MgSO_4$. Relative quantities of propylbenzenes and benzene as well as any residual chlorobenzene can now be analysed by glc using an OV-11 (or related silicone gum) column. Set the injection port and detector temperatures to 150 °C and the column oven to 90 °C. Adjust the mobile phase (nitrogen) gas pressure to 30–35 psi. Identify components by comparison of their retention times with those in a standard mixture (provided). Integrate the peak areas by cutting them from the chart paper and weighing them on an analytical balance. (Two chromatograms for each sample will be necessary so that one of each can be included in your report). Tabulate your results. (Assume that the detector response-factor for each of your components is 1.0; a flame ionisation detector responds to the *mass* of each solute present).

Exercises

1. Comment on the notable colour difference between the two catalysts. How could you verify your explanation?
2. Estimate a) the relative extent of overall coupling and b) the relative extent of regioretention for the two catalysts, noting any other reactions which are evident.

3. Suggest a plausible catalytic cycle for the cross-coupling reaction, explaining, by way of a suitable scheme, the nature of each step (the reference literature will help).
4. Suggest why the two catalysts differ in their activity, particularly with respect to the regioselectivity of the coupling reaction. Give a mechanistic explanation for the observed rearrangement.

References

M. Kumada, *et al.*, a) *J. Am. Chem. Soc.* **1972,** *94,* 4374; b) *idem., ibid.* **1972,** *94,* 9268; c) *idem., Pure and Applied Chem.* **1980,** *52,* 669.

H. Felkin, G. Swierczewski, *Tetrahedron* **1975,** *31,* 2735; R. J. P. Corrin, J. P. R. Masse, B. Mennier, *J. Organomet. Chem.* **1973,** *55,* 73.

3.6 Transition Metal Catalysis

Neil M. Boag

Transition metal complexes may be used as catalysts for a wide variety of organic syntheses. In this experiment, a catalyst precursor is prepared and used to prepare an alkyne.

Diarylacetylenes may be prepared by a variety of routes. In recent years, two methods based on transition metals have been developed. The first method devised by Castro involves the use of copper acetylides. These reagents are prepared by treatment of an aqueous ammonia solution of cuprous iodide with an ethanol solution of mono-substituted aryl acetylenes (Eqn. 1).

$$n\text{CuI} + n\text{ArC}{\equiv}\text{CH} \rightarrow (\text{CuC}{\equiv}\text{CAr})_n + n\text{HI} \qquad (1)$$

Treatment of these cuprous acetylides with aryl iodides in refluxing pyridine under nitrogen yields diaryl acetylenes stoichiometrically (Eqn. 2).

$$(\text{CuC}{\equiv}\text{CAr})_n + n\text{Ar'I} \rightarrow n\text{Ar'C}{\equiv}\text{CAr} + n\text{CuI} \qquad (2)$$

The second method, developed independently by Cassar and Heck, utilises a palladium complex to effect this reaction catalytically (Eqn. 3).

$$\text{ArC}{\equiv}\text{CH} + \text{Ar'X} + \text{NaOMe} \rightarrow \text{ArC}{\equiv}\text{CAr'} + \text{NaX} + \text{HOMe} \qquad (3)$$

The early procedure has been modified through the use of amines as solvents with CuI as a co-catalyst. These modifications allow the reaction to proceed at room temperature and has enabled the introduction of alkyl substituents. The mild conditions have resulted in this method being used for the introduction of carbon-carbon triple bonds during the synthesis of natural products.

In this experiment, a catalytic precursor is prepared which may be identified. This precursor is then be used to catalytically synthesise an alkyne using the conditions devised by Cassar. Under these conditions, the precursor generates the active species $Pd(PPh_3)_4$.

Experimental

a) The catalytic precursor

Note: It is important to determine the yield of product as accurately as possible (Exercise 1). Add water (5 cm^3) to a mixture of palladous chloride (0.15 g) and sodium chloride (0.12 g). Stir the slurry and slowly warm to about 50 °C. While the resultant solution is cooling, prepare a solution of triphenylphosphine (0.5 g) in ethanol (75 cm^3). Add the palladous chloride solution to the triphenylphosphine solution dropwise with stirring using a Pasteur pipette. Be careful not to add any residue which may remain. When the addition is complete, use an extra 1 cm^3 of water to ensure all the palladium salts are transferred. Warm the resultant slurry to about 50 °C over 30 minutes to coagulate the precipitate, cool and filter. Wash well with water, acetone and ether. Dry the product in a 50 °C oven for 30 minutes. Weigh the product.

b) The catalytic reaction

Add the following reagents to a B14 100 cm^3 round-bottomed flask.

4-bromoacetophenone (note 1)	2.00 g
palladium complex	0.42 g
sodium methoxide	0.60 g
triphenylphosphine	0.31 g
phenylacetylene	1.00 g
dimethylformamide (dmf)	40 cm^3
magnetic stirrer bar	

Using a Pasteur pipette, bubble nitrogen through the solution for ten minutes. Attach a condenser to the flask. To the top of the condenser attach a B14 T-piece. Connect one end of the T-piece to a nitrogen cylinder and the other to an outlet bubbler consisting of a Dreschel bottle 1/4 full of water Using a water bath, heat the reaction mixture with stirring to 100 °C and maintain this temperature for four hours. Cool and pour into 200 cm^3 of water. Extract with 3×40 cm^3 of diethyl ether. Wash the ether extracts with 2×30 cm^3 of water and dry with sodium sulphate. Filter and take to dryness on a rotary evaporator. Dissolve the residue in the minimum of hot methanol, add decolourising charcoal and filter. Cool in ice and filter off the product. Wash with a little cold methanol. Air dry.

Exercises

1. Deduce a structure for the catalytic precursor based on the yield of product (which is essentially quantitative) and the fact that only one palladium-chloride stretch is observed in the IR spectrum of this material. What physical measurement could be made to support the structural formulation?
2. Record an IR spectrum of the product of the catalytic reaction as a Nujol mull and identify any important absorption bands.
3. Record a proton NMR spectrum of the product from the catalytic reaction in $CDCl_3$. Integrate the spectrum. (A better result will be obtained if the sample is left overnight or dried in a vacuum dessicator to ensuure complete removal of the recrystallisation solvent which can interfere).
4. A $^{13}C\{^1H\}$ NMR spectrum of the catalytic product is reproduced in Figure **3.6-1.** Use all the spectroscopic information to confirm the identity of this material.
5. Complete the following catalytic cycle. Calculate the electron counts and oxidation states of the transition-metal intermediates.

PhC≡CAr
$Pd(PPh_3)_4$
ArX
a
a
C
A
B
$PhC{\equiv}C^- Na^+$ + MeOH
b
PhC≡CH + $MeO^- Na^+$

6. Based on the yield of product from the catalytic reaction, calculate the number of turnovers obtained.

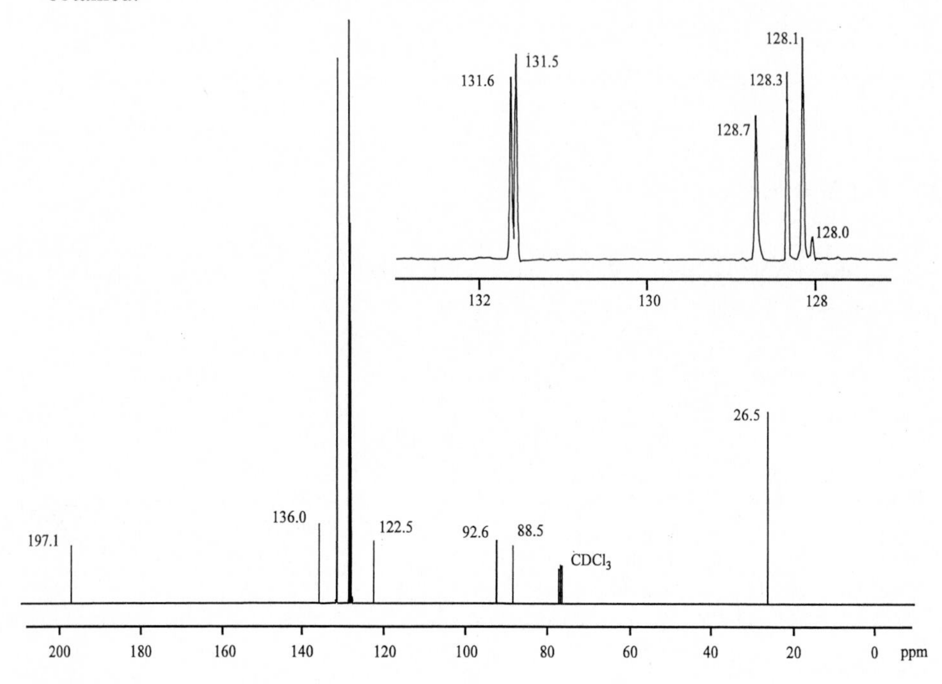

Figure 3.6-1. $^{13}C\{^{1}H\}$ spectrum of catalytic product.

References

R. D. Stephens, C. E. Castro, *J. Org. Chem.* **1963,** *28,* 3313.
L. Cassar, *J. Organomet. Chem.* **1975,** *93,* 253.
H. A. Dieck, R. F. Heck, *J. Organomet. Chem.* **1975,** *93,* 258.
Y. Abe, A. Ohsawa, H. Arai, H. Igeta, *Heterocycles* **1978,** *9,* 1397.

Notes

1. A large variety of alkynes can be prepared (see references by Cassar and Heck for details).
2. Copies of proton, carbon-13 and infrared spectra of the catalytic product are available from the author. These can be supplied as HPGL or Word for Windows files if required.

3.7 Transition Metal-Carbon Bonds in Chemistry and Biology

C. J. Jones

Special Safety Precautions

If any chemicals in this experiment should come in contact with your skin, wash them off immediately with copious amounts of water and then consult a demonstrator.
In the event of spillage in the fume cupboard, consult a demonstrator and, where permitted, flush all the spilled materials down the cupboard drain with copious amounts of water. In the event that formaldehyde solution, pyridine or methanol are spilled outside a fume cupboard, keep others away from the area of the spill and consult a demonstrator. The liquid should be diluted with water, collected and flushed to the drain in a fume cupbard with copious amounts of water. In the event that iodomethane or 2-methoxyethanol is spilled outside a fume cupboard, keep others away, consult a demonstrator and absorb the spillage on an inert absorbent (*e.g.* vermiculite) and remove it to a fume cupboard to be packaged for disposal. Consult a demonstrator about spillage of any other chemicals involved. Skin contamination by, or inhalation of, these materials must be avoided.

Material Hazards

Hexa-aquocobalt(II)chloride, $CoCl_2 \cdot 6H_2O$	Toxic, Irritant
Dimethylglyoxime	Harmful
Ethanol	Harmful, Flammable
Pyridine	Harmful, Flammable
Diethyl ether	Highly Flammable
Dichloromethane	Harmful
[Chloropyridinebis(dimethylglyoximato)cobalt(III)]	Properties unknown, Assume toxic
Sodium borohydride	Toxic, Flammable
Iodomethane	Highly toxic, Cancer suspect agent
Methanol	Toxic, Flammable
Petroleum ether 40/60	Irritant, Flammable
Ruthenium trichloride trihydrate	Harmful by skin absorption, Corrosive
2-Methoxyethanol	Harmful
Triphenylphosphine	Irritant
Potassium hydroxide	Corrosive
40% aqueous formaldehyde solution	Toxic, Cancer suspect agent

Disposal of Wastes

Waste solvents must be disposed of into the containers provided, in the fume cupboard. These must be removed before commencing the next reactions.

The 'classical' coordination chemistry which was developed in the first half of the 20th century was primarily concerned with compounds containing bonds between transition-metal ions and nitrogen, oxygen or halogen donor atoms. In the second half of this century, there has been increasing interest in the use of so-called 'soft ligands' containing donor atoms such as sulphur, phosphorus, arsenic or selenium. Such ligands tend to form complexes with metals in their lower oxidation states; phosphine ligands in particular may often be found in compounds containing carbon monoxide or unsaturated hydrocarbons as co-ligands. To some extent these developments occured in parallel with a large expansion of organometallic chemistry, which is concerned with compounds containing metal-carbon bonds and which has become a major field of research. The study of metal complexes containing alkyl, alkenyl, acyl, carbonyl and hydride ligands has led to the discovery of many novel reactions, some of which have important industrial applications in catalytic processes. These include alkene hydrogenation, alkene isomerisation, alkene metathesis, the hydroformylation of alkenes and the carbonylation of methanol to produce acetic acid.

Although organometallic chemistry may appear to be a rather recent human invention, for millions of years biology has exploited the reactivity of a metal-carbon bond in the vitamin B_{12} coenzyme. Among other things, B_{12} is involved in effecting reactions such as the group migration shown in Scheme **3.7-1.**

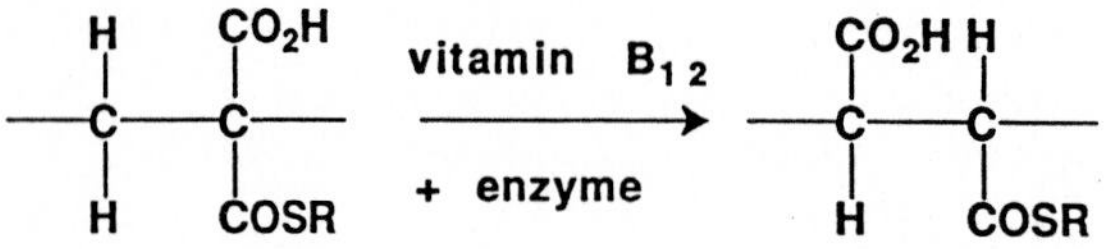

$$-\overset{\mathrm{H}}{\underset{\mathrm{H}}{\mathrm{C}}}-\overset{\mathrm{CO_2H}}{\underset{\mathrm{COSR}}{\mathrm{C}}}- \xrightarrow[\text{+ enzyme}]{\text{vitamin } B_{12}} -\overset{\mathrm{CO_2H}}{\underset{\mathrm{H}}{\mathrm{C}}}-\overset{\mathrm{H}}{\underset{\mathrm{COSR}}{\mathrm{C}}}-$$

Scheme 3.7-1.

This practical exercise is primarily intended to provide experience in compound synthesis and characterisation using spectroscopic methods. However, it also illustrates some chemical aspects of organometallic chemistry which relate to both biological inorganic chemistry and homogenous catalysis. In Part 1 a simple model compound is prepared which recreates the basic coordination environment of cobalt in vitamin B_{12}. In Part 2, two ruthenium carbonyl hydride complexes are prepared using a primary alcohol and formaldehyde as the source of the hydride and carbonyl ligands. In each case, the nature of the reaction products is determined using spectroscopic data.

Part 1

Methylcobaloxime: a model for vitamin B_{12} coenzyme

Although the average human contains only 2 to 5 mg of cobalt, concentrated in the liver, it is an essential trace element whose absence causes serious illness. The cobalt is present as vitamin B_{12}, a coenzyme which, among other roles, is associated with biosynthetic methylation reactions in organisms ranging from man to bacteria. Vitamin B_{12} is necessary for the proper development of red blood cells. It is also the agent responsible for the microbial conversion of environmetal Hg^{2+} to highly toxic CH_3Hg^+ derivatives. These concentrate in fish and have lead to serious outbreaks of poisoning among human populations eating fish from mercury contaminated waters.

Because vitamin B_{12} is present in animals in such small quantities, it proved extremely difficult to isolate. However, in 1957 Hodgkin and her associates reported a successful structure determination on cyanocobalamin, the cyanide derivative of B_{12}, using X-ray diffraction techniques. This structure is shown in Figure **3.7-1** and consists of two main components. The first is a planar tetradentate macrocyclic ligand, similar in some ways to the porphyrin unit found complexing the iron in haemoglobin. This highly substituted macrocyclic ring chelates the cobalt atom via four nitrogen atoms and, in the deprotonated form, is formally anionic. Attached to this ring via a sugar-phosphate linkage is the second component, a benzimidazole base, which also bonds to the cobalt. The sixth position in the octahedral coordination sphere of the cobalt is occupied by a cyanide ion. However, the active coenzyme would normally have either a water molecule or a CH_2 group from a sugar residue occupying this site.

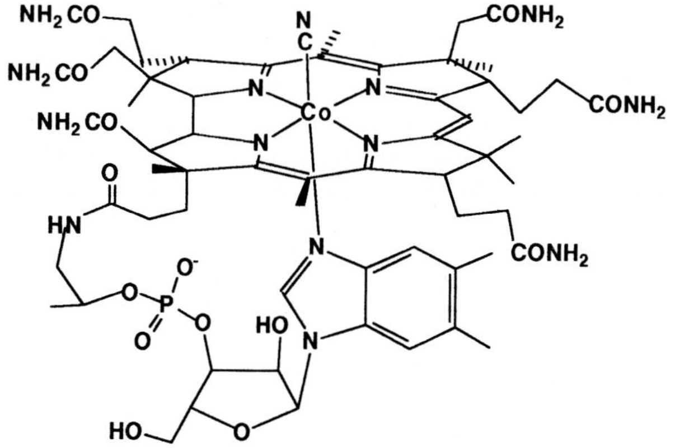

Figure 3.7-1. Cyanocobalamin – the cyano derivative of vitamin B_{12}.

The coordination sphere of cobalt in vitamin B_{12} is rather similar to that in cyanopyridine-bis(dimethylglyoximato)cobalt(III), *i.e.* cyanopyridine cobaloxime, whose structure is shown in Figure **3.7-2.** This much simpler molecule may easily be synthesised in the laboratory and provides a chemical model for vitamin B_{12}. Instead of working with small amounts of expensive coenzyme, it is possible to study the chemistry of cobalt in a similar, though not identical, environment using cobaloximes as models.

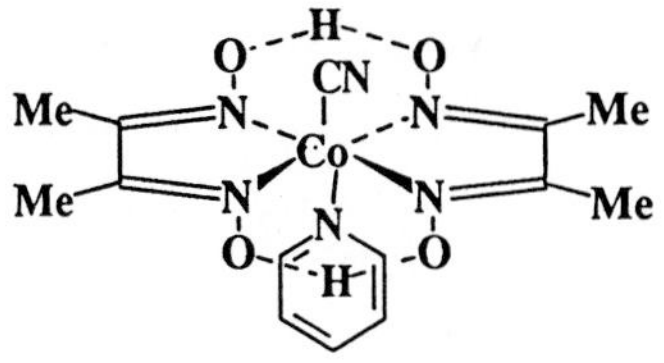

Figure 3.7-2. Cyanopyridinecobaloxime or cyanopyridinebis(dimethylglyoximato)cobalt(III).

Using this model system, it is possible to demonstrate that reduction of Co(III) to Co(I) produces a powerful nucleophilic metal centre, which may be methylated using CH_3I to give an analogue of the methylcobalamin involved in the biomethylation of mercury as mentioned earlier. The reactivity of the Co(I) model complex may be contrasted with the inert behaviour of the Co(III) complex, which may only be methylated with difficulty using CH_3MgI. The reactions of vitamin B_{12} (VB_{12}), copied using the cobaloxime model, may be summarised as follows:

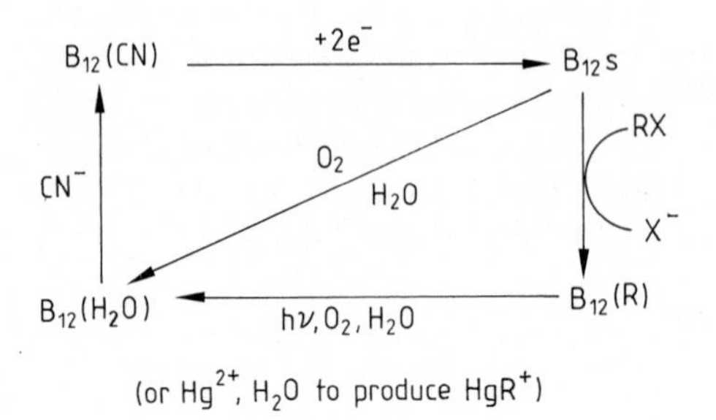

Following the procedure described in the experimental section, prepare a sample of chloropyridinecobaloxime. This material is then reduced and reacted with iodomethane to give methylpyridinecobaloxime, a model for the cobalt alkyl moiety in the vitamin B_{12} coenzyme. Infra-red and ^1H NMR spectra may be used to demonstrate the formation of the metal-alkyl complex.

Part 2

Formation of Metal-Hydrogen and Metal-Carbonyl Bonds

Ruthenium has a very extensive chemistry involving oxidation states -2 to $+8$ which serves to illustrate some of the chemistry found for the later 2nd and 3rd row *d*-block transition metals. Several ruthenium complexes show catalytic behaviour in hydrogenation and hydroformylation processes which are of commerical importance. The experiments described below illustrate decarbonylation reactions of alcohols and aldehydes with the concomitant formation of ruthenium carbonyl hydrides.

Chloro(hydrido)(carbonyl)tris(triphenylphosphine)ruthenium(II)

It has been known for many years that halides of the platinum metals can be reduced by alcohols or aldehydes in the presence of ligands (*e.g.* tertiary phosphines) to give carbonyl and/or hydrido complexes. Using isotopic labelling techniques, it is possible to show that the mechanism of metal-hydride formation from the alcohol involves a so-called *β*-elimination step (Eqn. 1).

$$M{-}Cl + RCH_2OH \xrightarrow{-HCl} M(O)(H)C(H)R \longrightarrow M{-}H + RCHO \quad (1)$$

It should be noted that in the addition of transition-metal hydrides to olefins the reverse of *β*-elimination takes place (Eqn. 2).

$$CH_3CH_2Mn(CO)_5 \underset{\text{insertion}}{\overset{\text{elimination}}{\rightleftharpoons}} HMn(CO)_5 + C_2H_4 \quad (2)$$

The complex $[Ru(H)(Cl)(CO)(Ph_3P)_3]$ is obtained by the reaction in Eqn. (3).

$$RuCl_3 + Ph_3P \xrightarrow[\text{excess HCHO}]{MeOCH_2CH_2OH} [Ru(H)(Cl)(CO)(Ph_3P)_3] \quad (3)$$

In this synthesis of $[Ru(H)(Cl)(CO)(Ph_3P)_3]$, the source of the hydride is the CH_2 group of the primary alcohol and the source of the CO ligand is the aldehyde. The steps of the mechanistic pathway for M–CO formation are uncertain, but are believed to involve "oxidative-addition" and "reductive-elimination" processes (Eqn. 4), *i.e.* to depend upon the ability of the metal to exist in various oxidation states differing relatively little in energy.

$$M\text{–}Cl + RCHO \xrightarrow{RCH_2OH} (H)(Cl)M\text{–}(O{=}C(H)R) \xrightarrow{-HCl} H\text{–}M\text{–}C(=O)\text{–}R \rightarrow H\text{–}M(R)\text{–}C{=}O \xrightarrow{-RH} M\text{–}CO \quad (4)$$

(R = H if HCHO is used)

The transformation [M · CO · R] into [M(R)(CO)] involves migration of the R group from carbon to the metal. This can be a reversible process and is well established both in the "hydroformylation" of olefins and in carbonyl insertion reactions (*e.g.* Eqn. 5).

$$(Me)Mn(CO)_5 + CO \longrightarrow (MeCO)Mn(CO)_4(CO) \quad (5)$$

A further example of this type of chemistry is provided by the synthesis of the iridium(I) complex *trans*-$[IrCl(CO)(Ph_3P)_2]$, commonly known as Vaska's compound (Eqn. 6). This material is important because it exhibits an extensive oxidative addition chemistry and is of historical importance in the development of this area of chemistry.

$$IrCl_3 \cdot 3H_2O + Ph_3P + Me_2NCHO \xrightarrow[Ph_3PHCl]{Me_2NH} [Ir(Cl)(CO))(Ph_3P)_2] \quad (6)$$

Following the procedures described in the experimental section, prepare samples of carbonylohydridotris(triphenylphosphine)ruthenium(II) and carbonyldihydridotris(triphenylphosphine)ruthenium(II). Infra-red, ^1H NMR and ^{31}P NMR spectra may be used to determine the formulae and structures of the products.

Experimental

Part 1

Avoid skin contact with, or inhalation of, any of the materials used. All operations should be performed in a fume cupboard. Wash all contaminated apparatus thoroughly with water before removing it from the fume cupboard. In the event of spillage, see the *Special Safety Precautions* section.

Chloropyridinebis(dimethylglyoximato)cobalt(III)

In a fume cupboard place 95% ethanol (200 cm^3) in a 400 cm^3 beaker and add dimethylglyoxime (5.5 g). Warm the mixture on a steam bath and stir the mixture to dissolve the solid then add $CoCl_2 \cdot 6H_2O$ (5.0 g). When this too has dissolved, add pyridine (4 cm^3) to the mixture and allow it to cool to room temperature. Transfer the solution to a 500 cm^3 Büchner flask and fit a rubber bung with a dip tube which projects to just below the surface of the liquid in the flask. Using a water pump draw a slow steam of air through the solution *via* the dip tube for about 30 minutes. Allow the solution to stand for 30 minutes then collect the deposited brown solid by filtration. Wash the brown product with aliquots (5 cm^3) of water, then ethanol and finally diethyl ether. Allow the product to air dry. Record the yield. Redissolve the product in the minimum volume of dichloromethane necessary, filter the solution and add an equal volume of 95% ethanol. Using a rotary evaporator, reduce the volume of solvent to about 1/3 the original volume. A brown crystalline product should be deposited and may be collected by filtration and dried in air. Record your yield of purified product and submit all of your unused product for assessment. Obtain IR (KBr disc) and ^1H NMR spectra on your product.

Methylpyridinebis(dimethylglyoximato)cobalt(III)

In a fume cupboard set up equipment for carrying out a reaction under nitrogen as shown in Figure **3.7-3.** Allow nitrogen to pass through the system for 5 minutes, escaping via a loosened stopper in the flask. Place methanol (10 cm^3) in the round-bottomed flask and bubble nitrogen slowly through the liquid for 30 seconds using a glass tube. Remove the tube, stopper the flask and allow nitrogen to pass over the apparatus and out through the bubbler.

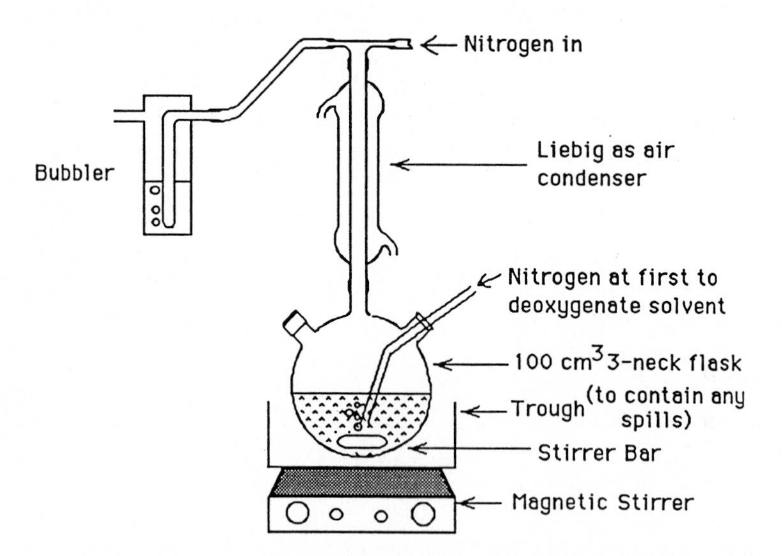

Figure 3.7-3. Apparatus for the methylation of chloropyridine cobaloxime.

Remove a stopper only briefly to add reagents, the nitrogen escaping when a stopper is removed should limit the entry of air into the flask. Add chloropyridine-bis(dimethylglyoximato)cobalt(III) (0.8 g) to the methanol followed by iodomethane (2 cm^3). While stirring the mixture, slowly add sodium borohydride (0.4 g) in portions. Stir the mixture for 10 minutes, allowing the evolved gas to vent through the bubbler. After this time, remove the condenser and other fittings and add water dropwise to the reaction mixture to precipitate the orange product. Do not add too much water. Collect the orange solid by filtration and wash it with aliquots (5 cm^3) of petroleum ether (boiling range 40–60 °C). Allow the product to air dry, then record the yield and submit all of your product in a tube wrapped in aluminum foil as the compound is light sensitive. Obtain IR (KBr disc) and ^1H NMR spectra on your product.

Part 2

Avoid skin contact with, or inhalation of, any of the materials used and wear protective gloves. Ruthenium trichloride is highly coloured and will stain skin and clothing. The solvents used are flammable and 2-methoxyethanol and formaldehyde are toxic so that all operations must be performed in a fume cupboard. Wash all contaminated apparatus thoroughly with water before removing it from the fume cupboard. In the event of spillage, see the *Special Safety Precautions* section.

Chlorohydridocarbonyltris(triphenylphosphine)ruthenium(II)

In a fume cupboard, set up a 100 cm^3 round-bottomed flask fitted with a reflux condenser and small magnetic stirrer bar above a stirrer hotplate and small oil bath. To the flask add triphenylphosphine (1.0 g) followed by 2-methoxyethanol (30 cm^3). While stirring, heat the mixture under reflux. Shake the sample of ruthenium trichloride hydrate (0.15 g) provided with 2-methoxyethanol (10 cm^3) in a stoppered 50 cm^3 flask to procduce a solution. Using a dropping pipette, add this solution to the reaction vessel *via* the condenser. Immediately afterwards add the 40% aqueous formaldehyde solution (10 cm^3), again added via the condenser using a dropping pipette. Continue to heat the mixture under reflux for 20 minutes. After this time, remove the flask from the oil bath and slowly add ethanol (25 cm^3) via the condenser. Remove the condenser and allow the flask to stand on a cork ring to cool to room temperature. A cream solid should be desposited. If not, cool the mixture further in ice and allow to stand longer. Collect the solid product by filtration and wash it with a little water (10 cm^3), followed by ethanol (20 cm^3). Place the filtrate in the bottle provided for Ru wastes in the fume cupboard. Dry the product in a vacuum dessicator, then record the yield and submit all of your product in a tube wrapped in aluminum foil as the compound is light sensitive. Obtain the IR (KBr disc) spectrum of your product.

Dihydridocarbonyltris(triphenylphosphine)ruthenium

In a fume cupboard, set up a 100 cm^3 round-bottomed flask fitted with a reflux condenser and small magnetic stirrer bar above a stirrer hotplate and small oil bath. Prepare the following three solutions in 25 cm^3 conical flasks: a) ruthenium trichloride trihydrate (0.15 g) in ethanol (10 cm^3), b) 40% aqueous formaldehyde (10 cm^3) (available in prepared form), c) postassium hydroxide (0.5 g) in ethanol (10 cm^3). To the round-bottomed flask add triphenylphosphine (1.0 g) followed by ethanol (20 cm^3). While stirring, heat the mixture to reflux, then quickly and successively add solutions a), b) and c) via the condenser. Continue to heat the mixture under reflux for 15 minutes. After this time, remove the flask from the oil bath and allow the reaction mixture to cool to room temperature. Collect the deposited white solid product by filtration and wash it with a little ethanol (5 cm^3), then water (5 cm^3) followed by ethanol (5 cm^3). Place the filtrate in the waste bottle provided. Redissolve the product in the minimum volume of dichloromethane necessary, filter the solution and add ethanol dropwise to induce precipitation of the purified material. Allow the mixture to stand to complete the crystallisation, then collect the product by filtration. Dry the product in a vacuum dessicator, then record the yield and submit all of your product in a tube wrapped in aluminum foil as the compound is light sensitive. Obtain IR (KBr disc), ^1H NMR and 31P-{^1H} NMR spectra on your product. [*Note:* ^{31}P-{^1H} NMR means the ^{31}P NMR spectrum is recorded with the coupling to ^1H removed by irradiation of the sample at the frequency of the ^1H signals].

Exercises

Examine the IR, ^1H NMR and ^{31}P-{^1H} NMR spectra obtained, list the bands or signals observed and identify which bands or signals arise from which groups in the molecules you have synthesised. Your report on this experiment should also contain the following information.

Part 1

1. The equations of the reactions carried out indicating the reference molecular masses of the compounds on which your yield calculation is based. Cite the mass of product obtained and calculate the % yield for each product.
2. A list of the principle infra-red bands for each product identifying which vibration they arise from. In each case, indicate how the spectral bands confirm the presence of particular groups in the molecule (*e.g.* Co – Cl, C=N).
3. A list of the shifts, integrations and, where appropriate, coupling constants for each signal in the ^1H NMR spectra, along with its assignment indicating how the ^1H NMR spectra confirm the formulations of your products.
4. A brief comment on the limitations of cobaloximes as models for B_{12} coenzyme.

Part 2

1. The equations of the reactions carried out indicating the reference molecular masses of the compounds on which your yield calculation is based. Cite the mass of product obtained and calculate the % yield for each product.
2. A list of the principle infra-red bands for each product indentifying which vibration they arise from. In each case, indicate how the spectral bands confirm the presence of particular groups in the molecule (*e.g.* Ru – H, Ru – CO, PPh_3).
3. A list of the shift and coupling constants for each hydride signal in the ^1H NMR spectrum and the signals in the ^{31}P-{^1H} NMR spectrum, along with its assignment indicating how the NMR spectra confirm the formulations of your products. You should measure the P – H and P – P coupling constants from the NMR spectra obtained. (The signals due to

the aryl protons in PPh_3 are usually complex and not well resolved, do not try to analyse these).

4. An explanation of why more than one carbonyl stretching frequency is observed in the IR spectrum of $Ru(CO)(H)(Cl)(PPh_3)_3$.
5. A suggestion about how you could determine which of the bands in the 2000 cm^{-1} region of the IR spectra of these carbonyl hydrides is due to Ru – H and which to C – O stretching vibrations.

References

F. A. Cotton, G. W. Wilkinson, *Advanced Inorganic Chemistry,* 5th ed., Wiley Interscience, New York, **1988,** Chpt. 27, 30 and p. 894–895.

M. N. Hughes, *The Inorganic Chemistry of Biological Processes,* 2nd ed., Wiley, Chichester, **1981,** p. 177–181.

J. E. Huheey, *Inorganic Chemistry,* 3rd ed., Harper and Row, Cambridge, **1983,** p. 878–880.

G. N. Schrauzer, *Acc. Chem. Res.* **1968,** *1,* 97.

G. N. Schrauzer, L. P. Lee, *J. Am. Chem. Soc.* **1968,** *90,* 6541

L. Vaska, E. M. Sloane, *J. Am. Chem. Soc.* **1960,** *82,* 1263.

B. N. Chaudret, D. J. Cole-Hamilton, R. S. Nohr, G. Wilkinson, *J. Chem. Soc., Dalton Trans.* **1977,** 1546.

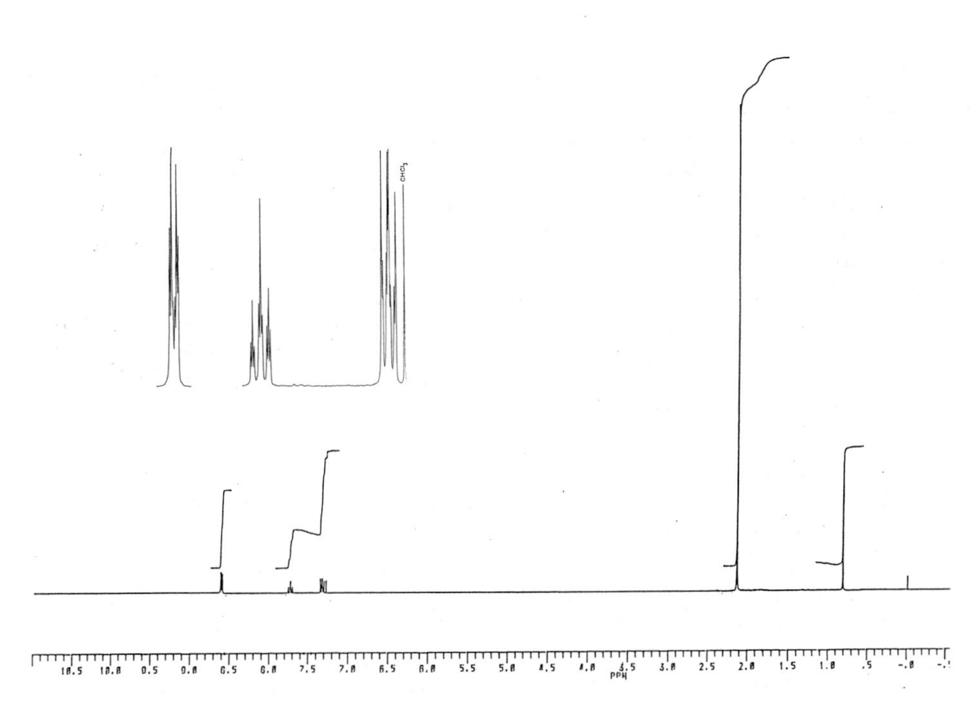

Figure 3.7-4. The 300 MHz ^1H NMR spectrum of methylpyridinecobaloxime in $CDCl_3$. The insert shows an expansion of the aromatic region.

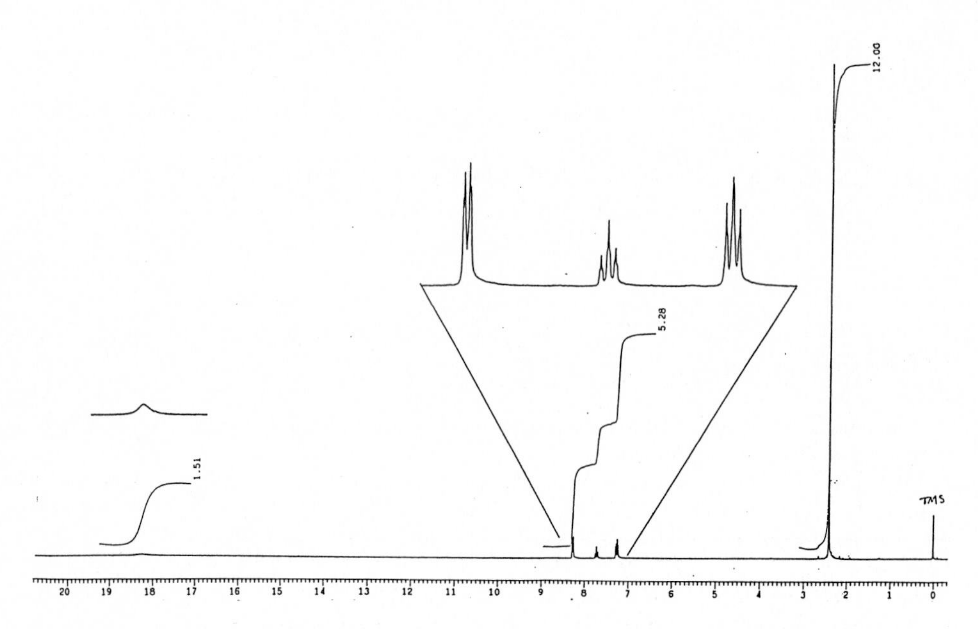

Figure 3.7-5. The 270 MHz ^1H NMR spectrum of chloropyridinecobaloxime recorded in $CDCl_3$ solution.

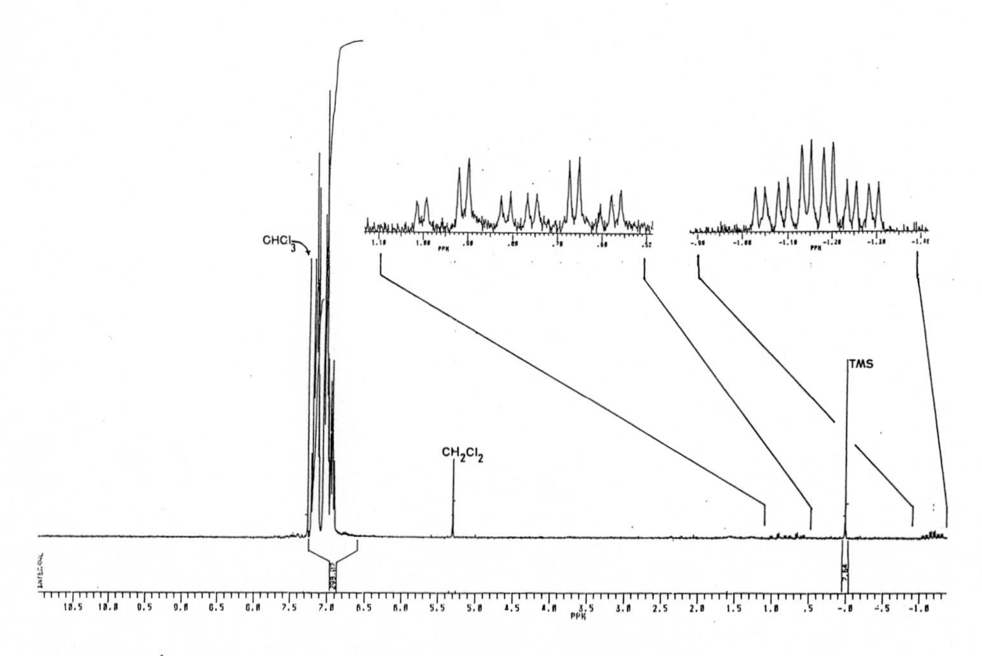

Figure 3.7-6. ^1H NMR spectrum of $[Ru(H)_2(CO)(PPh_3)_3]$ in $CDCl_3$ at 300 MHz.

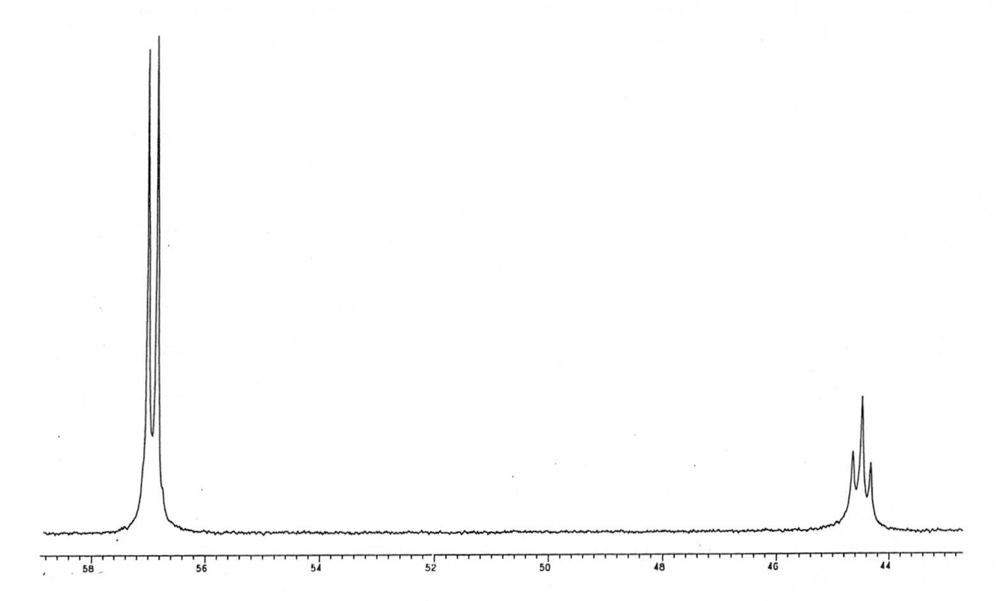

Figure 3.7-7. 109 MHz $^{31}P-\{^1H\}$ NMR spectrum of $[Ru(H)_2(CO)(PPh_3)_3]$ in $CDCl_3$.

3.8 Cobalt Complexes of Dioxygen

David T. Richens and Christopher Glidewell

Cobalt forms a large number of binuclear complexes of general type $[L_5CoXCoL_5]^{n+}$. This experiment investigates a pair of complexes having L = NH_3, X = O_2 and $n = 4$ or 5. When $n = 4$, the O_2 ligand is present as peroxide, O_2^{2-} thus, Co(III)(O_2^{-2})Co(III), having $r(O-O)$ = 148 pm. When $n = 5$, the O_2 ligand is present as superoxide O_2^- thus, Co(III)(O_2^-)Co(III), having $r(O-O)$ = 129 pm. (In ionic peroxides, $r(O-O)$ = 149 pm; in ionic superoxides, $r(O-O)$ = 128 pm). Some of these species have been of interest as models for biological oxygen containing metalloenzyme species.

In the experiment, you are asked to prepare samples of two compounds, one containing X = O_2^{2-} and other containing X = O_2^-, and then to identify them on chemical and physical grounds.

Experimental

Complex **A**

Dissolve 12 g of $Co(NO_3)_2 \cdot 6H_2O$ in 25 cm^3 water and filter the solution. Add 60 cm^3 of a slightly diluted concentrated aqueous ammonia solution (15 mol dm^{-3}; NB concentrated ammonia is 18 mol dm^{-3}), cool to below 10 °C and suck a current of air through the mixture for 3 hours by use of a water pump. Then add a solution of 5 g sodium nitrate in 10 cm^3 H_2O and suck air through for a further hour. Cool in ice and filter off the dark green crystals. Wash with concentrated ammonia solution and then with ethanol. This yields a dihydrate, which can be dehydrated by leaving overnight in a vacuum desiccator with silica gel. Record the yield of anhydrous product.

Complex **B**

At intervals of **10 seconds***, add the following solutions to a 1 litre conical flask, stirring vigorously throughout: 50 cm^3 of 1 mol dm^{-3} $(NH_4)_2SO_4$, 100 cm^3 of conc. NH_3, 100 cm^3 H_2O, 50 cm^3 of 1 mol dm^{-3} $CoSO_4 \cdot 7H_2O$, 50 cm^3 of 3% hydrogen peroxide and finally 50 cm^3 of 1 mol dm^{-3} $(NH_4)_2S_2O_8$. Then stir for a further 15 minutes, after which a fine precipitate forms. Filter this off, wash with a dilute NH_3 solution and then with ethanol. Dry the dark green crystals in a desiccator over silica gel (preferably in the dark). Record the yield.

***NB** It is important to add the reagents **quickly** and in the **correct** order.

Dissolve ~5.0 g of the crude complex **B** (scale down if a low yield was obtained) in a minimum volume of 1.0 mol dm^{-3} sulphuric acid at 80 °C. Filter while still hot and then leave for several hours to crystallise (preferably overnight if time allows). Filter off the dark green lustrous crystals, and wash once with the mother liquor. Dry in a desiccator and record the yield of recrystallised product.

Atomic Absorption Analysis

Prepare samples of your two complexes (recrystallised form of **B**) for AAS analysis. Approximate concentrations of Co in the final solutions for analysis should be around 5 ppm for most instruments.

Magnetic Measurements

Complex **A** can be shown to be **diamagnetic. B** however is **paramagnetic.** Measure the molar magnetic susceptibility of **B.**

Exercises

1. Given that: '**A**' contains 34.5% of nitrogen
   and that: '**B**' contains 14.5% of sulphur, and 21.1% of nitrogen;
   Calculate the atomic ratios, N : Co in **A** and S : N : Co in **B**. Use these atomic ratios, together with the magnetic data, to deduce the correct chemical identities of **A** and **B**. Note that sulphur may be present both as SO_4^{2-} and/or as HSO_4^-.
2. See what you can find out about the ESR spectra of transition metal complexes. Predict the number of lines and their relative intensities in the ESR spectrum of the superoxo complex (cobalt is monoisotopic and ^{59}Co has $I = 7/2$).
3. Observe the reaction of a sample of **A** with a dilute solution of sulphuric acid. What is the reaction occurring? Write a balanced equation. Compare with the behaviour of **B** which can be conveniently recrystallised from hot sulphuric acid as you have shown. Can you offer an explanation for the different chemical behaviour of **A** versus **B** here?
4. The Raman spectrum of **A** shows that it has an O–O stretching frequency at 851 cm^{-1} while that of **B** is at 1075 cm^{-1}. Explain this difference and why these bands are only very weak in the IR spectrum.
5. Consider the significance of the following observed magnetic moments (BM):
   $K_3[Mn(C_2O_4)_3] = 4.81$ $K_4[Mn(CN)_6] = 1.80$
   $K_3[Mn(CN)_6] = 3.18$ $K_3[Fe(C_2O_4)_3] = 5.75$
   $K_3[Fe(CN)_6] = 2.40$ $K_3[FeF_6] = 5.90$
   $K_3[CoF_6] = 4.70$

References

E. A. V. Ebsworth, *et al., Structural Methods in Inorganic Chemistry* **1987,** Chpt. 3, especially page 114 onwards.

R. Davies, M. Mori, A. G. Sykes, J. A. Weil, *Inorg. Synth.* **1970,** *12,* 199.

A. G. Sykes, J. A. Weil, *Prog. Inorg. Chem.* **1970,** *13,* 1

D. T. Richens, A. G. Sykes, *J. Chem. Soc., Dalton Trans.* **1982,** 1621.

3.9 Preparation of Nitrosyl Complexes of Iron and Nickel

Andrew W. G. Platt and John F. Gibson

Special Safety Precautions

The toxicity of nitrosyl complexes is unknown. It is therefore prudent to assume that they are highly toxic. Avoid all contact with skin and wash throroughly in the event of contamination.

Nitric oxide is capable of forming many complexes with transition metals. There is some similarity between the NO molecule and carbon monoxide and nitric oxide does form a series of binary nitrosyls analogous to the binary carbonyls of the first row transition metals. Formally, nitric oxide can be considered as either a neutral three electron donor, a cationic two electron donor (isoelectronic with carbon monoxide) or even an anionic four electron donor.

Unlike carbon monoxide, nitric oxide is not stable in air as it is rapidly oxidised to NO_2. Nitric oxide can be used directly to generate complexes but only if atmospheric oxygen is excluded. This experiment illustrates two methods of forming nitrosyl complexes without the use of free nitric oxide. The first method generates NO in the presence of the metal, whilst the second method involves the reduction of nitrite ion already coordinated to the metal.

Experimental

a) Nitrosylbis(diethyldithiocarbamato)iron, $Fe(NO)(S_2CNEt_2)_2$

Carry out all manipulations in the fume cupboard. Weigh 2.5 g iron(II) sulphate, 0.75 g sodium nitrite and 5 g sodium diethyldithiocarbamate. Dissolve the iron sulphate and sodium nitrite in 15 cm^3 of dilute sulphuric acid and immediately add the sodium diethyldithiocarbamate. Stir vigorously for 5 minutes. Transfer the dark slurry to a separatory funnel and extract with small volumes of chloroform until the extracts are only lightly coloured. During the initial extractions, the boundary between the layers may be difficult to observe so proceed with care.

Combine the chloroform extracts, dry over anhydrous magnesium sulphate, filter and evaporate to about 10 cm^3. Slowly add about 50 cm^3 of petrol, filter the product and dry at the pump. If difficulty is experienced in crystallising the product in this way, decant the petrol from the dark oil and treat with fresh petrol. Record the weight obtained and calculate the percentage yield.

b) Bromonitrosylbis(triphenylphosphine)nickel, $NiBr(NO)(PPh_3)_2$

This preparation requires the use of $NiBr_2\,(PPh_3)_2$. Either use a provided sample or synthesise your own by adding the stoichiometric amount of nickel bromide in ethanol to a refluxing solution of triphenylphosphine in propan-2-ol.

Place 8 g of finely powdered, dry sodium nitrite in a flask with 5 g of $NiBr_2(PPh_3)_2$, 1.8 g triphenylphosphine and 50 cm^3 tetrahydrofuran. Stir under reflux for about 35 minutes. Cool and filter the solution and reduce the volume to about 25 cm^3 by evaporation on a steam bath (fume cupboard). Slowly add 25 cm^3 of petrol to the warm solution with stirring. Allow to cool to room temperature, filter the purple product and dry at the pump. Record the weight obtained and calculate the percentage yield.

Record the infrared spectra of both compounds.

Exercises

1. Assign the bands in your IR spectra which are due to the coordinated NO group. What can be deduced about the mode of bonding of the NO group from your spectra (free NO absorbs at 1878 cm^{-1}).
2. Discuss the possible oxidation states of the metal in the two complexes.
3. Electron spin resonance is a form of magnetic resonance in which an unpaired electron in a magnetic field is excited from one spin to another by a quantum of energy in the microwave region of the electromagnetic spectrum. It is entirely analogous to nuclear magnetic resonance but the appearance of the spectrum is usually rather different because it is normally presented as the first derivative of the absorption with respect to field, plotted against field. In practical terms, what this means is that a peak position is precisely measured by the field value at which the derivative *crosses* the baseline. The ESR spectrum of the iron complex is shown in Figure **3.9-1.** The spectrum is characterised by two parameters, *g* and *a*.

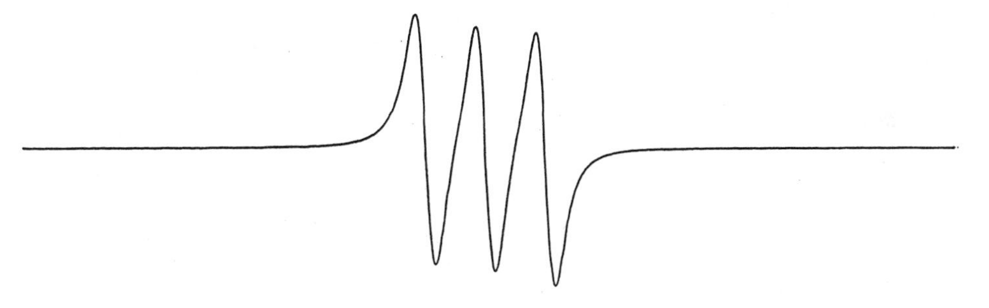

Figure 3.9-1. Electron Spin Resonance spectrum of Fe(NO) $(dtc)_x$ in $CHCl_3$, frequency = 9.515 GHz, scale = 0.1 mT nm^{-1}, centre line = 0.333 T.

The *g* value: this is analogous to the chemical shift in NMR. The free electron value is 2.0023; the value for a free radical in which the unpaired electron is centred on carbon is usually around 2.003; and if centred on nitrogen around 2.006. Metal ions can have very different *g* values (in general between about 1 and 8) but if there is only one unpaired electron the shift away from 2.0023 is often no more than about 0.2.

Calculate the g value from the field position of the central line using the formula $H\nu = g\beta\boldsymbol{B}$, which gives $g = 0.0714484\ \nu(\text{GHz})/\boldsymbol{B}(\text{tesla})$.

The *a* value: this is the symbol for the isotropic hyderfine coupling constant which arises when a nearby nucleus interacts with the electron; it is analogous to the spin-spin coupling constant in NMR. In your molecule, the unpaired electron clearly interacts with a nitrogen nucleus which has spin $I = 1$, having three projections along the magnetic field characterised by $M_I = +1$, O, -1. The isotropic hyperfine coupling constant a_N may be measured in field units by measuring the separation between two adjacent lines. To convert to frequency units use a (MHz) $= g \times 13.99626 \times a$(mT).
What can you say about your molecule from its g value?
Given that an unpaired electron situated entirely on a free nitrogen atom has an isotropic hyperfine coupling constant a_0 of 1540 MHz, calculate by direct ratio of the coupling constants the percent probability of finding an unpaired electron on nitrogen in your molecule. What does this value tell you about the electronic structure?

References

R. D. Feltham, *Inorg. Chem.* **1964,** *3,* 116.
D. M. Adams, *Metal Ligand and Related Vibrations*
K. Nakamoto, *Infrared and Raman Spectra of Inorganic and Coordination Compounds,* 3rd ed., John Wiley and Sons, New York, **1978.**

3.10 Preparation of an Iron Dinitrogen Complex

Andrew W. G. Platt

Special Safety Precautions

Sodium azide is toxic and liberates a harmful gas on treatment with acids. The compound must be handled with care and used only in the manner described.
Concentrated ammonia solution is irritating to eyes and respiratory system and should only be used in the fume cupboard.

Since the initial report on the synthesis of $[Ru(NH_3)_5N_2]X_2$ dinitrogen complexes, there has been considerable interest in coordination complexes of dinitrogen, partly due to the possible relationship between such complexes and the nitrogen fixation process. Whilst many dinitrogen complexes are sensitive to air and moisture, the compound synthesised here, $Na_2[Fe(EDTA)N_2] \cdot 2H_2O$, is stable to moisture and air over extended periods.

Experimental

a) $[Fe(HY)H_2O] \cdot H_2O$ (H_4Y = EDTA)

Note that this experiment will take more than one laboratory session. The reaction can be left at places marked* without loss of yield.

This complex is prepared by reacting freshly precipitated iron(III) hydroxide with EDTA. Dissolve 13.5 g of hydrated iron(III) chloride in 15 cm^3 of water with gentle warming. Allow to cool to room temperature and slowly add 20 cm^3 of SG 0.880 ammonia solution with constant stirring (carry out this operation in the fume cupboard). Heat the mixture on the steam bath for 15–20 minutes filter the iron(III) hydroxide by suction and wash well with water. Suspend 17.5 g of EDTA in 40 cm^3 water and add the moist iron(III) hydroxide. Make the volume up to about 100 cm^3 with water and heat on a steam bath for 2 hours.* Allow the mixture to cool and filter the yellow solution from any unreacted EDTA.

Reduce the volume to about 60 cm^3, cool and slowly add acetone to precipitate the product. The exact amount of acetone will depend on the yield of $[Fe(HY)H_2O] \cdot H_2O$, but 100 cm^3 is usually sufficient. If the product precipitates as an oil, decant the supernatant and stir the residue with small quantities of acetone to induce crystallisation. Filter and dry at the pump.* Record the weight obtained and calculate the percentage yield. Record the infra-red spectrum of the product.

b) The dinitrogen complex

Note the final product is reported to be thermally unstable. You should make sure that there is sufficient time to complete the preparation and the characterisation in one laboratory session.

Dissolve 0.5 g of the iron EDTA complex in 20 cm^3 nitrogen degassed water in a flask equipped with a magnetic stirrer bar and a nitrogen inlet. Add 0.5 g sodium azide and stir the solution at about 70 °C for 1 hour whilst maintaining a steady flow of nitrogen through the solution. Reduce the volume to about 10 cm^3. This can be done under reduced pressure or by evaporation under a flow of nitrogen. Cool to room temperature and add 50 cm^3 of ethanol. If the product precipitates as an oil, decant the supernatant and stir the residual oil with more ethanol to induce crystallisation. Record the weight obtained and calculate the percentage yield. Record the infra-red spectrum as either a KBr disc or as a Nujol mull and identify the bands due to coordinated N_2 and the CO groups of the EDTA.

Exercises

1. What is the structure of the $[Fe(HY)\,H_2O] \cdot H_2O$ complex?
2. Discuss the bonding of dinitrogen to metals.

References

R. A. Henderson, *Adv. Inorg. Chem. Radio Chem.* **1983,** *27,* 198.
M. Hidai, *J. Amer. Chem. Soc.* **1983,** *105,* 1680.
R. J. H. Clark, *J. Chem. Soc. Dalton Trans.* **1983,** 2005.
M. Garcia Basallote, J. M. Lopez Alcala, M. C. Puerta Vizcaino, F. Gonzalez Vilchez, *Inorg. Synth., 24,* 207.

3.11 Nickel Dihalide Phosphine Complexes

Ivan P. Parkin
Adapted from the Open University CHEM777 Summer School as prepared by the S343 course team

Special Safety Precautions

1. Nickel salts are toxic. They are also irritants. Wear gloves to handle all nickel salts.
2. The carbon disulphide generated in the experimental is highly toxic and flammable. This preparation must be carried out in a fume cupboard.
3. Phosphines are irritants and toxic.
4. Carry out all reactions in a fume cupboard.

The physical and chemical properties of a coordination complex are dependent on its geometry. This distinction becomes important when different geometric isomers are possible. For example, *cis* and *trans* isomers and square planar or tetrahedral isomers can have different IR spectra, magnetic moment values and UV spectra.

In this experiment, you will synthesise three nickel halide bis-phosphine complexes and determine their geometry using spectroscopic methods. From this data, you will be in a position to predict the geometry of the complexes and suggest reasons for a specific geometric isomer.

Experimental

a) $NiCl_2(PPh_3)_2$

Triphenylphosphine (2.8 g), propan-2-ol (30 cm^3, dried) and two antibumping granules are placed in a dry 100 ml B24 neck round-bottomed flask. The flask is fitted with a B24 reflux condenser and the mixture brought to reflux.

Nickel dichloride (1.2 g $NiCl_2 \cdot 6H_2O$) is dissolved in ethanol (15 ml) in a 50 ml conical flask and the solution warmed to approximately 40 °C. The warm solution of nickel dichloride is added with care down the condenser to the refluxing triphenylphosphine solution. Normally, the precipitate of $NiCl_2(PPh_3)_2$ will form immediately upon addition. The reaction mixture is allowed to cool to room temperature and the product filtered off using a sintered glass filter and Büchner funnel. Wash the precipitate on the filter with cold ethanol (15 ml) and cold dry ether (15 ml). Dry by suction.

Record the yield, melting point, IR (4000–200 cm^{-1}), UV (in toluene or dry $CHCl_3$) and magnetic moment.

b) [Ni(NCS)$_2$(PPh$_3$)$_2$]

Charge a B24 100 ml "Quickfit" round-bottomed flask with nickel nitrate, 1.5 g $Ni(NO_3) \cdot 6H_2O$, ethanol (25 ml) and two antibumping granules or glass beads. Stir or swirl to dissolve, heating gently if necessary. Then add the *finely ground* sodium thiocyanate (0.8 g), top the flask with a B24 reflux condenser and reflux for 20 minutes (during which time you should prepare a phosphine solution as described in the first paragraph of the $NiCl_2(PPh_3)_2$ preparation). Cool the thiocyanate solution in an ice bath, while scratching the inside of the flask with a glass rod to precipitate the sodium nitrate and any unreacted sodium thiocyanate.

Filter the nickel thiocyanate solution through a sintered glass funnel into a Büchner flask. Always remove the tubing from your Büchner flask *before* you turn off the water tap, so that water does not suck back. The thiocyanate solution should be transferred to a conical flask and heated on a hotplate with one or two (not more) antibumping granules. The hot nickel thiocyanate solution is than carefully added down the reflux condenser to the refluxing phosphine solution. The reaction mixture is allowed to cool to room temperature and the product filtered off using a sintered glass filter and Büchner funnel. Wash the precipitate on the filter with cold ethanol (15 ml) and cold dry ether (15 ml). Dry by suction.

Record the yield, melting point, IR (4000–200 cm^{-1}), UV (in toluene or dry $CHCl_3$) and magnetic moment.

c) [NiCl$_2$(PCy$_3$)$_2$]

A dinitrogen atmosphere is essential to avoid oxidation of the phosphine. Dinitrogen is bubbled through the solution of the adduct $PCy_3 \cdot CS_2$ to remove the carbon disulphide, which is highly flammable and toxic. *This preparation must therefore be carried out in a fume cupboard.*

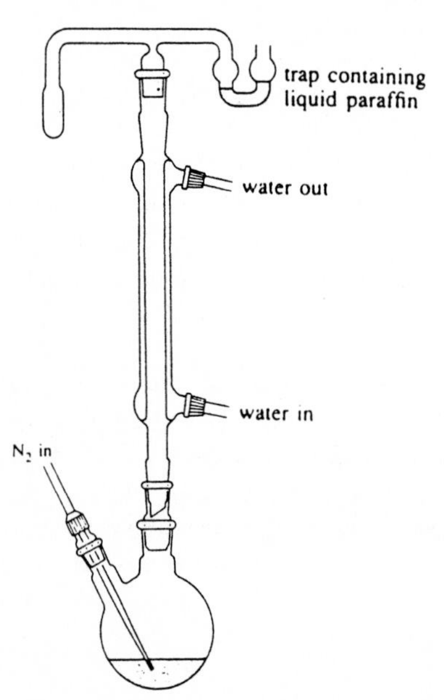

Figure 3.11-1. Reflux apparatus for the preparation of $[NiX_2(PR_3)_2]$.

Assemble the reflux apparatus as shown in Figure **3.11-1.** There should be a small amount of liquid paraffin in the dinitrogen trap to prevent air entering the apparatus, and to allow dinitrogen to escape from the apparatus.

Flush the system with dinitrogen by passing a fairly rapid stream of dinitrogen through the apparatus for a few minutes. The pressure should not exceed 25 kPa, which gives a flow rate of about 4 bubbles per second.

Weigh 1.9 g of the tricyclohexylphosphine-carbon disulphide adduct into the flask and add dry propan-2-ol (20 ml) together with antibumping granules. Attach the condenser and pass dinitrogen through the solution for five minutes. Make sure that the dinitrogen inlet is well below the liquid surface.

Maintain the dinitrogen supply and heat the flask to reflux. If the flow is sufficient, the CS_2 will be driven off within fifteen minutes. The solution will then be clear and colourless (or a very pale straw colour), but you should continue until you are certain that there is no further colour change, and until you are ready to add the warm nickel salt. (You may have to top up with degassed solvent, since this evaporates during the reflux.)

Nickel dichloride (0.6 g $NiCl_2 \cdot 6H_2O$) is dissolved in dried ethanol (15 ml) in a 50 ml conical flask and the solution warmed to 40 °C. The warm nickel dihalide solution is added to the refluxing phosphine carefully down the condenser and the product obtained as previously described for $NiCl_2(PPh_3)_2$,

Record the yield, melting point, IR (4000–200 cm^{-1}), UV (in toluene or dry $CHCl_3$) and magnetic moment.

Exercises

1. Four coordinate nickel complexes can be either square planar or tetrahedral. Draw three diagrams for $NiCl_2(PPh_3)_2$ showing the possible geometric isomers.
2. Metal chloride stretches occur in the infra-red between 400–250 cm^{-1}. How many Ni–Cl stretches are seen for $NiCl_2(PPh_3)_2$ and $NiCl_2(PCy_3)_2$? Does this suggest a particular geometric isomer for each complex?
3. Calculate the extinction coefficient ε for the bands in the UV spectra of the three nickel complexes. Does this value suggest tetrahedral or square planar coordination?
4. Draw a simplified splitting diagram for the *d*-orbitals in a tetrahedral and square planar crystal field. Fill in the appropriate number of electrons for Ni^{2+} in both cases. Compare the number of unpaired electrons with the magnetic moments you measured. Does this predict a certain geometry?
5. Compare the ligand field properties of the thiocyanate and chloride ions and use this to explain the differences in the experimental results for $NiCl_2(PPh_3)_2$ and $Ni(NCS)_2(PPh_3)_2$. What are the steric and electronic differences between PPh_3 and PCy_3 which account for your findings?
6. Can you suggest any further methods we could use to differentiate between square planar and tetrahedral geometries.

3.12 Influence of Ligand Field Tetragonality on the Ground-State Spin of Nickel(II) Complexes

David M. L. Goodgame

Special Safety Precautions

N,N-Diethylethylenediamine should be stored and used in a fume cupboard and any skin contact with it avoided; wear rubber gloves when using it. Also wear rubber gloves when working with solutions of the nickel salts and their complexes, to avoid the possibility of heavy metal skin allergy.

Regularly octahedral, first series transition-metal complexes of electronic configuration d^n ($n = 4-7$) may be either high- or low-spin depending, primarily, on the strength, Δ or $10Dq$, of the ligand field. For distorted ligand fields, a wider range of possibilities exists for changes in the ground-state spin. Changes in spin can have important effects, *e. g.* high-spin octahedral iron(II) complexes are much more labile than their low-spin counterparts. Moreover, the haem complexes involved in haemoglobin, catalase, cytochrome C, etc., contain the iron atom in either a high- or low-spin state, depending on the axial ligand.

The effects of changes in the axial ligands, X, on the electronic properties of a tetragonally distorted complex *trans*-ML_4X_2 are well illustrated by the magnetic properties and *d-d* spectra of a series of nickel(II) complexes of formula $Ni(Et_2en)_2X_2$, where Et_2en = *N,N*-diethylethylenediamine ($Et_2NCH_2CH_2NH_2$) and X = Cl^-, Br^-, I^-, NCS^-, etc.

In the following experiment, a number of compounds of this type can be made and their spin-states determined by magnetic susceptibility measurements, supplemented, if possible, by their solid state (reflectance) *d-d* spectra.

Experimental

a) $Ni(Et_2en)_2(NCS)_2$

Prepare an ethanol solution of nickel thiocyanate by dissolving nickel nitrate hexahydrate (2.0 g) and finely powdered potassium thiocyanate (1.3 g) separately in hot absolute ethanol and mixing the two solutions. Keep the volume of solution to a minimum and cool the mixture thoroughly before filtering off the precipitated potassium nitrate. (For your guidance, the solubility of nickel nitrate in hot ethanol is *ca.* 50 g/100 ml, and that of *powdered* potassium thiocyanate is *ca.* 10 g/100 ml).

To the resulting solution of nickel thiocyanate, add 2 ml of Et_2en by means of a syringe, and stir or shake the solution during the addition. Filter off the precipitated complex, wash it with a little cold ethanol, and dry it (air dry or vacuum desiccator). Record the yield of crude product and then recrystallise it from methanol. Record the yield of pure, crystalline product and take its m. p.

b) Ni(Et$_2$en)$_2$I$_2$

Prepare an ethanol solution of nickel iodide containing *ca.* 2 g NiI_2 from ethanol solutions of the appropriate amounts of nickel nitrate hexahydrate and sodium iodide (solubility of NaI in hot ethanol *ca.* 16 g/100 ml). (Again, cool the solution well before filtering off the precipitated sodium nitrate). Add Et$_2$en (2 ml) using a syringe and filter off, wash (ethanol) and dry the precipitated complex. Do not attempt to recrystallise it. (Alternatively, or additionally, the analogous nickel nitrate complex can be made by simple addition of Et$_2$en to a concentrated ethanol solution of nickel nitrate).

c) Physical measurements

Measure the magnetic susceptibilities of the complexes and, in the case of the thiocyanate complex, also calculate the magnetic moment. If the thiocyanate complex has formed large crystals, it may be necessary to powder some of the sample to achieve good packing in the Gouy (or Evans) tube used for the magnetic measurements.

Obtain the IR spectrum of the thiocyanate complex and identify the bands due to the NCS group, which coordinates to the nickel via the nitrogen atom, *i. e.* it is an isothiocyanato complex. If you have access to a spectrometer capable of measuring the reflectance spectra of solid samples in the UV-vis region, record the spectra of the complexes and relate the spectra to the colours of the compounds.

d) Other complexes

The analogous complexes with nickel chloride and nickel bromide exist as dihydrates as well as anhydrous compounds. Investigate the formation of these compounds by dissolving 1 g of nickel chloride or bromide respectively, in ethanol and adding Et$_2$en (1 ml) to each of these solutions. Observe the colours of the initial precipitates and any colour changes that take place when the products are filtered off, washed with ethanol and air dried. Put a small amount of each compound in a melting-point tube, record the melting point or decomposition point and note any colour changes that occur prior to that point.

If you have isolated both anhydrous and hydrated forms of each complex, record their infrared spectra (using *dry* Nujol) and identify additional bands from the water molecules. If time permits, the weight loss when one of the dihydrates is converted to its anhydrous analogue can be determined, for example, by heating in an oven or a drying pistol at *ca.* 100 °C, or on prolonged storage in an evacuated desiccator over a good drying agent.

Exercises

Draw the crystal field splitting diagram expected for a tetragonally elongated octahedron and use it to explain why some of the compounds you have made are diamagnetic and the others are paramagnetic with magnetic moments indicative of two unpaired electrons per nickel atom.

References

D. M. L. Goodgame, L. M. Venanzi, *J. Chem. Soc.* **1963**, 616, 5909.

A simple method for measuring magnetic susceptibilities may be found in: S. S. Eaton, G. R. Eaton, *J. Chem. Educ.* **1979**, *56*, 170.

3.13 Preparation of Trans-$PtHCl(PPh_3)_2$ and Measurement of its ^1H, ^{31}P and ^{195}Pt NMR Spectra

William P. Griffith

Special Safety Precautions

1. Platinum salts can, in rare cases, cause skin disorders and asthma. Wash your hands after using them.
2. Hydrazine hydrate, potassium hydroxide solution and acetic acid are corrosive. If any is spilled on the skin wash it off with copious quantities of water.
3. Triphenylphosphine is toxic; avoid inhaling its dust.

The development of Fourier transform techniques for NMR spectroscopy has greatly stimulated the study of many nuclei including those with quite low magnetic moments and low sensitivities. In this experiment, you will make *cis*-$PtCl_2(PPh_3)_2$, convert it to *trans*-$PtHCl(PPh_3)_2$ and measure the ^1H resonance spectrum of the latter. The ^{31}P and ^{195}Pt spectra are provided in Figures **3.13-1** and **3.13-2.** From these and your ^1H spectrum you will be able to derive a number of chemical shifts and coupling constants for the hydrido complex.

Experimental

a) Cis-$PtCl_2(PPh_3)_2$

Dissolve finely powdered $K_2[PtCl_4]$ (0.6 g) in 10 cm^3 of water with stirring. In a 100 cm^3 three-necked flask equipped with a reflux condenser, stirrer bar, N_2 inlet and dropping funnel, dissolve finely powdered triphenylphosphine (0.85 g) in degassed ethanol (10 cm^3), passing N_2 over the solution to avoid formation of Ph_3PO. Boil the solution gently and add the $K_2[PtCl_4]$ solution dropwise with continuous stirring. The solution should go pale yellow, followed by precipitation of the white complex.

Allow the solution to cool and continue stirring under a slow nitrogen stream. Centrifuge off the compound in air, wash with a little alcohol, then ether and allow to dry. Take the melting point, record your yield, and run the infra-red spectrum (4000–200 cm^{-1}, pressed KBr disc). Compare your spectrum with that in the literature. Save a small (0.05 g) sample before proceeding to the preparation of *trans*-$PtHCl(PPh_3)_2$.

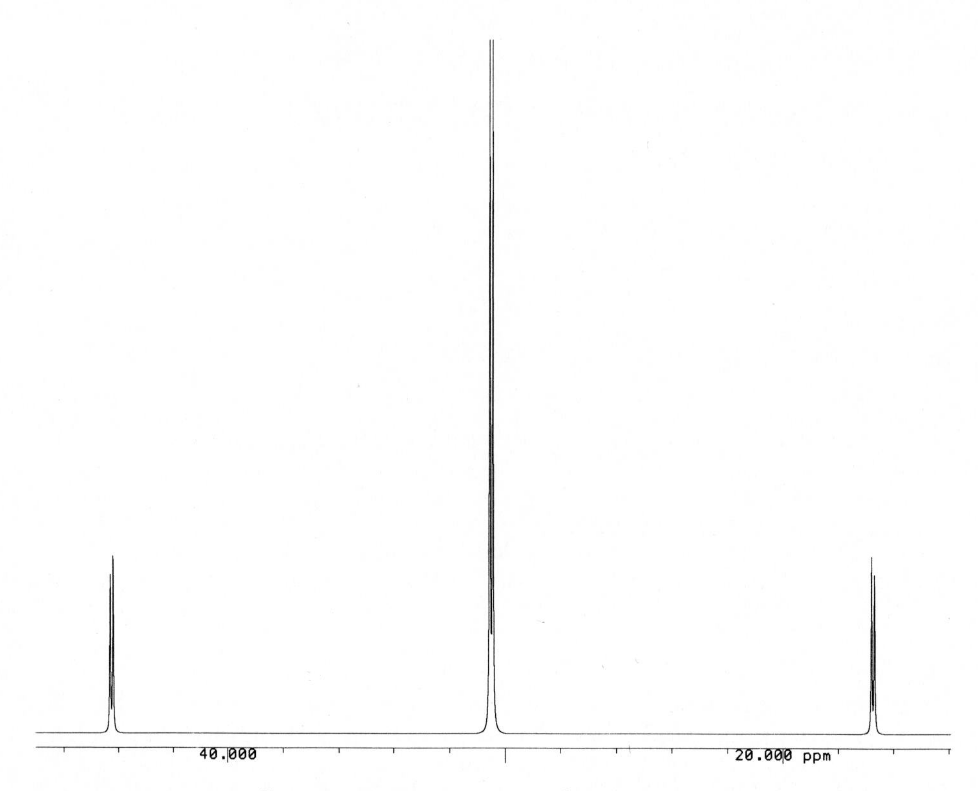

Figure 3.13-1. 109.3 MHz ^{31}P NMR of *trans*-$PtHCl(PPh_3)_2$ ^{1}H Decoupled at 7 ppm.

b) Trans-$PtHCl(PPh_3)_2$

The quantities given here are based on the use of 1.0 g of *cis*-$PtCl_2(PPh_3)_2$. You should, however, use all of your product apart from the amount needed for the IR spectra, etc. in Part a) above, and adjust quantities according to the amount you have available. Make a suspension of *cis*-$PtCl_2(PPh_3)_2$ (1.0 g) in ethanol (30 cm^3) under N_2 using a three-necked flask equipped as before. With stirring, add 1 cm^3 from syringe of concentrated hydrazine hydrate (Corrosive!), and then quickly bring the mixture to reflux with a preheated heating mantle. Continue refluxing for five minutes, swirling the flask or stirring the contents. As clear pale-yellow solution should result.* To the clear stirred solution or supernatant, still under nitrogen, add a mixture of 2 cm^3 of glacial acetic acid in 15 cm^3 of ethanol dropwise. Allow the solution to cool overnight. If little or no precipitate appears, evaporate off about half the solvent and leave the remainder in the refrigerator.

Centrifuge off the white crystals of the hydride, record the yield and melting point, and measure the infra-red spectrum (4000–200 cm^{-1}, pressed KBr disc). Compare your spectrum with that in the literature. Save a small (0.05 g) sample before running the ^{1}H NMR spectrum.

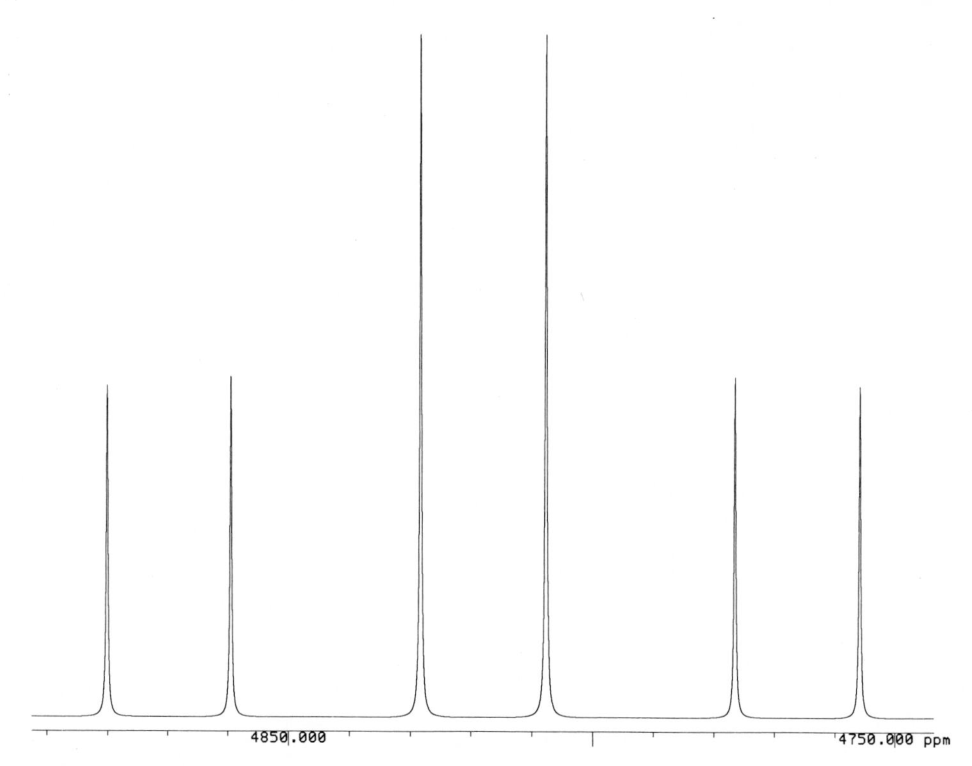

Figure 3.13-2. 58.05 MHz ^{195}Pt NMR of *trans*-PtHCl(PPh$_3$)$_2$.

c) ^{1}H NMR spectrum of trans-PtHCl(PPh$_3$)$_2$

Dissolve as much of your sample as necessary to make a saturated solution in the CH_2Cl_2 (1 cm^3); you may need to decant away from any undissolved solid. Measure the ^{1}H resonance spectrum. Since measurements will necessarily be made at high sensitivity levels, make sure that the phase trim is correct.

Interpret the proton resonance spectrum and calculate *chemical shifts* and *coupling constants* using your spectrum and using the ^{195}Pt and ^{31}P spectra given here.

References

S. H. Mastin, *Inorg. Chem.* **1974,** *13,* 1003.
J. C. Bailar, H. Itatami, *Inorg. Chem.* **1965,** *4,* 1618.

* If a yellow solid is formed, this is likely to be *trans*-Pt(OH)Cl(PPh$_3$)$_2$. In this event decant off the supernatant and convert it to the desired product (*trans*-PtHCl(PPh$_3$)$_2$) by dissolving the solid in the minimum quantity of 3M HCl (30 cm^3), refluxing under nitrogen for 15 minutes and filtering off the white hydride. Continue with the product as in c).

3.14 Identification of Stereochemical (Geometrical) Isomers of $[Mo(CO)_4(L)_2]$ by Infra-Red Spectroscopy

Michael A. Beckett

Special Safety Precautions

Molybdenum hexacarbonyl and its derivatives are highly toxic, volatile materials. Chlorinated hydrocarbons, including CH_2Cl_2 and tetrachloroethene, are toxic and may be carcinogenic. Piperidene is a highly toxic, flammable liquid. Toluene, methanol and 60–80 °C petroleum ether are toxic and flammable. Triphenylphosphine is an irritant. The reactions given in the experimental are on a small scale and should present no special hazards provided reactions and manipulations (*e. g.* weighing, making up Nujol mulls, etc.) are carried out in a fume cupboard. Eye protection and rubber gloves should be worn at all times. Time allowed for the experiment should be *ca.* 6 h.

Physical techniques are widely used to obtain structural information at a molecular level in both organic and inorganic chemistry. A quick and convenient laboratory bench-top spectroscopic analysis of a sample will often allow chemists insight into a structural problem without recourse to single crystal X-ray diffraction study. Vibrational spectroscopy is one such spectroscopic technique. Functional groups within a molecule vibrate at characteristic frequencies (group frequencies) and in doing so absorb radiation in the infra-red (IR) region of the electromagnetic spectrum, 4000–200 cm^{-1}. Mononuclear metal carbonyl complexes are well suited to study by IR spectroscopy since intense absorptions due to the CO oscillations usually occur in the range 2100–1750 cm^{-1}. Furthermore, this region is also generally free from interference from absorptions due to other functional groups. *

Separable geometrical isomers of complexes are common in ligand substituted mononuclear metal carbonyl chemistry. Group theory can be used to predict the number of IR active CO absorption bands to be expected for any particular isomer and so if the structural formula of a metal carbonyl complex is known, it is often possible to correctly identify which particular isomer is present by examining the CO stretching region of its IR spectrum. Thus, four carbonyl absorption bands are to be expected from a *cis*-$[M(CO)_4(L)_2]$ complex whereas only one band is expected from the *trans* isomer.

In this experiment, you will prepare, according to convenient literature methods, one isomer of the molybdenum carbonyl complex $[Mo(CO)_4(pip)_2]$ and both possible isomers of $[Mo(CO)_4(PPh_3)_2]$. These isomers will be identified by examining the CO stretching region of their IR spectra.

---

* The N–N and N–O stretches in M–N_2 and M–NO complexes and M–H stretches also occur in this region. The method used for the preparation of the compounds should give some indication as to whether such functional groups might be present.

Experimental

a) An isomer of [Mo(CO)$_4$(pip)$_2$] (1) (pip = piperidene, HNC_5H_{10})

[Mo(CO)$_6$] (1.0 g) is suspended in dry toluene (40 cm^3) under N_2 and piperidene (10 cm^3) is added. The mixture is heated at reflux for 2 hours and the [Mo(CO)$_6$] should fully dissolve to give a yellow-orange solution. This solution should slowly become opaque as a yellow precipitate of the product **(1)** is produced. The reaction mixture is filtered *hot* using a Büchner flask/pump set up and the bright yellow product, remaining on the filter paper, is washed with cold 60–80 °C petroleum ether (2 × 10 cm^3 portions, previously cooled in an ice bath for 15 min). The product is conveniently dried at the pump. Record the weight of the product and calculate the percentage yield. Obtain the melting point of the complex and its IR spectrum as a Nujol mull.

b) An isomer of [Mo(CO)$_4$(PPh$_3$)$_2$] (2)

[Mo(CO)$_4$(pip)$_2$] **(1)** (0.5 g, 1.32 mmol) is partially dissolved in dry CH_2Cl_2 (20 cm^3) under N_2 and PPh_3 (0.75 g, 2.86 mmol) is added as a solid. The reaction mixture is heated to reflux (whereupon **1** should fully dissolve) and reflux is maintained for 15 min. The reaction solution is allowed to cool to room temperature and the orange solution is filtered. The filtrate is reduced in volume to *ca.* 8 cm^3 (rotary evaporator or vacuum pump) and methanol (15 cm^3) is added. The solution is cooled in a freezer (<0 °C) for 15 min and the pale yellow product should crystallise out. The product is collected by filtration using a Büchner flask/pump set up and dried at the pump. Record the weight of the product and calculate the percentage yield. Obtain the melting point of the complex and its IR spectrum in tetrachloroethene solution and as a Nujol mull.

*c) Thermal isomerisation of **2** to give **3***

[Mo(CO)$_4$(PPh$_3$)$_2$] **(2)** (0.5 g, 0.68 mmol) is dissolved in dry toluene (10 cm^3) under N_2 to give a pale yellow solution. This solution is heated at reflux for 30 min and then allowed to cool to room temperature. The solution may darken upon heating. The toluene is removed on a rotary evaporator (or vacuum pump) to yield the product as a brownish off-white residue **(3)**. The product may be purified by 'dissolving' in CH_2Cl_2 (15 cm^3). The coloured insoluble material can be removed by filtration and the pure product can be obtained from the filtrate by removal of the solvent. Record the weight of the residue and calculate the percentage yield. Obtain the melting point of the complex and its IR spectrum in tetrachloroethene solution and as a Nujol mull.

Exercises

1. Decide, by inspection of the CO stretching region of your IR spectra, the stereochemistry (*cis* or *trans* geometrical isomers) of your products.
2. Draw a reaction scheme which clearly shows the stereochemistry of **1, 2,** and **3,** and conditions for conversions.
3. Compare the melting point of your complexes with those quoted in the literature.

4. Do these metal carbonyl substitution reactions go *via* a dissociative or an associative mechanism?
5. Explain why the mechanisms for the isomerization of $[Mo(CO)_4(L)_2]$ with $L = P^nBu_3$ and $L = PPh_3$ are different.
6. Are the *cis* or the *trans* isomers of $[Mo(CO)_4(PR_3)_2]$ thermodynamically more stable?

References

R. M. Silverstein, G. C. Bassler, T. C. Morrill, *Spectrometric Identification of Organic Compounds,* John Wiley and Sons, 5th ed. **1991.**

E. A. V. Ebsworth, D. W. H. Rankin, S. Cradock, *Structural Methods in Inorganic Chemistry,* Blackwell Scientific, 2nd ed., **1991.**

K. Nakamoto, *Infrared and Raman Spectra of Inorganic and Coordination Compounds,* John Wiley and Sons, 3rd ed., New York, **1978.**

M. Y. Darensbourg, D. J. Darensbourg, *J. Chem. Ed.* **1970,** *47,* 33.

D. J. Darensbourg, R. L. Kump, *Inorg. Chem.* **1978,** *17,* 2680.

W. Strohmeier, K. Gerlach, D. v. Hobe, *Chem. Ber.* **1961,** *94,* 164.

W. Hieber, J. Peterhans, *Z. Naturforsch.* **1959,** *14B,* 462.

A. D. Allen, P. F. Barrett, *Can. J. Chem.* **1968,** *46,* 1649.

D. J. Darensbourg, *Inorg. Chem.* **1979,** *18,* 14.

3.15 Five Coordinate Complexes: $[VO(acac)_2]$ and $[Cr(NH_3)_6][CuCl_5]$

Christopher P. Morley

Special Safety Precautions

1. Anhydrous liquid ammonia has a boiling point of $-33\,°C$, and is corrosive. All operations involving liquid ammonia should therefore be conducted in a fume cupboard, whilst wearing protective gloves.
2. Sodium reacts violently with water. Any waste sodium scraps should be destroyed by adding them to a small beaker of anhydrous ethanol. When reaction is complete, the solution may be washed cautiously down the drain.
3. Chloroform is toxic by inhalation or contact.
4. Sulphur dioxide is a severe respiratory irritant. All reactions which may result in its evolution should be carried out in a fume cupboard.

Five coordinate complexes are rare when compared to the abundance of four and six coordinate species. They commonly exhibit either square pyramidal or trigonal bipyramidal geometry. In this experiment, you will prepare and analyse an example of each type: vanadyl acetylacetonate, $[VO(acac)_2]$, and the pentachlorocuprate(II) anion as its hexaamminechromium(III) salt, $[Cr(NH_3)_6][CuCl_5]$.

Experimental

a) $VO(acac)_2$

Water (50 cm^3) and concentrated hydrochloric acid (25 cm^3) are added to vanadium (V) oxide (3.3 g) in a 600 cm^3 beaker. The mixture is heated on a hotplate (in the fume cupboard) and sodium sulphite (5 g) is added in 1 g quantities. After a few minutes at the boiling point, all the V_2O_5 should be reduced. After cooling, the solution is filtered. Acetylacetone (10 cm^3) is added, followed by saturated sodium carbonate solution until no further effervescence occurs. The precipitate is collected by filtration on a Büchner funnel and dried thoroughly in a vacuum dessicator. The product is purified by dissolving it in a small volume of chloroform, filtering if necessary, and reprecipitating by slow addition of petroleum ether. Dry the product as before and record the yield.

Record the melting point, infra-red spectrum (Nujol mull, NaCl plates), and visible spectrum ($CHCl_3$ solution, *ca.* 0.1 g in 20 cm^3) of your product. Rerun the visible spectrum using tetrahydrofuran as solvent and comment on any differences between the two spectra. Measure the magnetic susceptibility of your sample, and calculate the number of unpaired electrons using the spin-only formula.

Vanadium may be estimated by titration with permanganate. The acetylacetonate groups are, however, oxidised only relatively slowly by permanganate and must therefore first be destroyed. Add 25 cm^3 dilute sulphuric acid to a weighed sample of complex (ca. 0.25 g) and bring to the boiling point. From a burette, slowly add concentrated (approx. 0.2 M) $KMnO_4$ solution, reheating when necessary. Add a slight excess of permanganate and boil for a few minutes. Cool, reduce with an excess of sodium sulphite, and boil off the excess SO_2 (in a fume cupboard): this will take 10–15 minutes. A blue solution of V(IV) will be obtained. Allow this to cool to 50–70 °C and then titrate with standard (ca. 0.02 M) potassium permanganate until the first permanent pink colour is observed. After recording the endpoint, add a slight excess of $KMnO_4$ solution and boil. Repeat the reduction and titration to check that oxidation of the acetylacetonate groups was complete. Repeat if necessary until constant titres are achieved. Calculate the percentage vanadium in the complex, compare with the theoretical value and comment on the purity of your sample.

b) $[Cr(NH_3)_6]Cl_3$

A slush bath at −78 °C is prepared by slowly adding solid carbon dioxide to acetone in a Dewar vessel until the solid remains at the bottom of the vessel without vapourising. The reaction vessel is clamped so that about half of it is cooled in the slush bath (see diagram), and ammonia is then passed into the reaction vessel from a cylinder. Allow about 40 cm^3 of NH_3 to condense. A small piece (<0.1 g) of freshly cut sodium is then added to the liquid ammonia, followed by a small crystal of ferric nitrate (or other Fe(III) salt) to discharge the blue colour. The solution is stirred if necessary.

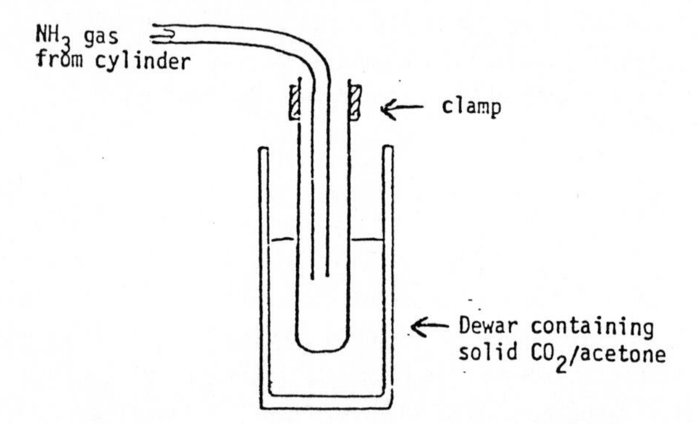

The dark solution of sodium amide thus produced, which also contains finely divided iron, is treated in small portions with anhydrous $CrCl_3$ (2.5 g in total), previously ground to a fine powder using a pestle and mortar. Only small portions of the relatively warm $CrCl_3$ are added, in order to avoid boiling the solution over the sides of the reaction vessel.

The contents of the reaction vessel are poured into an evaporating dish and allowed to dry. 1 g of the crude product thus obtained should be used in the preparation of the pentachlorocuprate(II) salt (see below). The remainder may be recrystallised as follows. Dissolve it in a small volume of 1 M hydrochloric acid at room temperature, filter to remove any unreacted $CrCl_3$ and iron containing impurities, and cool the solution to 0 °C in an ice bath. Add concentrated hydrochloric acid slowly until the initial volume is approximately doubled. The crystals obtained are collected by filtration, washed with ethanol, then dried.

Record the infra-red (KBr disk) and visible spectra (aqueous solution, *ca.* 0.1 g in 20 cm^3) of your product and assign the absorptions observed. Measure the magnetic susceptibility of the sample and calculate the number of unpaired electrons using the spin-only formula.

c) $[Cr(NH_3)_6][CuCl_5]$

Hexaamminechromium(III) chloride (1 g, see above) is dissolved in water (15 cm^3). A solution of copper(II) chloride dihydrate (1 g) in water (15 cm^3) is added. The mixture is filtered if necessary, then heated to 60 °C. Concentrated hydrochloric acid (10 cm^3) is added: crystals form as the solution cools to room temperature. Filter off the product, wash with ethanol and dry in a desiccator. Calculate your yield.

Record the infra-red spectrum (KBr disk) of your product and compare it with the spectrum of $[Cr(NH_3)_6]Cl_3$ obtained above.

The chromium and copper content of the sample should be determined using atomic absorption spectroscopy. The complex may be brought into solution using a small quantity of a mixture of concentrated nitric and sulphuric acids, followed by dilution with water to a known volume. Use sufficient complex to ensure that the final concentration of each metal is in the range 10–20 ppm. Calculate the percentage chromium and copper, compare with the theoretical value and comment on the purity of your sample.

References

R. A. Rowe, M. M. Jones, *Inorg. Synth.* **1967,** *5,* 114.
J. Selbin, H. R. Manning, G. Cessac, *J. Inorg. Nucl. Chem.* **1963,** *25,* 1253.
K. Nakamoto, Y. Morimoto, A. E. Martell, *J. Am. Chem. Soc.* **1961,** *83,* 4533.
A. L. Oppegard, J. C. Bailar Jr., *Inorg. Synth.* **1950,** *3,* 153.
G. C. Allen, N. S. Hush, *Inorg. Chem.* **1967,** *6,* 4.

3.16 Metal Acetylacetonate Complexes: Preparation and Characterisation

Christopher Glidewell

Special Safety Precautions

1. Vanadium pentoxide, V_2O_5, is toxic.
2. Cyclohexane is flammable.
3. All organic liquid waste from recrystallisations must be placed in the waste bottles provided.

Acetylacetone (2,4-pentanedione), $CH_3COCH_2COCH_3$, is a typical β-diketone which can ionise in aqueous solution as a weak acid (Eqn. 1).

$$CH_3COCH_2COCH_3 \rightleftharpoons H^+ + CH_3COCHCOCH_3^- \tag{1}$$

The resulting anion can act as a ligand towards metal ions, forming complexes in which the ligand is usually bonded to the metal through both oxygen atoms, hence forming a six-membered ring (Eqn. 2).

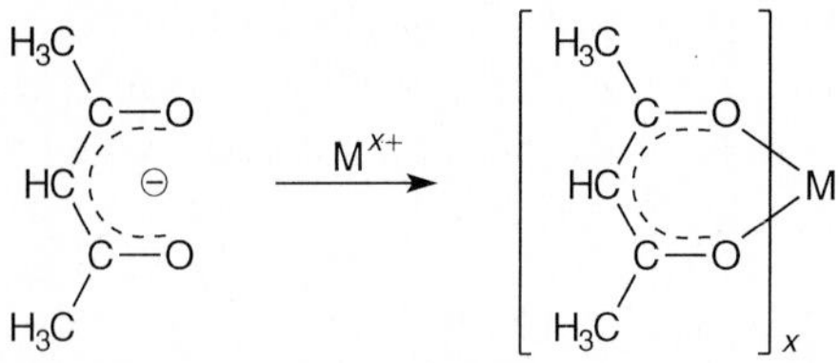

H₃C–C(–O)–HC(⊖)–C(–O)–CH₃ $\xrightarrow{M^{x+}}$ $[M(O{-}C(CH_3){-}CH{-}C(CH_3){-}O)]_x$ (2)

In general, the complexes isolated as crystalline solids are neutral so that a metal ion M^{x+} forms a complex having the stoichiometry $M(CH_3COCHCOCH_3)_x$ (Eqn. 3).

$$xCH_3COCH_2COCH_3 + M^{x+} \rightleftharpoons xH^+ + M(CH_3COCHCOCH_3)_x \tag{3}$$

In the complexes, since the MO_2C_3 six-membered rings are planar and contain 6 π-electrons, they may be regarded as weakly aromatic. In $M(CH_3COCHCOCH_3)_3$ complexes the MO_6 array is octahedral, in $Cu(CH_3COCHCOCH_3)_2$ the CuO_4 group is square planar and in $VO(CH_3COCHCOCH_3)_2$, the VO_5 group is square pyramidal.

In pure acetylacetone, or in its solution in non polar organic solvents, the diketo form is in equilibrium with a cyclic enol-like form (Eqn. 4).

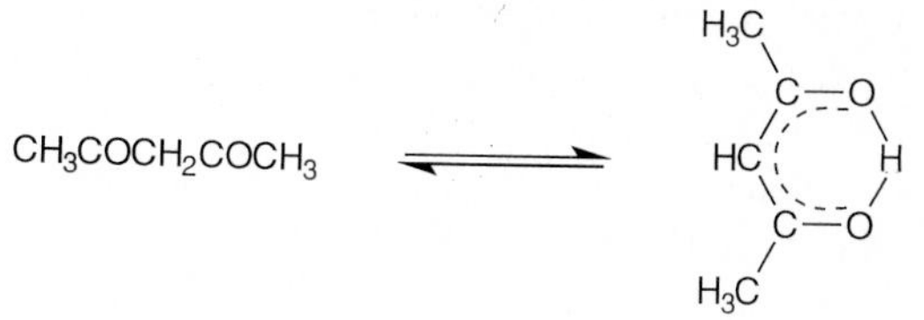

$$CH_3COCH_2COCH_3 \rightleftharpoons \text{(chelated enol form: } H_3C-C(=O)-CH=C(CH_3)-O\cdots H \text{)} \quad (4)$$

This second tautomer may be regarded as a complex in which the proton H^+ takes the role of the metal ion M^{x+}.

In this experiment, you will prepare and purify one metal complex (consult a demonstrator about this), and identify the components in a mixture of complexes using thin layer chromatography. You will also interpret ^1H and ^{13}C NMR spectra and mass spectra of representative examples.

Experimental

a) Metal acetylacetonates

Al^{3+} complex

Weigh into a 100 cm^3 conical flask 3 g (3 cm^3; *ca.* 0.03 mol) of acetylacetone, using a dropping pipette to transfer the liquid into the flask. Add 40 cm^3 of distilled water followed by 8 cm^3 of 5 mol dm^{-3} ammonia solution (dil. NH_4OH). Dissolve 3 g (*ca.* 0.005 mol) of aluminum sulphate $[Al_2(SO_4)_3 \cdot 16H_2O]$ in 30 cm^3 cold distilled water. To the almost clear solution add the ammoniacal acetylacetone solution in portions with shaking. After complete addition of the acetylacetone, check the pH using blue litmus paper (or pH paper) and, if the solution is still acidic, add further small portions of 5 mol dm^{-3} ammonia solution until it is neutral to litmus. Allow to stand for 15 minutes. Filter off the cream coloured product at the water pump, wash with 100 cm^3 of cold distilled water and suck dry for 10 minutes. Transfer the dry product to a weighed sample tube and dry in a vacuum desiccator over anhydrous $CaCl_2$. Weigh the dry product and calculate the percentage yield (Eqn. 5).

$$Al^{3+} + 3CH_3COCH_2COCH_3 \rightarrow 3H^+ + Al(CH_3COCHCOCH_3)_3 \quad (5)$$

Recrystallise a small sample (*ca.* 0.5 g) from cyclohexane. The reslulting needles should be filtered at the water pump, washed with a little cold cyclohexane, and sucked dry for 15 minutes. Record the melting point.

$(VO)^{2+}$ complex

Place 5 cm^3 distilled water in a 50 cm^3 "Quickfit" round-bottomed flask and slowly add to it an equal volume of concentrated H_2SO_4. Then add 12 cm^3 of ethanol followed by 2.5 g (*ca.* 0.014 mol) vanadium pentoxide, V_2O_5. Add a water cooled vertical condenser and reflux this mixture for about 1.5 hours over a small bunsen flame. The solution will turn a dark blue-green colour. Cool the mixture under the tap and filter using cotton wool, discarding any solid residue. Add 6 cm^3 of acetylacetone dropwise to the filtrate with shaking. Neutralise the mixture by adding it carefully to a solution of 20 g of anhydrous Na_2CO_3 in 150 cm^3 distilled water, contained in a 500 cm^3 conical flask, while stirring the mixture using a magnetic stirrer. The resulting mixture should then be cooled in ice water for 15 minutes before filtering at the pump. Wash the dark green product with cold distilled water (2 × 15 cm^3). Suck dry at the pump for 15 minutes, then dry in a vacuum desiccator over anhydrous $CaCl_2$. Weigh the dried product and calculate the percentage yield (Eqns. 6 and 7).

$$V_2O_5 + 4H^2 \rightarrow 2(VO)^{2+} + 2H_2O + 1/2O_2 \quad (6)$$

$$(VO)^{2+} + 2CH_3COCH_2COCH_3 \rightarrow 2H^+ + VO(CH_3COCHCOCH_3)_2 \quad (7)$$

Dissolve a small portion (*ca.* 0.5 g) of the crude, dry product in 6 cm^3 dichloromethane; carefully decant from any residue. Add 20 cm^3 of light petroleum (b.p. 40/60 °C), swirl the mixture and allow to stand for 10 minutes. Filter off the product at the water pump and wash with cold petroleum (2 × 20 cm^3). Suck dry and record the melting point.

Cr^{3+} complex

Weigh directly into a 100 cm^3 conical flask 1.4 g (*ca.* 0.005 mol) of chromium(III) chloride hexahydrate ($CrCl_3 \cdot 6H_2O$) and dissolve it in 50 cm^3 distilled water. Weigh out 10 g urea and add it in 3 or 4 portions to the deep green chromium solution, shaking well after each addition. Then add 3 g (*ca.* 0.03 mol) of acetylacetone, using a dropping pipette. Shake the resulting mixture, cover it with a watch glass and heat rapidly to 80–90 °C on a hot plate. Monitor the temperature with a short stem thermometer. The solution is initially very dark and almost black in appearance, but as the reaction proceeds, deep maroon plate-like crystals form. After 1.5 hours heating, cool the reaction mixture and filter off the product at the water pump. Do not wash the product: dry it in air. Record the percentage yield (Eqns. 8 and 9).

$$CO(NH_2)_2 + H_2O \rightarrow 2NH_3 + CO_2 \quad (8)$$

$$Cr^{3+} + 3CH_3COCH_2COCH_3 + 3NH_3 \rightarrow 3NH_4^+ + Cr(CH_3COCHCOCH_3)_3 \quad (9)$$

The urea undergoes slow hydrolysis, liberating ammonia, which then controls the pH of the reaction mixture.

Dissolve a small sample (*ca.* 0.4 g) in approx. 70 cm^3 of boiling cyclohexane, decant the solution, reduce the volume to about half (do this by adding 2–3 anti-bumping granules and heating on a steam bath in a fume cupboard), and then allow to cool. Filter off the deep red needles at the water pump. Suck dry, then determine the melting point.

Mn^{3+} complex

Dissolve 2.6 g (*ca.* 0.013 mol) of manganese(II) chloride tetrahydrate ($MnCl_2 \cdot 4H_2O$) and 6.8 g (*ca.* 0.05 mol) of sodium acetate in 100 cm^3 of distilled water. Add 10 g (*ca.* 0.1 mol) of acetylacetone and shake the mixture before adding, portionwise with magnetic stirring, a solution of 0.52 g (*ca.* 0.003 mol) of potassium permanganate in 25 cm^3 of distilled water (ensure that the $KMnO_4$ is completely dissolved and use a dropping pipette for this addition, which should take 10–15 minutes).

Stir for a further 10 minutes and then add, in a similar manner, a solution of 6.3 g (*ca.* 0.046 mol) of sodium acetate in 25 cm^3 distilled water. While still stirring, heat the resulting dark mixture on a hot plate to between 60 ° and 70 °C for 15 minutes (monitor the temperature using a short stem thermometer) and then cool to room temperature. Filter off, at the water pump, the very dark, almost black product and wash it with 60 cm^3 of cold distilled water. Suck dry for 15 minutes and dry it in a vacuum desiccator over anhydrous $CaCl_2$. Weigh the dry product and record the percentage yield.

The stoichiometry of this preparation is complicated. First the Mn(II) complex is formed according to Eqn. (10). The manganese(II) then reacts with the manganese(VII) of the permanganate to give manganese(III) (Eqn. 11) and, in outline, the overall stoichiometry is in Eqn. (12).

$$Mn^{2+} + 2CH_2COCH_2COCH_3 \rightarrow 2H^+ + Mn(CH_3COCHCOCH_3)_2 \quad (10)$$

$$Mn(VII) + 4Mn(II) \rightarrow 5Mn(III) \quad (11)$$

$$5Mn^{2+} + MnO_4^- + 15CH_3COCH_2COCH_3 \rightarrow 4H_2O + 7H^+ + 5Mn(CH_3COCHCOCH_3)_3 \quad (12)$$

The purpose of the sodium acetate is to neutralise the acid released, since acetic acid is a *weak* acid (Eqn. 13).

$$7H^+ + 7CH_3COO^- \rightarrow 7CH_3COOH \quad (13)$$

In this preparation, the potassium permanganate is the limiting component on which percentage yields should be based, *i.e.* 1 mol $KMnO_4$ gives 5 mol product.

To about 0.2 g of the dry crude product contained in a 25 cm^3 conical flask, add 12 cm^3 of cyclohexane and boil on a steam bath for 1 minute. Place a small filter funnel in the neck of the flask to act as a reflux condenser. Allow the mixture to settle for 30 seconds before **carefully decanting** from any solid residue into a clean 100 cm^3 conical flask. Reheat for 1 minute to ensure complete solution before adding 40 cm^3 of light petroleum (b.p. 40–60 °C). Cool slowly to room temperature and, when cool, further in an ice water bath for 15 minutes. Filter off at the water pump the black lustrous needles, wash with 10 cm^3 of cold light petroleum and suck dry for 15 minutes. Record the melting point.

Fe^{3+} complex

Dissolve 3.3 g (*ca.* 0.012 mol) of finely-ground iron(II) chloride hexahydrate ($FeCl_3 \cdot 6H_2O$) in 25 cm^3 of distilled water. Add, over a period of 15 minutes, a solution of 3.8 g (*ca.* 0.038 mol) of acetylacetone in 10 cm^3 methanol: stir throughout the addition using a **large** magnetic stirrer. Add to the resulting blood red mixture, over a period of 5 minutes, a solution of 5.1 g of sodium acetate in 15 cm^3 of distilled water, maintaining the stirring throughout. At this point, a red crystalline solid should precipitate. Heat the whole rapidly to about 80 °C using a hot plate (monitor the temperature with a short stem thermometer) and keep at this temperature for 15 minutes, still maintaining rapid stirring. Cool to room temperature under the tap, and then in an ice water bath. Filter off the product at the water pump, wash with 100 cm^3 of cold distilled water and suck dry for 15 minutes. Then dry in a vacuum desiccator over anhydrous $CaCl_2$. Weigh the dry product and calculate the percentage yield (Eqn. 14). The sodium acatate is added to neutralise the acid released: acetic acid is a *weak* acid (Eqn. 15).

$$Fe^{3+} + 3\,CH_3COCH_2COCH_3 \rightarrow 3\,H^+ + Fe(CH_3COCHCOCH_3)_3 \quad (14)$$

$$3\,H^+ + 3\,CH_3COO^- \rightarrow 3\,CH_3COOH \quad (15)$$

Weigh *ca.* 0.2 g of the dried crude product into a 25 cm^3 conical flask and add 3 cm^3 distilled water. Warm on a steam bath and add methanol dropwise, maintaining gentle heating, until the crude product *just* dissolves. Cool in an ice water bath for 15–30 minutes. Filter at the water pump, suck dry for 15 minutes and finally dry in a vacuum desiccator over anhydrous $CaCl_2$. Record the melting point

Co^{3+} complex

Weigh into a 100 cm^3 conical flask 2.5 g (*ca.* 0.021 mol) of cobalt carbonate and add 20 cm^3 (*ca.* 0.20 mol) of acetylacetone. Heat the mixture to about 90 °C on a hotplate with continuous stirring. Monitor the temperature with a short stem thermometer. While maintaining the temperature around 90 °C, add dropwise 30 cm^3 of a 10% hydrogen peroxide solution using a dropping pipette. Cover the flask with a watch glass between H_2O_2 additions. The whole addition of the H_2O_2 solution should occupy about half an hour. Stirring should be maintained throughout the addition, and then for a further 15 minutes. Cool in an ice-water/salt bath for 30 minutes. Filter the dark green product at the water pump, suck dry for 15 minutes and then dry in the oven at 110 °C. Weigh the dry product and record the percentage yield (Eqns. 16, 17 and 18).

In this preparation, the Co(II) complex is formed first according to Eqn. (16). This Co(II) complex is then oxidised by the hydrogen peroxede (Eqn. 17) and the overall stoichiometry may be written as in Eqn. (18).

$$CoCO_3 + 2\,CH_3COCH_2COCH_3 \rightarrow CO_2 + H_2O + Co(CH_3COCHCOCH_3)_2 \quad (16)$$

$$2\,Co^{2+} + H_2O_2 \rightarrow 2\,Co^{3+} + 2\,OH^- \quad (17)$$

$$2CoCO_3 + 6CH_3COCH_2COCH_3 + H_2O_2 \rightarrow 2Co(CH_3COCHCOCH_3)_3 + 2CO_2 + 4H_2O \quad (18)$$

Add *ca.* 0.3 g of the dried product to 10 cm^3 of toluene and heat for about 5 minutes on a steam bath. Allow to settle for about a minute and carefully decant the very dark solution into a 100 cm^3 flask through a small plug of cotton wool contained in a filter funnel. Reheat the solution and then add 20 cm^3 of boiling light petroleum (b.p. 40/60 °C). Cool to room temperature, and then in an ice water bath for 15 minutes. Filter at the water pump and wash the crystals with 50 cm^3 cold light petroleum. Suck dry and record the melting point.

Cu^{2+} complex

To a solution of 4 g (*ca.* 0.025 mol) copper(II) chloride dihydrate ($CuCl_2 \cdot 2\,H_2O$) in 25 cm^3 of distilled water, contained in a 250 cm^3 conical flask, add dropwise over a period of 20 min a solution of 5 cm^3 (*ca.* 0.05 mol) of acetylacetone in 10 cm^3 methanol, while maintaining constant stirring. Add to the resulting mixture 6.8 g of sodium acetate in 15 cm^3 distilled water over a period of 5 min. Heat the mixture to *ca.* 80 °C on a hot plate for 15 minutes, still maintaining rapid stirring. (Monitor the temperature with a short stem thermometer). Cool to room temperature and then in an ice water bath. Filter off the blue-grey product at the water pump, wash with 100 cm^3 of cold distilled water and suck dry for 15 minutes before drying in an oven at 110 °C. Weigh the dry product and record the percentage yield (Eqn. 19).

$$Cu^{2+} + 2CH_3COCH_2COCH_3 \rightarrow 2H^+ + Cu(CH_3COCHCOCH_3)_2 \quad (19)$$

The sodium acetate is added to neutralise the H^+ liberated: acetic acid is a *weak* acid (Eqn. 20).

$$H^+ + CH_3COO^- \rightarrow CH_3COOH \quad (20)$$

To about 0.2 g of the dried crude product contained in a 100 cm^3 conical flask, add 25 cm^3 of methanol and 2 anti-bumping granules. Place a small filter funnel in the neck of the flask to act as a reflux condenser and boil on a steam bath for 5 minutes. Carefully decant from any solid residue the blue solution into a clean 100 cm^3 flask containing about 5 cm^3 of hot methanol. Cool to room temperature before filtering off the fine blue-grey needles at the water pump. Wash with a little ice-cold methanol and suck dry at the pump.

Submit labelled samples of your crude and purified products for inspection

b) Thin-layer chromatography (t.l.c.) of metal acetylacetonate complexes

T.l.c. is a useful technique for establishing either the number of components in a mixture or, alternatively, that a given substance is pure. For mixtures, it is also possible, by comparison with known substances, to establish the identity of the individual components. The stationary phase is usually silica (SiO_2) or alumina (Al_2O_3) spread in a thin layer of uniform thickness on a glass plate or an aluminum backed sheet. The sheet can incorporate a fluorescent dye for identification of colourless compounds by exposure to ultra-violet light. The mobile phase can be any suitable solvent. The unknown is introduced as a solution in the form of a **very** small spot, usually applied using a drawn-out melting-point tube. In this experiment, a mixture of metal ions already converted to the acetylacetonate complexes is subjected to t.l.c. and identified by comparison with samples of the pure metal acetylacetonates. It is the neutral character of these complexes, and their solubility in organic solvents, which enables t.l.c. to be done: the chromatography of the uncomplexed ions would be much less easy.

You are provided with seven solutions. One contains a mixture of three metal acetylacetonate complexes, selected from $(VO)^{2+}$, Cr^{3+}, Mn^{3+}, Fe^{3+}, Co^{3+} and Cu^{2+} complexes dissolved in dichloromethane. Each of the other six solutions contains one of the pure complexes, also dissolved in dichloromethane. Draw out 7 melting-point tubes for use in spotting the t.l.c. plates (consult a demonstrator here).

Spot the unknown mixture and the pure substances onto the plate, using **very small** spots (less than 2 mm in diameter). The spots should be in a line about 1 cm from the short edge of the plate. Each solution should be spotted, dried in air and respotted several times: the $(VO)^{2+}$, Cr^{3+} and Cu^{2+} can be respotted up to six times.

Carefully place the plate in the jar containing *ca.* 0.5 cm depth of a 1% solution of methanol in CH_2Cl_2 (provided in a Winchester bottle) ensuring that the top of the solvent is below the line of spots. Allow to run for *ca.* 30 minutes, after which time the solvent front should be at least 3/4 of the way up the plate. Measure the R_f values for each spot and tabulate these with the spot colours. Deduce the composition of the unknown mixture.

c) Nuclear magnetic resonance spectra of acetylacetonates

In this section, the ^1H and ^{13}C NMR spectra of one diamagnetic metal acetylacetonate, $Al(CH_3COCHCOCH_3)_3$, and of acetylacetone itself are examined and assigned.

For the ^{13}C spectra, several different versions of the spectrum were obtained using different pulse sequences. Note that in all the ^{13}C spectra, however, the H–C coupling is suppressed and not displayed. In the normal spectrum, all types of carbon environment (CH_3, CH_3, CH, and quaternary C) are displayed. In the DEPT spectrum, the resonances due to quarternary C are suppressed and those due to CH_2 are inverted, while in the DEPT-90 spectrum, only the resonances due to CH appear. Hence all four types of carbon environment can be readily identified.

Examine first the ^{13}C spectrum of $Al(CH_3COCHCOCH_3)_3$ (Fig. **3.16-1**) and, bearing in mind that the three ligands are all symmetrical and equivalent (D_3 molecular symmetry), assign the resonances. Next assign the ^1H spectrum of this complex (Fig. **3.16-2**).

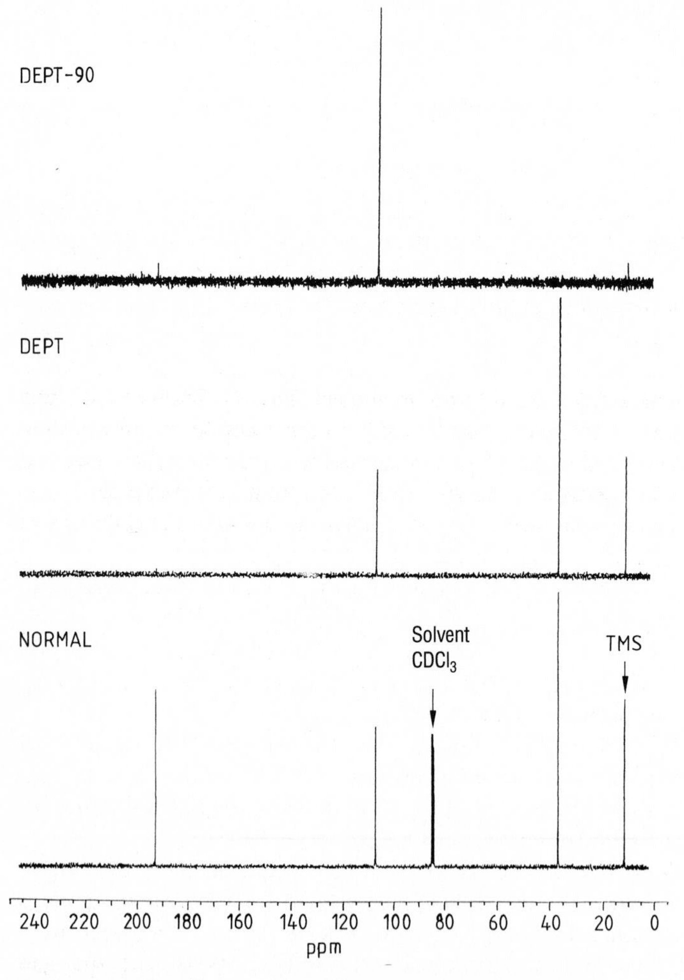

Figure 3.16-1. The ^{13}C nmr spectrum of $Al(CH_3COCHCOCH_3)_3$.

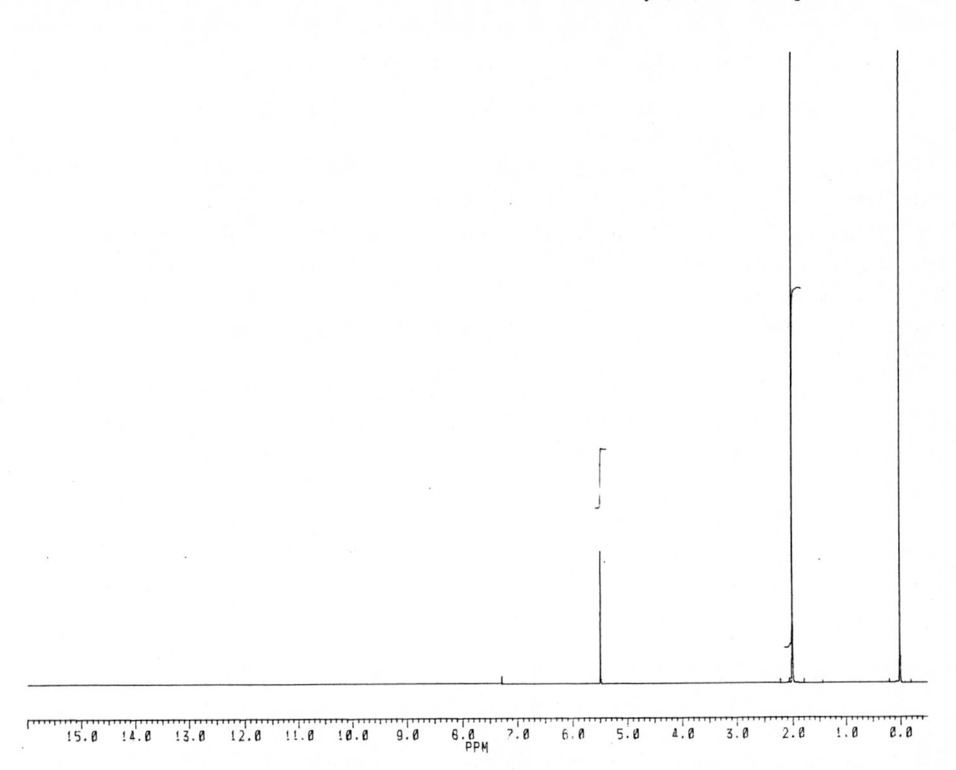

Figure 3.16-2. The ^{1}H nmr spectrum of $Al(CH_3COCHCOCH_3)_3$.

In acetylacetone itself, both keto and enol forms are present (Eqn. 4). Since the enol form can be regarded as a complex of H^+, its ^{1}H and ^{13}C NMR spectra may be expected to show similarities with those of the aluminum complex. Examine first the ^{13}C spectrum (Fig. **3.16-3**) and assign the resonances due to the enol tautomer by comparison with Figure **3.16-1**; then assign the keto resonances. Finally, assign the ^{1}H spectrum (Fig. **3.16-4**), again by comparison with Fig. **3.16-2**, and calculate the keto: enol ratio.

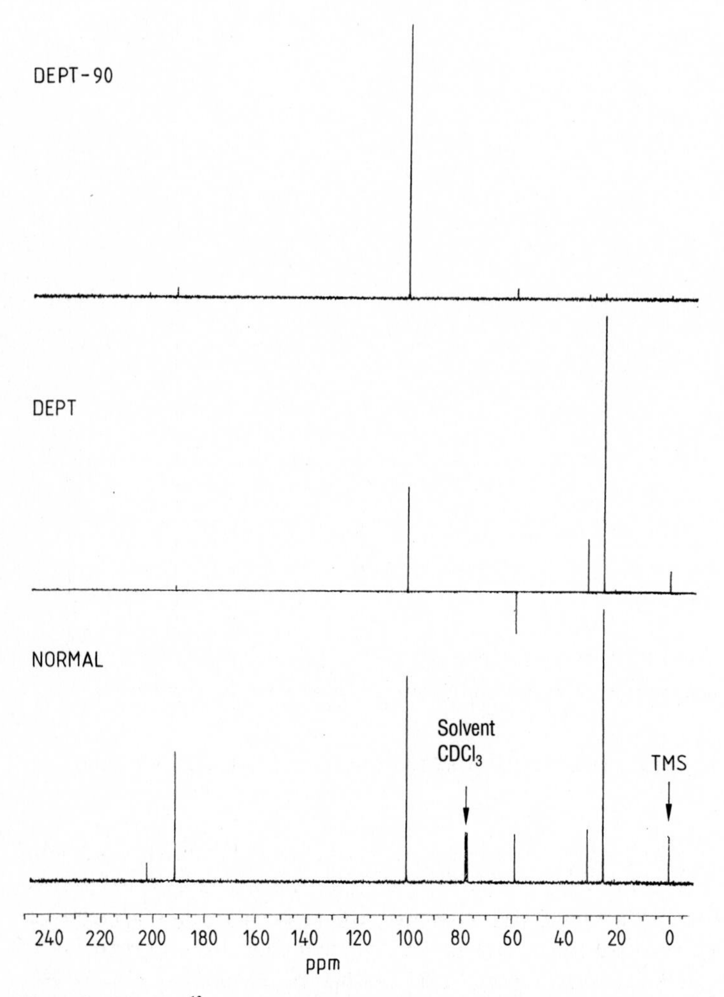

Figure 3.16-3. The ^{13}C nmr spectrum of $CH_3COCH_2COCH_3$containing both keto and enol forms.

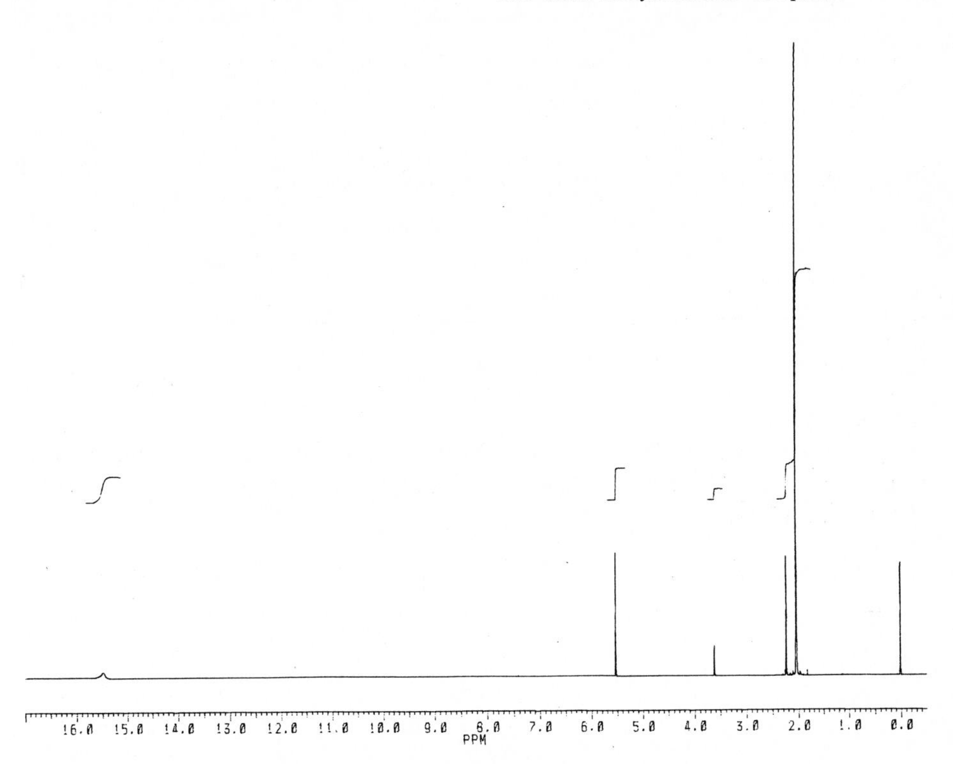

Figure 3.16-4. The ^1H nmr spectrum of $CH_3COCH_2COCH_3$containing both keto and enol forms.

d) Mass spectra of acetylacetone and its aluminum complex

The mass spectrum of acetylacetone contains the following peaks (with relative abundances scaled to the most intense as 100%):

$^m/_z$	**RI** (/%)
100	67
85	89
72	8
59	8
43	100

Identify the composition of the ions at $^m/_z$ = 100, 85, 72 and 43, and suggest structures for these ions. Suggest also how they might arise. Speculate on the structure of the ion having $^m/_z$ = 59.

The mass spectrum of the aluminum complex $Al(acac)_3$ contains the following peaks:

$^m/_z$	RI (/%)
324	4
226	17
225	100
143	10
141	17
127	5
126	6
43	18

Identify the ions having $^m/_z = 324$, 225, 126 and 43. Suggest compositions for the ions having $^m/_z = 226$, 141 and 127. Suggest a structure for the ion having $^m/_z = 143$, whose composition is $(C_5H_8AlO_3)^+$.

3.17 Nickel(II) Complexes of Some Schiff Base Ligands

C. J. Jones

Special Safety Precautions

If any chemicals in this experiment should come in contact with your skin, wash them off immediately with copious amounts of water, then consult a demonstrator. In the event of spillage in the fume cupboard, consult a demonstrator and, where permitted, flush all the spilled materials down the cupboard drain with copious amounts of water. In the event that a diamine is spilled outside a fume cupboard, keep others away from the area of the spill and consult a demonstrator. Absorb the spillage on an inert absorbent (*e. g.* vermiculite) and remove it to a fume cupboard to be packaged for disposal. Consult a demonstrator about spillage of any othe chemicals involved. Skin contamination by, or inhalation of, these materials must be avoided.

Material Hazards

Nickel(II)ethanoate, $Ni(O_2CCH_3)_2 \cdot 6H_2O$	Toxic
1,2-Diaminopropane	Corrosive
1,3-Diaminopropane	Toxic, Corrosive
Ethanol	Harmful, Flammable
Dichloromethane	Harmful
Sodium carbonate	Irritant
Petroleum ether 40/60	Irritant, Flammable
Sodium hydroxide	Corrosive

Disposal of Wastes

Waste solvents must be disposed of into the containers provided.

For further information consult May & Baker Data sheet numbers 416 (nickel(II)ethanoate), 395 (dichloromethane).

The condensation of an aldehyde or ketone with a primary amine leads to the formation of an azomethine (imine) linkage with the liberation of one molecule of water as shown in Eqn. (1) where R, R^1 and R^2 are hydrocarbyl substituents. The nitrogen atom in the product carries a lone pair of electrons and can function as a Lewis base, forming complexes with transition-metal ions. The first recorded example of such a complex was reported as early as 1840 by Ettling, who isolated a copper complex of the product formed in the reaction between salicylaldehyde and ammonia. However, it was Schiff who, in 1869, established the 1 : 2 metal : ligand stoichiometry of this complex and lent his name to the compounds containing the azomethine fragment.

$$R^1R^2C{=}O + RNH_2 \longrightarrow R^1R^2C{=}N{-}R + H_2O \qquad (1)$$

Since these early discoveries, a wide range of complexes derived from Schiff base ligands have been isolated. These compounds have played a major role in the development of modern coordination chemistry, providing examples of macrocyclic ligand systems and the effects of steric interactions on coordination geometries. Schiff base complexes have also been used to model biological systems, such as the haem group and vitamin B_{12} coenzyme, which contain transition metals.

Two general methods may be used to prepare Schiff base complexes. The first entails the prior formation and isolation of the ligand system, followed by its reaction with the metal to form a complex. The second method does not entail prior isolation of the ligand but instead, the condensation and complexation reactions are performed together during a single synthetic process. In fact, there are some ligands which will only form in the presence of a metal ion in a so-called 'template' reaction. In some reactions of this type, it is thought that the metal ion complexes one or both of the ligand precursors before the condensation reaction occurs in what has been called a kinetic template effect. Thus, the metal ion can act as a template orienting the reacting species and controlling the product formed. One example of such a system is provided by the reaction between 1,8-diaminonapthalene and pyrrole-2-aldehyde. The Normal product of the reaction between 1,8-diaminonapthalene and an aldehyde in air is a heterocycle (Eqn. 2). With pyrrole-2-aldehyde in the presence of nickel(II) ions, however, a different product can be obtained as its nickel complex (Eqn. 3).

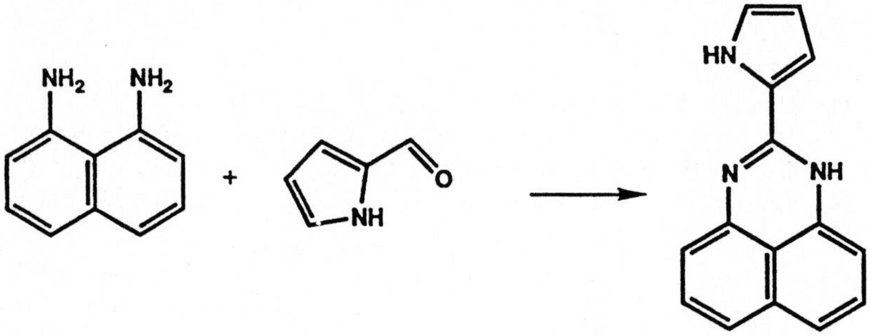

NH2 NH2 + O NH → HN N NH (2)

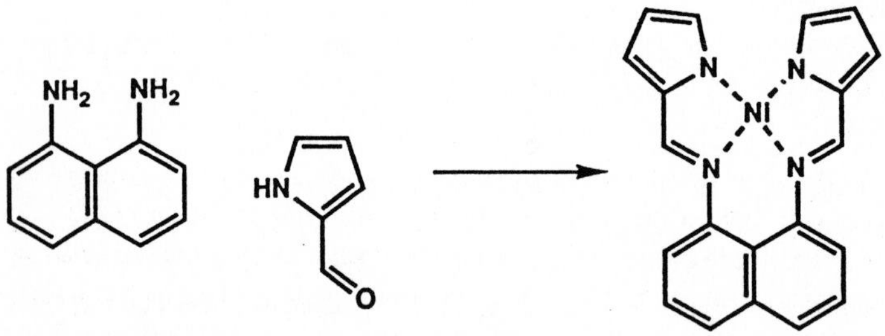

NH2 NH2 HN O → N N Ni N N (3)

In the following experiments, nickel complexes of two isomeric ligands derived from diaminopropane and pyrrole-2-aldehyde are prepared using different methods. ^1H NMR, MS and IR spectral measurements can be used to investigate the structures of these compounds.

Experimental

a) A Schiff base ligand from 1,3-diaminopropane and pyrrole-2-aldehyde

This reaction is to be carried out in the designated fume cupboard. Dissolve pyrrole-2-aldehyde (0.95 g) in ethanol (5 cm^3) in a round-bottomed "Quickfit" flask (100 cm^3). Using a graduated pipette, add 1,3-diaminopropane (0.40 cm^3) to the solution and mix the liquids. Fit a reflux condenser and warm the flask and contents on a steam (or boiling water bath) to boiling for 3–4 min and then stand in an ice bath for 2 hours. The mixture may solidify to a crystalline mass or remain liquid. If a solid has deposited on cooling, collect this by filtration and wash it with a few cm^3 of diethyl ether. The combined filtrate and washings may deposit more product, as crystalline needles, on standing. If the mixture remains liquid, reduce the volume on a rotary evaporatory, until solid starts to appear and then stand the flask in an ice bath to complete the crystallisation. Proceed to collect the product as described above. Allow the product to air dry and record your yield. Obtain the IR (KBr disc), ^1H NMR and mass spectra of your product.

b) A Ni(II) complex from the Schiff base ligand

Dissolve a portion (0.5 g) of the ligand prepared above in a) in warm ethanol (10 cm^3). Slowly add a solution of nickel ethanoate (nickel acetate, $Ni(OCOCH_3)_2 \cdot 4H_2O$, 0.5 g) in water (10 cm^3) to produce a turbid, brick red mixture. Next add a solution of sodium carbonate (0.2 g) in water (5 cm^3) and stir the mixture for 20 minutes. After this time, collect the crude product by filtration and wash it with a little ethanol-water mixture (1 : 1, a few cm^3). Redissolve the red product in dichloromethane (*ca.* 40 cm^3) and dry the solution over a little magnesium sulphate. Remove the magnesium sulphate by filtration and wash it with a little dichloromethane. Add light-petroleum (80–100 °C, 40 cm^3) to the combined washings and filtrate, then remove the dichloromethane using a rotary evaporator (use a room temperature water bath, do not heat the flask). The red product will precipitate from the light-petroleum as the dichloromethane is removed. The product may be collected by filtration and air dried. Record your yield and obtain the IR (KBr disc), ^1H NMR and mass spectra of your product.

c) A nickel(II) complex with 1,2-diaminopropane and pyrrole-2-aldehyde

In the designated fume cupboard, set up a round-bottomed "Quickfit" flask (100 cm^3) equipped with twin-necked adaptor, reflux condenser and dropping funnel. Place an ethanol-water mixture (1 : 1 v/v, 50 cm^3) in the flask along with pyrrole-2-aldehyde (0.95 g), nickel ethanoate (nickel acetate, $Ni(OCOCH_3)_2 \cdot 4H_2O$, 1.25 g) and 3 or 4 anti-bumping granules. Heat the flask to dissolve the nickel ethanoate (a turbid rather than clear solution will form) and then add an aqueous solution of NaOH (10% w/v, 4 cm^3). Dissolve 1,2-diaminopropane (0.4 cm^3) in water (20 cm^3) in the dropping funnel. Add this diamine solution dropwise, over a period of about 20 minutes, to the refluxing suspension of nickel hydroxide and aldehyde. Next add water (10 cm^3) and allow the mixture to cool. Collect the crude orange product by filtration and wash it with a little ethanol-water (1 : 1). Redissolve the product in dichloromethane (*ca.* 40 cm^3) while still in the filter funnel and allow the orange dichloromethane solution to filter into a clean conical flask (100 cm^3). Dry this solution with a little magnesium sulphate, remove the magnesium sulphate by filtration and wash the magnesium sulphate with a little dichloromethane. Add light-petroleum (80–100 °C) to the combined filtrate and washings and remove the dichloromethane using a rotary evaporator. (Use a room temperature water bath and do not heat the flask). The orange product will precipitate from the light-petroleum as the dichloromethane is removed and may then be collected by filtration and allowed to dry in air. Record your yield and obtain the IR (KBr disc), ^{1}H NMR and mass spectra of your product.

Exercises

1. Consult the IR spectra obtained. Given that >N–H bonds give rise to IR bands (νNH) in the region 3000 to 3400 cm^{-1} and that >C=N^{-} bonds give rise to IR bands (νC=N) in the region 1550 to 1600 cm^{-1}, what evidence do the spectra provide for the formation of a nickel complex from the ligand in parts a) and b).
2. Consult the mass spectra and list the major ions observed. Comment on the appearance of the molecular ion peaks in the nickel complexes and explain why prominent ions are observed at $^{m}/_{z}$ = 284 and 286 in each case? How do the spectra of the two isomers differ.
3. Draw the structures of the complexes prepared in Parts b) and c). Consult the ^{1}H NMR spectra and list the shifts and, where appropriate, coupling constants of the signals in these spectra. Show how these spectra are consistent with the structures of the compounds prepared and how the spectra of the two isomeric nickel complexes differ. Explain the appearance of the signals in the region δ_{TMS} 3 to 4 in the NMR spectrum of the compound derived from 1,2-diaminopropane.
4. Square planar complexes of nickel(II) are diamagnetic and exhibit NMR spectra. Use a crystal field splitting diagram to explain why this should be so when octahedral nickel(II) complexes are paramagnetic with 2 unpaired electrons.

References

R. H. Holm, *Prog. Inorg. Chem.* **1966**, 7, 83.
N. F. Curtis, *Coord. Chem. Rev.* **1968**, 3, 3.
M. Hobday, T. D. Smith, *Coord. Chem. Rev.* **1972**, 9, 311.
J. H. Weber, *Inorg. Chem.* **1967**, 6, 258.

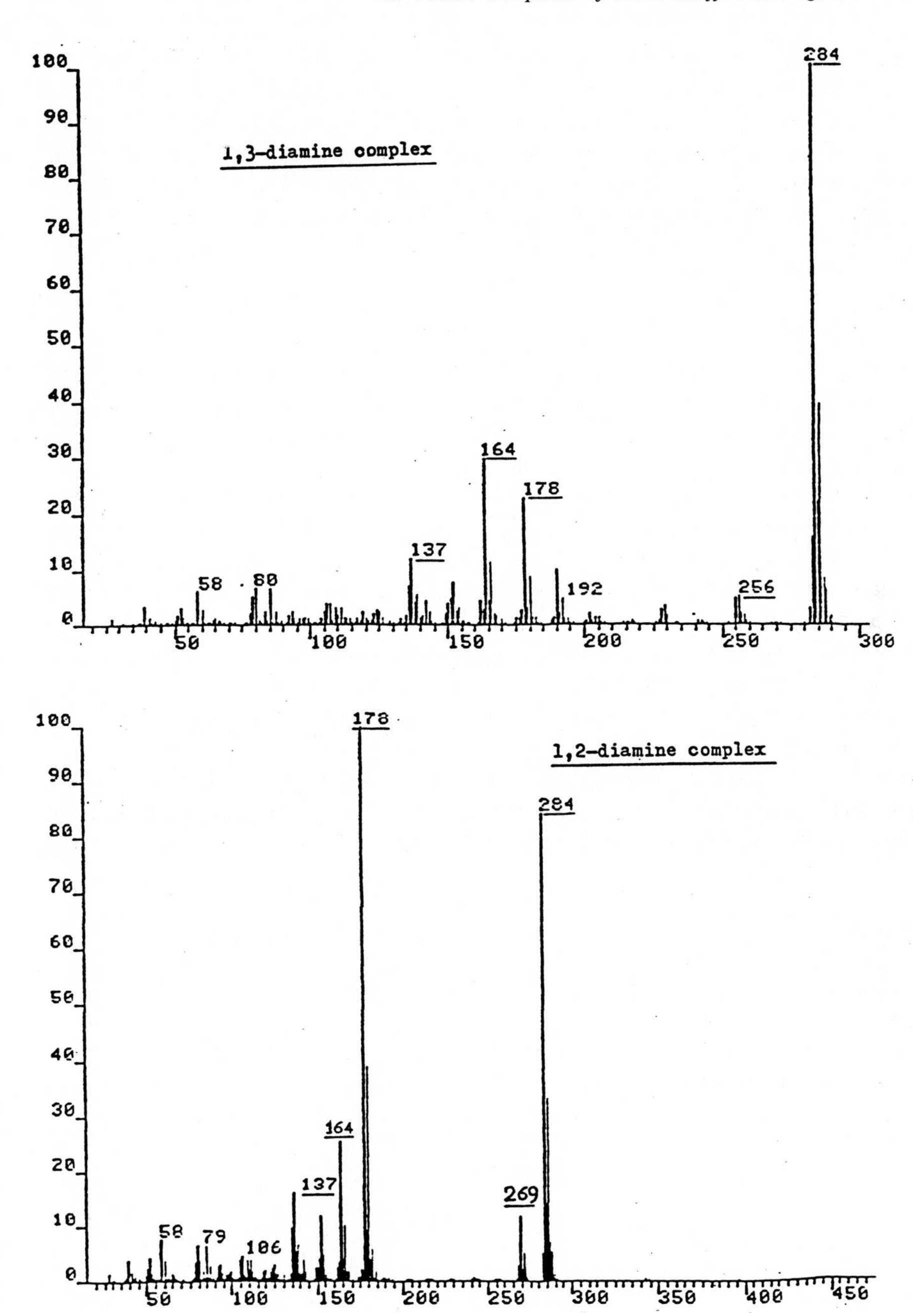

Figure 3.17-1. Mass spectra of the nickel Schiff base complexes

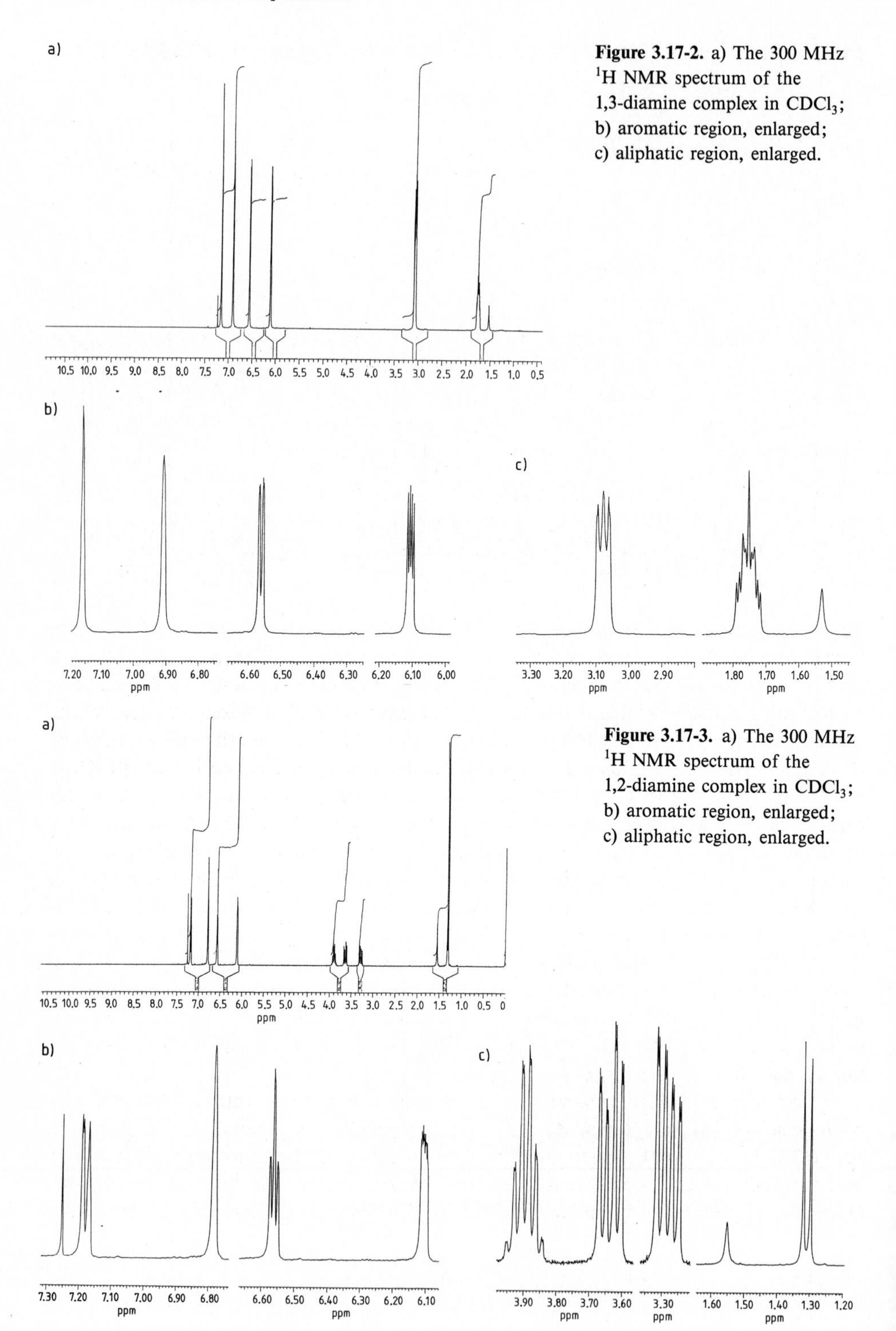

Figure 3.17-2. a) The 300 MHz ^{1}H NMR spectrum of the 1,3-diamine complex in $CDCl_3$; b) aromatic region, enlarged; c) aliphatic region, enlarged.

Figure 3.17-3. a) The 300 MHz ^{1}H NMR spectrum of the 1,2-diamine complex in $CDCl_3$; b) aromatic region, enlarged; c) aliphatic region, enlarged.

3.18 The Synthesis and Coordination Chemistry of Macrocyclic Complexes

Martin Schröder

Special Safety Precautions

1. All perchlorate (ClO_4^-) salts are potentially explosive, particularly when heated to dryness, and must be handled with care at all time. ***Perchlorates should not be heated.*** In all the experiment below, $NaClO_4$ may be replaced by $NaBF_4$ or NH_4PF_6, although the resultant metal complex BF_4^- and PF_6^- salts are often more difficult to isolate from solution than their corresponding ClO_4^- salts.
2. Nickel salts are potential carcinogens. Avoid skin contact.
3. Hydrogen peroxide is highly oxidising and corrosive. Spills should be cleaned and washed off immediately.

Macrocyclic ligands are ligands in which the donor atoms are contained within the same ring. Common examples of macrocyclic ligands include porphyrins, which are present at the active site of *haem* proteins, and phthalocyanins. The beginnings of current interest in synthetic macrocyclic ligands and their complexes can be traced to the synthesis of the nickel(II) complex of the tetraaza macrocycle, L^1 Fig. **3.18-1.** * The original paper by Curtis describes the synthesis of $[Ni(L^1)]^{2+}$ by the condensation of *tris*(ethylenediamine)nickel(II)perchlorate, $[Ni(en)_3](ClO_4)_2$, with acetone. The synthesis of $[Ni(L^1)]^{2+}$ represents a common method of macrocyclic ligand preparation, namely, Schiff-base template condensation of an amine with a ketone around a metal ion. However, this synthesis is not suitable as an undergraduate experiment because the condensation reaction occurs rather slowly over several days and affords a mixture of nickel(II) complexes of L^1 and L^2** (Fig. **3.18-1**), which are related by the position of the imino C=N bonds within the ring. Complexes of L^1 can, however, be conveniently prepared by metal insertion into the preformed macrocyclic ligand. Curtis and Hay first reported the preparation of the metal-free ligand as its diprotonated salt, and this synthetic procedure was subsequently adapted by Tait and Busch. The preparation of the metal-free ligand and its corresponding copper(II), nickel(II) and cobalt(III) complexes via metal insertion reactions provides an excellent introduction to macrocyclic chemistry for undergraduate students.

As a result of the restricted inversion of the chiral secondary amine centres, metal complexes of L^1 can exist as the N-*meso* (**I**) or the N-*racemic* (**II**) diastereoisomers as shown in Figure **3.18-1.** These can be interconverted in basic solution. The N-*meso* diastereoisomers can be distinguished in the case of the diamagnetic nickel(II) complexes by ^1H NMR studies. The N-*racemic* diastereoisomer (thiocyanate salt) displays three equally intense resonances at δ

* The full name for L^1 is: 5,5,7,12,12,14-hexamethyl-1,4,8,11-tetra-azacyclotetradeca-7,14-diene.

** The full name for L^2 is: 5,7,7,12,12,14-hexamethyl-1,4,8,11-tetra-azacyclotetradeca-4,14-diene.

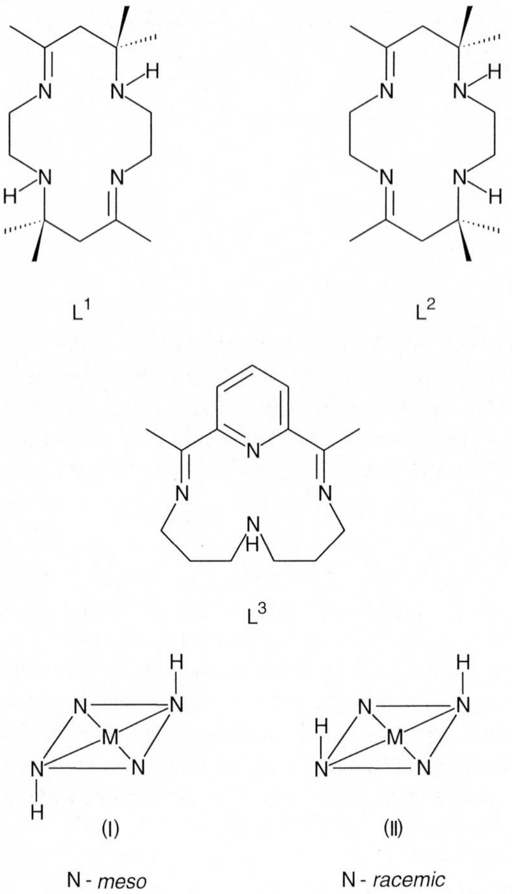

L^1 L^2

L^3

(I) (II)

N - *meso* N - *racemic*

Figure 3.18-1.

2.67, 2.52 and 1.75, while the N-*meso* diastereoisomer (thiocyanate salt) shows these signals at δ 2.69, 2.21 and 1.75 in D_2O. These methyl signals are assigned to the imine methyls (equatorial) and to the axial and equatorial geminal methyls groups respectively. The infra-red bands near 3200 and 1650 cm^{-1} are assignable to the ν(N–H) and ν(C=N) vibrations respectively. Ionic perchlorate (T_d symmetry) leads to a broad band at 1100 cm^{-1} and a sharp band at 625 cm^{-1}. The ligand field strength of L^1 has been calculated to be 15690 cm^{-1}, making the ligand one of the most strongly coordinating synthetic macrocyclic quadridentates on nickel(II). The NCS^-, PF_6^- and BF_4^- salts have been prepared and characterised.

A further set of experiments can be undertaken based upon a convenient *direct* Schiff-base condensation of 2,6-diacetylpyridine with an appropriate diamine, in this case, 4-aza-heptane-1,7-diamine, in the presence of nickel(II) to form $[Ni(L^3)]^{2+}$.* In the absence of nickel(II) ion, the Schiff-base condensation reaction affords intractable polymeric materials.

* The full name for L^3 is: 2,12-dimethyl-3,7,11,17-tetraazabicyclo[11.3.1]septadec-1(17),2,11-13,15-pentaene.

The complex $[Ni(L^3)]^{2+}$ is particularly suitable for electrochemical investigation by cyclic voltammetry and coulometry. The complex shows one one-electron oxidation to afford a stable nickel(III) species, and two one-electron reductions, the first corresponding to the formation of a nickel(II)-ligand radical species. Both one-electron redox products can be detected by ESR spectroscopy.

Experimental

a) 1,2-Diaminoethane dihydrobromide

Dissolve 1,2-diaminoethane (30 g, 33.5 cm^3, 0.5 mol) in methanol (100 cm^3). Cool the solution in an ice bath and carefully add concentrated (49%) hydrobromic acid (175 cm^3) dropwise with stirring. Collect the colourless dihydrobromide salt by filtration on a glass sinter, wash the product with diethyl ether and dry *in vacuo.* A further crop of dihydrobromide salt can be obtained by reducing the volume of the filrate or by adding ether to the filtrate.

b) $L^1 \cdot HBr \cdot 2H_2O$

To 1,2-diaminoethane dihydrobromide (11.1 g, 0.05 mol) prepared above, add acetone (100 cm^3) and 1,2-diaminoethane (3.0 g, 0.05 mol). Heat the mixture on a water bath using a reflux condenser for *ca.* 30 minutes, during which time a copious white precipitate of the crystalline macrocycle dihydrobromide forms. Cool the solution, collect the product by filtration and wash the precipitate with ice-cold acetone, then diethyl ether and dry *in vacuo.* The yield is *ca.* 15.5 g (80% based on ethylenediamine). Measure the melting point of the sample [m.p.: 107–108 (dec.)], and run an IR spectrum as a KBr disk to identify the C=N stretching vibration [ν_{max}: 1670 cm^{-1}]. The dihydrobromide salt can be stored for 2–3 months without appreciable decomposition.

c) N-meso and N-racemic nickel(II) complexes of L^1, $[Ni(L^1)]^{2+}$

Dissolve $L^1 \cdot 2HBr \cdot 2H_2O$ (5 g, 0.01 mol) in the minimum volume of water (*ca.* 30 cm^3) and add an excess of nickel(II) carbonate. Heat the resultant slurry on a water bath for *ca.* 30 minutes. Remove any unreacted nickel(II) carbonate by filtration and add a saturated aqueous solution of sodium perchlorate, $NaClO_4$ (3 cm^3).* Cooling in an ice bath affords yellow crystals of the perchlorate salt of $[Ni(L^1)]^{2+}$. The sample obtained is a mixture of N-*meso* (**I**) and N-*racemic* (**II**) diastereoisomers. These diastereoisomers can be distinguished by IR and ^1H NMR spectroscopy. The pure N-meso diastereoisomer can be obtained by rapid recrystallisation of the sample from hot methanol.

d) N-meso and N-racemic copper(II) complexes of L^1, $[Cu(L^1)]^{2+}$

A mixture of N-*meso* and N-*racemic* diastereoisomers of $[Cu(L^1)]^{2+}$ can be prepared using the same procedure as for the nickel(II) complex, but replacing nickel(II) carbonate with copper(II) carbonate. Addition of a saturated aqueous solution of sodium perchlorate, $NaClO_4$

---

* $NaClO_4$ may be replaced by $NaBF_4$ or NH_4PF_6 in the preparation although the resultant BF_4^- and PF_6^- salts are often more difficult to isolate from solution than the corresponding ClO_4^-.

(3 cm^3) leads to rapid crystallisation of the perchlorate salt. The N-*meso* diastereoisomer is orange and the N-*racemic* diastereoisomer red. The N-*meso* diastereoisomer can be obtained by recrystallisation from water-ethanol, and the N-*racemic* diastereoisomer by extraction into boiling dioxane.

e) Cobalt(III) complexes

The N-*racemic* cobalt(III) complex of L^1 can be prepared by reacting the ligand dihydrobromide salt with $Na_3[CO_3)_3 \cdot 3]H_2O$ in the presence of hydrochloric acid.

Sodium tris(carbonata)cobalt(III)trihydrate, $Na_3[Co(CO_3)_3] \cdot 3H_2O$

This complex can be readily prepared by the method of Bauer and Drinkard. Dissolve $[Co(NO_3)_2] \cdot 6H_2O$ (29.1 g, 0.10 mol) in water (40 cm^3) and slowly add hydrogen peroxide (30%) (10 cm^3). Add this solution dropwise with stirring to a cold slurry of sodium hydrogen carbonate, $NaHCO_3$, (42.0 g, 0.50 mol) in water (50 cm^3). Allow the mixture to stand at 0 °C for 1 h with continuous stirring. Collect the olive green product by filtration, wash the precipitate with 3 × 10 cm^3 portions of cold water, then thoroughly wash with absolute alcohol and diethyl ether and dry *in vacuo*.

Trans-$[CoCl_2(L^1)]ClO_4$

Heat a slurry of $Na_3[Co(CO_3)_3] \cdot 3H_2O$ (7.24 g, 0.02 mol) and the ligand dihydrobromide (12 g, 0.025 mol) in methanol-water (1 : 1 v/v, 100 cm^3) on a water bath until effervescence ceases (*ca.* 20 min). Filter the resultant red solution hot and add concentrated hydrochloric acid (20 cm^3) to the filtrate, which is then heated on a water bath and the volume reduced to *ca.* 20 cm^3. Allow the solution to cool and add a concentrated aqueous solution of sodium perchlorate, $NaClO_4$, (*ca.* 5 cm^3). Green crystals of the required complex crystallise on cooling.

Discussion

Measure the electronic spectra of the complexes and obtain the energy and intensities (extinction coefficients) of the $d-d$ transitions. Measure the molar conductivities of the complexes. Measure the IR, NMR and ESR spectra of the products as appropriate.

The metal complexes of L^1 display many interesting features not normally seen with classical polyamine complexes. The copper(II) complex is quite stable to strong mineral acids and dissociates only slowly ($t_{1/2} \approx 9.6$ min in 6.1 M HCl at 25 °C). The slow dissociation can be readily monitored spectrophotometrically at 500 nm in 6.1 M HCl at 25 °C. A plot of log $(A_t - A_\infty)$ versus time is linear and the observed first order rate constant (k_{obs}) can be readily derived from such plots.

Molecular models can be constructed to illustrate the detailed stereochemistry of these complexes. Discuss the formation of L^1 from the condensation of diamine with acetone.

f) Template Schiff-base condensation to form $[Ni(L^3)]^{2+}$

Dissolve $[Ni(H_2O)_6]Cl_2$ (2.4 g, 0.1 mol) and 2,6-diacetylpyridine (1.6 g, 0.1 mol) in 50% aqueous ethanol (200 cm^3) and bring the solution to reflux. Add 4-aza-heptane-1,7-diamine (1.3 g, 0.1 mol) dropwise with stirring to the solution and reflux the resultant solution for 12 h. Allow the reaction solution to cool and add a concentrated solution of sodium perchlorate,

$NaClO_4$. Cool the solution to 0 °C to allow full precipitation of the product, which should be collected by filtration and dried *in vacuo.*

References

The nomenclature of macrocyclic ligands and the basis for abbreviations to represent ligands are discussed in *J. Amer. Chem. Soc.* **1972,** *94,* 3397 and in *Inorg. Chem.* **1972,** 11, 1979.

N. F. Curtis, *J. Chem. Soc.* **1960,** 4409.

N. F. Curtis, D. A. House, *Chem. Ind. (London)* **1961,** *42,* 1708.

N. F. Curtis, R. W. Hay, *J. Chem. Soc., Chem. Commun.* **1966,** 524.

R. W. Hay, G. A. Lawrence, N. F. Curtis, *J. Chem. Soc., Perkin Transactions I,* **1975,** 591.

A. M. Tait, D. H. Busch, *Inorg. Synth.* (Ed.: B. E. Douglas), Vols. 18, 2.

D. H. Busch, J. L. Karn, *Inorg. Chem.* **1969,** *8,* 1149.

J. Lewis, M. Schröder, *J. Chem. Soc. Dalton Trans.* **1982,** 1085.

L. G. Warner, N. J. Rose, D. H. Busch, *J. Amer. Chem. Soc.* **1968,** *90,* 6938; **1967,** *89,* 703.

N. F. Curtis, Y. M. Curtis, H. J. K. Powell, *J. Chem. Soc. (A)* **1966,** 1015.

H. F. Bauer, W. C. Drinkard, *J. Amer. Chem. Soc.* **1960,** *82,* 5031.

D. K. Cabbiness, D. W. Margerum, *J. Amer. Chem. Soc.* **1970,** *92,* 2151.

F. V. Lovecchio, E. S. Gore, D. H. Busch, *J. Amer. Chem. Soc.* **1974,** *96,* 3109.

3.19 Synthesis and Reactivity of Tertiary Phosphines

Neil M. Boag

Special Safety Precautions

Phosphine halides are colourless fuming liquids with a penetrating, obnoxious odour. They react violently with water, liberating hydrochloric acid. When using, calculate quantities as volumes using the appropriate densities (Aldrich or Merck catalogue) and **only** use in a fume cupboard.

The common oxidation states of phosphorus are P(III) and P(V). A particularly important class of complexes are the tertiary phosphines PR_3 and phosphites $P(OR)_3$. These species have an extensive chemistry in their own right, but are often encountered in their common role as $2e^-$ Lewis bases.

Organophosphine complexes are generally unstable towards oxidation and are notoriously malodorous. However, the triarylphosphines are relatively air-stable compounds with no discernible smell. In this experiment, you will synthesise an example of this type of species and investigate its chemistry.

An important feature of phosphorus is that it exists as a single isotope, ^{31}P, with a nuclear spin of 1/2. The nucleus may be easily detected by NMR spectroscopy and $^{31}P\{^1H\}$ NMR spectroscopy will be used as a diagnostic tool.

Triarylphosphines are generally synthesised from halophosphines by metathesis using aryl Grignard reagents.

$$PCl_3 + 3\,ArMgCl \rightarrow PAr_3 + 3\,MgCl_3$$

$$PhPCl_2 + 2\,ArMgCl \rightarrow PhPAr_2 + 2\,MgCl_2$$

$$Ph_2PCl + ArMgCl \rightarrow Ph_2PAr + MgCl_2$$

The experiment makes use of this reaction to synthesise an example of a triarylphosphine and explores the chemistry of these species.

Experimental

Important: This method is applicable to PPh_2Ar, $PPhAr_2$ and PAr_3. The reaction conditions given below are for the synthesis of PPh_2Ar and uses one equivalent of ArMgBr. If $PhPAr_2$ is to be synthesised, then the scale of the Grignard synthesis must be doubled. If PAr_3 is to be synthesised, the scale of the Grignard reaction should be tripled.

Perform this synthesis in a fume cupboard

a) Grignard (scale as appropriate)

Set up an apparatus consisting of a 250 cm^3 3-neck flask containing a magnetic stirring bar, a dropping funnel and a condenser equipped with a $CaCl_2$ drying tube. Add magnesium turnings (1.2 g) to the flask and just cover the magnesium with dry THF (see Note 1). Place 0.05 moles of the appropriate aryl bromide dissolved in 25 cm^3 of dry THF in the dropping funnel. Add a little of this solution to the magnesium followed by one crystal of iodine. Do not stir at this stage. Surround the reaction flask with a warm water bath (about 40 °C). After a few minutes, the iodine crystal should dissolve and lose colour and the reaction mixture start to reflux. Start the magnetic stirrer and drop the aryl bromide solution into the reaction mixture at *ca.* 1 drop per second. After addition, remove the warm water bath and stir the solution for a further hour.

b) Phosphine

Add a solution of 0.045 moles of the appropriate phosphine halide in 10 cm^3 of dry THF to the dropping funnel. Add the solution of phosphine halide dropwise to the Grignard solution with stirring. When the addition is complete, surround the flask with a hot water bath and reflux the solution for 3 hours. Cool the solution and pour onto 100 cm^3 of ice. Neutralise the resultant solution with dilute hydrochloric acid and remove the THF on a rotary evaporator. Extract the solution with 3×40 cm^3 of toluene and then extract the toluene solution with 4×40 cm^3 on conc. hydrochloric acid (**care**!)

Carefully (**dropwise – strong acid/strong base neutralisation generates a lot of heat**) neutralise the acid with 0.880 ammonia (more than 100 cm^3 required). An oil should result which, upon cooling, will solidify (Note 2). Filter the solid, wash well with water (break up the lumps) and air dry. Recrystallise from hot methanol.

Measure the infra-red spectrum of the product as a Nujol mull and record its melting point. Use Chemical Abstracts to obtain a literature value of the melting point.

c) Phosphine oxide

Prepare a solution/suspension of the phosphine (1 g) in acetone (10 cm^3). Dissolve sufficient hydrogen peroxide to furnish 1.1 moles of peroxide per mole of phosphine in a few cm^3 of water and slowly add this solution to the acetone solution of phosphine. After five minutes, remove the acetone on a rotary evaporator and extract the phosphine oxide with 2×20 cm^3 of toluene. Wash the toluene extracts with ferrous ammonium sulphate solution and dry with anhydrous magnesium sulphate. Remove the toluene on a rotary evaporator. Recrystallise the product (try methanol first, if this is not successful try toluene/petroleum ether).

Record the infra-red spectrum of the product as a Nujol mull. Identify the P=O stretching frequency by comparing the infra-red spectrum with that of the phosphine.

d) Phosphine plus methyl iodide reaction

Add approx. 0.2 g of the phosphine to a small ignition tube (which should fit in a centrifuge). Add toluene to the tube until it is 3/4 full and dissolve the phosphine by warming gently (water bath). Using a Pasteur pipette, add six drops of methyl iodide and seal the tube with a stopper.

Heat in a 40 °C water bath for thirty minutes. A white solid should form. Decant off the supernatant liquid (centrifuge if necessary) and wash the solid three times with diethyl ether. Break up any lumps to ensure the material is well washed. Air dry and store in a desiccator (see Note 3).

Record proton NMR spectra of the three phosphorus compounds you have synthesised using approx. 50 mg of sample in $CDCl_3$. It is better if the products are left to dry for at least several hours (or *in vacuo*) before the NMR samples are prepared so as to ensure there is no residual solvent which would complicate the spectra.

Record $^{31}P\{^1H\}$ spectra of the three phosphorus compounds (spectra of a typical phosphine, $P(4\text{-}MePh)_3$, and its products are reproduced in Fig. **3.19-1**).

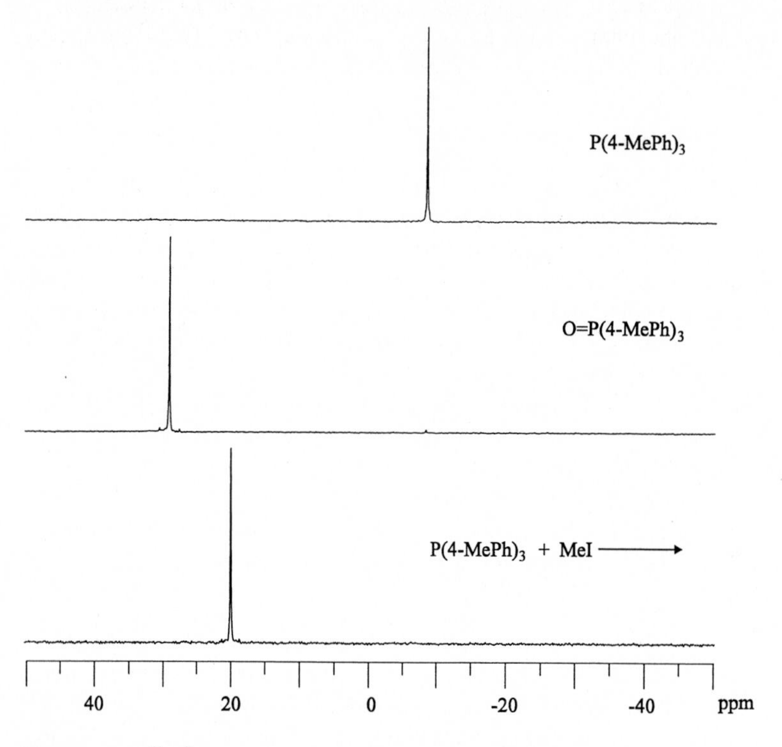

Figure 3.19-1. $^{31}P\{^1H\}$ spectra of $P(4\text{-}MePh)_3$ and its derivatives.

Exercises

1. Assign the resonances in the proton NMR spectra of the phosphine and phosphine oxide. Use the NMR spectra to identify the product formed by the reaction of the phosphine with methyl iodide. What is the oxidation state of the phosphorus in this complex? Are the $^{31}P\{^1H\}$ spectra useful in assigning oxidation states?
2. What is the chemistry behind the *conc.* HCl extraction and why is it undertaken?
3. How would a phosphite, $P(OR)_3$, be prepared? A particularly important reaction of phosphites is the Arbusov reaction. What is the Arbusov reaction?

References

F. A. Cotton, G. Wilkinson, Advanced Inorganic Chemistry 5th ed. pp. 404–415, 421–423.
Inorganic Chemistry 4th ed. pp. 459–473.

Notes

1. Tetrahydrofuran may be dried and stored over molecular sieves.
2. The oil can sometimes take several hours to solidify, particularly if it is hot.
3. The product can be very hygroscopic and oily. It may be prepared *in situ* by the addition of a drop of methyl iodide to a $CDCl_3$ solution of the phosphine in an NMR tube.
4. Copies of proton and phosphorus NMR as well as infra-red spectra of $P(4\text{-}MePh)_3$ and its derivatives are available from the author. These can be supplied as HPGL or Word for Windows files if required.

3.20 Inorganic Heterocycles: Cyclophosphazenes

Josef Novosad and Milan Alberti

Special Safety Precautions

Aniline: highly toxic, cancer suspect agent.
Phosphorus Pentachloride: violently decomposed by water with formation of hydrochloric acid and phosphoric acid; corrosive, moisture-sensitive.
Phosphorus Oxychloride: violently decomposed by water with formation of hydrochloric acid and phosphoric acid; corrosive, moisture-sensitive.
1,1,2,2 Tetrachloroethane: highly toxic, cancer suspect agent.
Dichloromethane: toxic, irritant.
Methanol: flammable liquid, toxic.

Cyclophosphazenes constitute an important class of inorganic heterocyclic ring systems. Hexachlorocyclotriphosphazene. $(NPCl_2)_3$, is the earliest reported inorganic heterocycle, dating back to 1834. The ring system is made up of alternating phosphorus and nitrogen atoms. Phosphorus is pentavalent and tetracoordinate while nitrogen is trivalent and dicoordinate. The phosphorus atom has two exocyclic substitutents but the ring nitrogen atom has none. The best studied examples are hexachlorocyclophosphazene, $(NPCl_2)_3$ (**1**), and octachlorocyclotetraphosphazene, $(NPCl_2)_4$ (**2**).

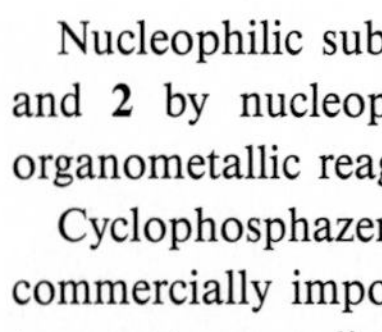

1 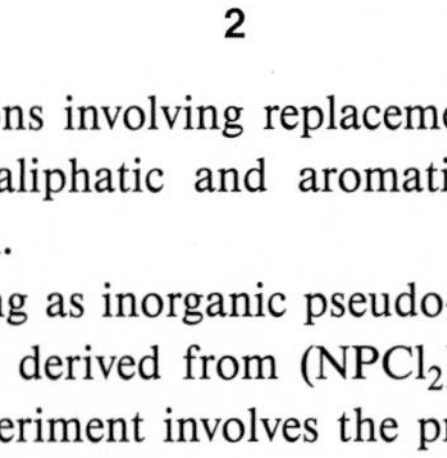 **2**

Nucleophilic substitution reactions involving replacement of the labile P – Cl bonds in **1** and **2** by nucleophiles such as aliphatic and aromatic amines, alcohols, phenols and organometallic reagents are known.

Cyclophosphazenes are interesting as inorganic pseudo-aromatic systems, as well as being commercially important. Polymers derived from $(NPCl_2)_3$ are useful, for example, in high temperature applications. This experiment involves the preparation of two cyclotriphosphazenes: $(NPCl_2)_3$ and its anilino-analogue.

Experimental

a) Hexachlorocyclotriphosphazene, [1,3,5,-triaza-2,4,6,-triphosphorin-2,2,4,4,6,6-hexachloride]

Caution: Carry out this procedure in a fume cupboard. PCl_5 is corrosive and HCl is liberated during the reaction.

The reaction of phosphorus pentachloride with ammonium chloride produces a mixture of cyclic and linear chlorophosphazenes (Eqn. 1).

$$nPCl_5 + nNH_4Cl \rightarrow (NPCl_2)_n + 4nHCl \qquad (1)$$

Equip a 250 ml three necked round-bottomed flask with a reflux condenser, a heating mantle, a scrubbing train consisting of one empty Dreschel bottle (reversed) and one half-filled with 2 M aqueous NaOH solution. Place in the flask 150 ml of 1,1,2,2-tetrachloroethane, *powdered* ammonium chloride NH_4Cl (75 mmol, 4 g), phosphorus pentachloride PCl_5 (75 mmol, 15.6 g) and phosphorus oxychloride $POCl_3$ (0.75 mmol, 1.15 g). After addition, heat the mixture at reflux for 3.5 hours until no more HCl is evolved. Disconnect the scrubbing assembly, allow the mixture to cool and filter off any excess NH_4Cl. Evaporate the filtrate using a rotary evaporator. Cool the crude product (in a fridge) and extract the solid with 60–80 °C petroleum ether (3 × 25 ml). Combine the extracts and reduce in volume (rotary evaporator) to *ca.* 10 ml. Cool this solution to recover the product as an off-white solid. Filter and vacuum dry the product. Record its yield, melting point and infra-red spectrum. The mass spectrum of a pure sample of $(NPCl_2)_3$ is given in Fig. **3.20-1.** Assign the major features of the IR and mass spectra.

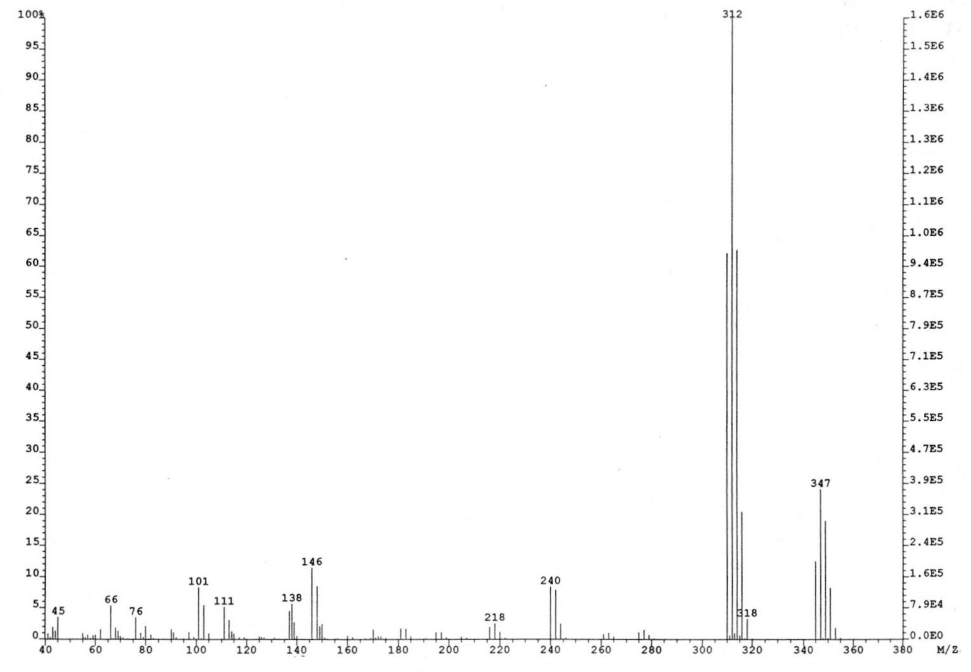

Figure 3.20-1.

b) Hexakis(anilino)cyclotriphosphazene, [1,3,5-triaza-2,4,6-triphosphorin-2,2,4,4,6,6-hexa-aniline]

$(NPCl_2)_3$ reacts with $C_6H_5NH_2$ to give hexakis(anilino)cyclotriphosphazene and anilino hydrochloride (Eqn. 2).

$$(NPCl_2)_3 + 12\,C_6H_5NH_2 \rightarrow N_3P_3(C_6H_5NH)_6 + 6\,C_6H_5NH_2 \cdot HCl \quad (2)$$

In a 100 ml round-bottomed flask equipped with a reflux condenser, place 2.5 g (7.2 mmol) of $(NPCl_2)_3$ and 32 ml (0.35 mol) aniline. Using an oil bath, heat the mixture at reflux, with stirring, for 6 hours. The resulting red-purple semicrystalline mass is allowed to cool, then washed successively with water (3 × 30 ml) and ethanol (3 × 30 ml), and air dried. Recrystallise part of this crude product by dissolving in a minimum volume of hot dichloromethane and adding up to an equivalent volume of methanol, cooling subsequently if necessary. Record the yields and melting points of both the crude and purified products and the infra-red spectrum of the pure phosphazene. Assign characteristic infra-red frequencies.

References

M. A. Armour, *Hazardous Laboratory Chemicals Disposal Guide,* CRC Press, **1991.**

H. R. Allcock, *Phosphorus-Nitrogen Compounds,* Academic Press, New York, **1972**

H. G. Heal, *The Inorganic Heterocyclic Chemistry of Phosphorus, Sulphur and Nitrogen,* Academic Press, London, **1980.**

C. W. Allen, *Chem. Rev.* **1991,** *91,* 119.

V. Chandrasekhar, M. G. Muralidhara, I. I. Selvaraj, *Heterocycles* **1990,** *31,* 2231.

J. D. Woollins, *Non-Metal Rings, Cages and Clusters,* J. Wiley, Chichester, **1988.**

J. E. Mark, H. R. Allcock, R. West, *Inorganic Polymers,* Prentice Hall, **1992.**

3.21 Inorganic (Carbon-Free) Chelate Rings: A Dithioimidodiphosphinato Ligand and Some of its Metal Complexes

Ionel Haiduc

Special Safety Precautions

1. Chlorodiphenylphosphine, Ph_2PCl, is corrosive, irritating and very toxic by inhalation. *Always* use it in a good fume cupboard. Spillages should be treated with solid sodium carbonate and washed with large amounts of water.
2. Carbon disulfide is a foul smelling, toxic and very flammable liquid. Petroleum ether and diethyl ether are also very flammable, volatile liquids. Keep these properties in mind when handling these solvents. Don't keep large amounts around.
3. Grease the glass joints with silicone grease.

Every coordination chemistry textbook describes [1, 2] the very stable chelate ring complexes of β-diketones, of which acetylacetone (pentane-2,4-dione, abbreviated *acac*) is the best known. This ligand forms a uninegative anion, **a**, with delocalized π-electrons, which is able to form neutral chelate rings, **b**, in complexes of the type $M(acac)_n$ (n = the oxidation state of the metal). These six-membered rings are very stable and even exhibit some aromatic character. Replacement of oxygen by sulphur results in the formation of a dithio analogue (abbreviated *sacsac*), which is also able to form similar sulphur based chelate rings, **c**.

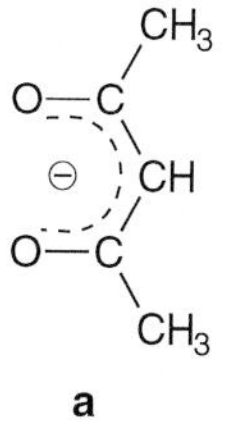

a 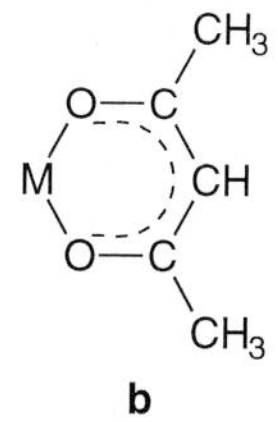 **b** 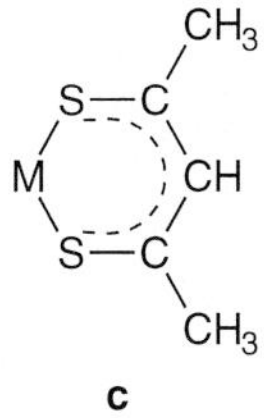 **c**

Whilst the oxygen/sulphur replacement is known by almost any student of inorganic chemistry, fewer people are aware that in the six-membered chelate rings **b**, and **c**, the carbon atoms can also be replaced, leading to fully inorganic, carbon-free chelate rings. Thus, the C⋯C⋯C sequence can be replaced [3] by a P⋯N⋯P sequence to produce rings **d**, and **e**.

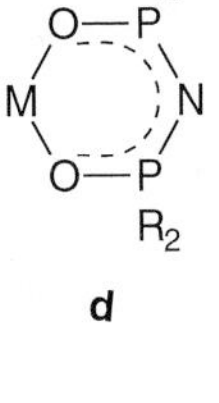

d 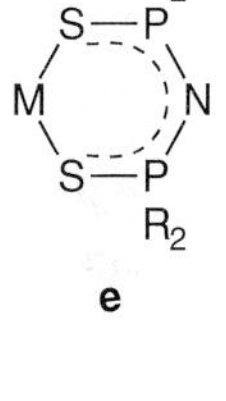 **e**

In this case, the phosphorus atoms will bear organic groups (exocyclic substituents), *e.g.* R = CH_3 or C_6H_5.

In this experiment, you will prepare the free ligand as the neutral compound **f**, tetraphenyldithiodiphosphinylimide, which is readily converted to a potassium or ammonium salt as the delocalized anion **g**, and from this, the chelate ring complexes **e** (m = Pb, Fe, Co, Ni).

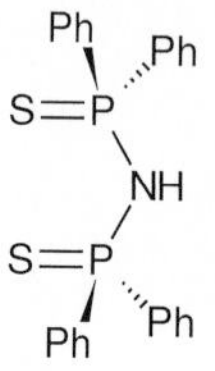

f 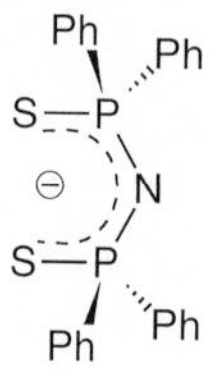 **g**

The structures of the neutral ligand **f**, and of the potassium salt, have been confirmed by X-ray diffraction.

The ligand described here forms the spirobicyclic complexes **h** with divalent metals, but other types of compounds are also known.

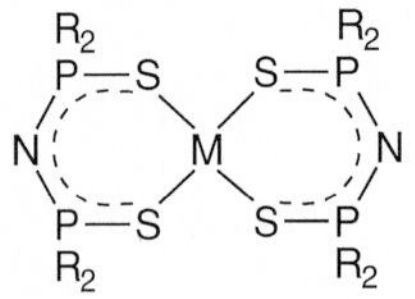

h

Experimental

a) $HN(PPh_2S)_2$

$$2\,Ph_2PCl + Me_3SiNHSiMe_3 \rightarrow Ph_2P-NH-PPh_2 + 2\,Me_3SiCl$$

$$Ph_2P-NH-PPh_2 + 2\,S \rightarrow Ph_2\underset{\underset{S}{\|}}{P}-NH-\underset{\underset{S}{\|}}{P}Ph_2$$

Refer to the set up in Figure **3.21-1.** A solution of hexamethyldisilazane (8.4 g) in 75 cm^3 of toluene is placed in a two-neck round-bottomed flask (250 cm^3) provided with a pressure-equalised dropping funnel on one neck and a condenser adapted for distillation on the second neck. The distillate (see further) will be collected in a round-bottomed flask (200 cm^3) and the system is closed with a calcium chloride tube. The flask is heated with a heating mantle and the contents stirred with a teflon-coated magnetic stirrer.

A solution of chlorodiphenylphosphine (23 g) in toluene (75 cm^3) is placed in the dropping funnel. This is added dropwise to the solution of hexamethyldisilazane while heating. As the chlorotrimethylsilane forms, it distills out. The toluene solvent boils at 110 °C and the heating is maintained below this temperature (preferably at 80–90 °C). The heating at

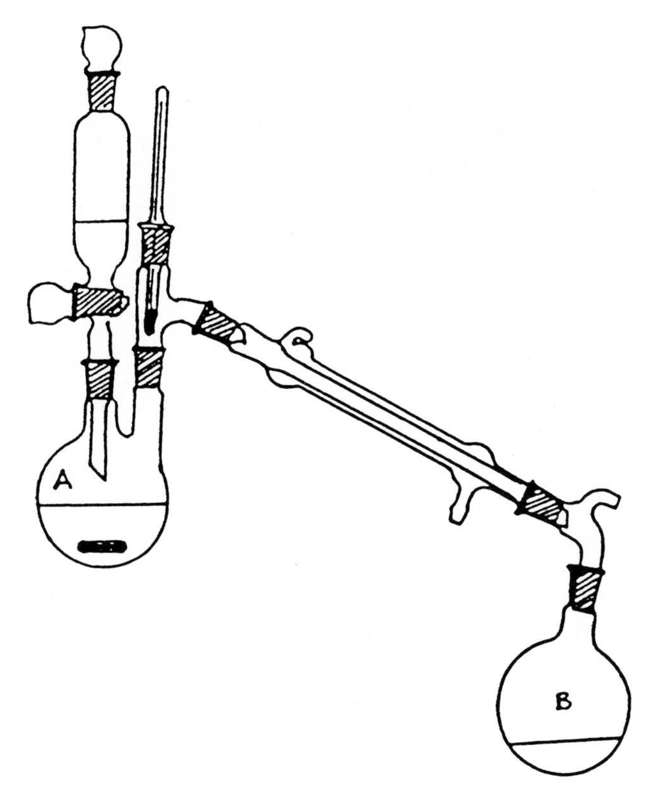

Figure 3.21-1. Apparatus for $HN(PPh_2S)_2$ synthesis.

80–90 °C is maintained for *ca.* 3h, to completely remove chlorotrimethylsilane. Then, the mixture is allowed to warm up to the boiling temperature of toluene and *ca.* 70–80 cm^3 of toluene is distilled out. The mixture is cooled to room temperature, the dropping funnel is removed and replaced with a stopper. Now 3.3 g elemental sulphur are added to flask A and the mixture is heated again to 80–90 °C and stirred for another 3h, until almost all sulphur is dissolved. Upon cooling, the precipitated solid is filtered on a glass frit, thoroughly washed with cold toluene, carbon disulfide (to remove unreacted sulphur from the product) and petroleum ether, and dried in air.

Calculate the yield of the first crop and compare the melting point of your product with the literature value (m.p.: 213.5–214.5 °C). Record and interpret the infra-red spectrum, identifying the P–N, P–S and N–H vibration bands.

b) $K^+[N(PPh_2S)_2]^-$

$$Ph_2(S)PNHP(S)Ph_2 + KO^tBu \rightarrow K^+[N(PPh_2S)_2]^- + {}^tBuOH$$

In a round-bottomed flask (200 ml), 10 g of $HN(PPh_2S)_2$ in 100 ml methanol is treated with 2.5 g potassium *tert*-butoxide. The solution is evaporated to dryness and the product is recrystallised from methanol to give the colorless crystalline potassium salt (m.p.: 363–366 °C).

c) $NH_4^+[N(PPh_2S)_2]^-$

$$Ph_2(S)PNHP(S)Ph_2 + NH_3 \rightarrow NH_4^+[N(PPh_2S)_2]^-$$

Dry ammonia gas is bubbled through a solution of 1 g $HN(PPh_2S)_2$ in 100 ml CH_2Cl_2. The ammonium salt precipitates as a needle-like crystalline solid, which decomposes on heating with evolution of ammonia and regeneration of the neutral ligand. Therefore, the melting point will be that of the ligand.

Some metal complexes

Lead

Lead acetate (0.4 g) is treated with the potassium salt (1 g) in 20 cm^3 methanol. The bright yellow precipitate which is formed can be recrystallised from a methylene chloride/diethyl ether mixture (1 : 1). M.p.: 239–241 °C.

Iron

A pale green precipitate is formed by adding an excess of $[Et_4N][FeCl_4]$ (*ca.* 1 g) to a solution of the potassium salt (0.9 g) in 100 cm^3 methanol. The solid is filtered, washed with methanol and dried. The complex $Fe[(SPPh_2)_2N]_2$ can be recrystallised from a mixture of CH_2Cl_2 and petroleum ether to give pale green crystals (m.p.: 285–286 °C).

Cobalt

The cobalt complex can be prepared similarly, from $CoCl_2 \cdot 6H_2O$ and the potassium salt of the ligand, in the form of green, air-stable crystals (m.p.: 304–309 °C).

Nickel

This brown complex (m.p. 265–266 °C (dec)) can be obtained as above, from $[Et_4N][NiCl_4]$ and the potassium salt. Calculate yourself the amounts of reagents required.

Other metal complexes described in the literature contain manganese, palladium, platinum, copper, gold, silver, indium, zinc, bismuth and tellurium. The structures of some of these metal compounds have been confirmed by X-ray diffraction. Organometallic derivatives **i**, with R_nM = PhTe, R_2Au, Me_3Sn and Me_2Sn have also been reported.

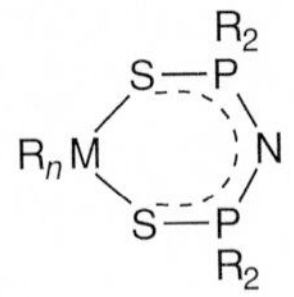

i

References

[1] F. A. Cotton, G. Wilkinson, *Advanced Inorganic Chemistry,* 5th ed., John Wiley & Sons, **1988,** p. 477.

[2] S. Kawasuchi, *Coord. Chem. Revs.* **1986,** *70,* 51.

[3] For a review of inorganic (carbon-free) chelate rings see I. Haiduc, I. Silaghi-Dumitrescu, *Coord. Chem. Revs.* **1986,** *74,* 127.

3.22 Amidosulfuric Acid, Trisilveramidosulfate Monohydrate and $SO_3 \cdot N(CH_3)_3$ Complexes

Antonín Růžička and Zdirad Žák

Special Safety Precautions

1 a) Fuming sulphuric acid (30% SO_3, oleum) is a very corrosive substance and reacts violently with water. Extensive care must be taken to avoid contact with skin – it can cause severe burns.
   b) You must work in a fume cupboard.
   c) Protective clothing, gloves and goggles should be worn at all times, do not inhale the vapours.
   d) For measuring, use only a dry graduated cylinder.

2. Clamp securely the Erlenmeyer flask with the reaction mixture in the water bath to prevent overturning.

3. The residue of oleum after the isolation of HSO_3NH_2 may be destroyed by adding it *dropwise* to a large amount of cold water.

4. CH_3I and HCOH are highly toxic and mutagenic. All manipulations should be done in a fume cupboard, do not inhale the vapours.

5. Solutions of $AgNO_3$ leave black stains of elemental silver on contact.

Amidosulphuric acid, HSO_3NH_2, is a white non-hygroscopic crystalline solid. It is a strong acid and serves *inter alia* as a titrimetric standard substance for acidimetry. In strongly alkaline solutions, it is possible to exchange up to all three H atoms in its molecule by metal ions. Thus, various salts have been prepared *e.g.* $KSO_3NHAg \cdot H_2O$, $NaSO_3NHA_g$, KSO_3NHAg and $AgSO_3NAg_2 \cdot H_2O$.

This experiment involves the preparation of HSO_3NH_2, $AgSO_3NAg_2 \cdot H_2O$ and, by the reaction of the trisilver salt with CH_3I, $SO_3 \cdot N(CH_3)_3$. This donor-acceptor complex is usually prepared by the direct reaction of SO_3 with $N(CH_3)_3$ and its formation during the action of CH_3I on $AgSO_3NAg_2 \cdot H_2O$ indicates that all three Ag ions of the trisilver salt are bonded to the nitrogen atom as has been confirmed by X-ray structure analysis. A large excess of $AgNO_3$ is required for the preparation of $AgSO_3NAg_2 \cdot H_2O$, but silver can be recovered from the mother liquid in elemental form.

Experimental

a) HSO_3NH_2

The formation of the amidosulphuric acid can be expressed by the overall Eqn. (1).

$$OC(NH_2)_2 + H_2SO_4 \cdot SO_3 \rightarrow 2HSO_3NH_2 + CO_2 \quad (1)$$

15 g of urea is dissolved in 25 ml of concentrated H_2SO_4 (*caution* – the mixture warms up strongly!) in a 500 ml Erlenmeyer flask. The flask is immersed in a boiling water bath (clamp securely!) and 60–70 ml of 30% oleum is added in small quantities (5–10 ml). After approx. 30 ml of oleum has been added to the flask, it is necessary to stop the addition of oleum and wait until the reaction starts (CO_2 evolves), otherwise a violent evolution of CO_2 can occur and the reaction mixture will be lost in the water bath. When the reaction proceeds, the addition of oleum is continued in the same manner. After all oleum has been added, leave the flask in the boiling water bath for a further 30 minutes. Then remove it from the bath and cool it down in an ice bath (ice water bath). HSO_3NH_2 crystallises out of the solution. Filter the product off through a sintered glass funnel, placing a trap between the filtering medium and the suction pump, and press out all the mother liquid you can. This raw amidosulphuric acid is contaminated by sulphuric acid which can be removed by recrystallisation from water as follows.

Dissolve the crude HSO_3NH_2 at 70 °C in as little water as possible (see Note), filter if necessary through a sintered glass funnel and cool in an ice bath to 0–2 °C. Vacuum filter the crystallised acid using a sintered glass funnel and a suction flask, wash with ethanol (15 ml), ether (15 ml) and allow to dry in air.

Note: The solubility of HSO_3NH_2 at 70 °C is 40 g of HSO_3NH_2 in 100 g of H_2O.

Test for SO_4^{2-} *ions:* Dissolve a pea-sized quantity of dry HSO_3NH_2 in 2–3 ml water in a test tube. Add a few drops of a 5% solution of $BaCl_2$. The presence of SO_4^{2-} ions is manifested by the formation of a white precipitate of $BaSO_4$ (barium amidosulphate is soluble in water).

Acidimetric titration of HSO_3NH_2 *with a 0.1 M solution of NaOH:* Use standard procedure and calculate the purity of the prepared acid.

b) $AgSO_3NAg_2 \cdot H_2O$

The preparation proceeds according to Eqns. (2) and (3).

$$2HSO_3NH_2 \rightarrow K_2CO_3 \rightarrow 2KSO_3NH_2 + H_2O + CO_2 \quad (2)$$

$$KSO_3NH_2 + 3AgNO_3 + 2NH_4OH \rightarrow AgSO_3NAg_2 \cdot H_2O + KNO_3 + 2NH_4NO_3 + H_2O \quad (3)$$

Dissolve 3.0 g HSO_3NH_2 in 50 ml water in a beaker and neutralise the solution with solid K_2CO_3. Add 5 ml of conc. NH_4OH solution and then, under continuous stirring with a magnetic stirrer, add slowly the solution of 25 g $AgNO_3$ in 100 ml of water. A yellow precipitate of an amorphous form of $AgSO_3NAg_2 \cdot H_2O$ is formed immediately. Filter off approx. 1/2 of the precipitate using a sintered glass funnel. Return the filtrate to the remaining unfiltered precipitate, cover the beaker with a watch glass and allow to stand for several days in a dark place. The bright yellow amorphous precipitate will recrystallise into a yellow-green crystalline product.

Wash the amorphous salt on the filter with a small amount of water (25 ml), ethanol (15 ml) and ether (15 ml), and allow to dry in air. Isolate the crystalline product in the same way. *Note:* Before washing the crystalline product with ethanol and ether, save the mother liquid for silver revovery!

c) $SO_3 \cdot N(CH_3)_3$

The reaction scheme is given by Eqn. (4).

$$AgSO_3NAg_2 \cdot H_2O + 3\,CH_3I \rightarrow SO_3 \cdot N(CH_3)_3 + 3\,AgI + H_2O \qquad (4)$$

Transfer 2.0 g of the dry amorphous salt to a 25 ml distillation round-bottomed flask with a ground joint. Attach a reflux condenser to the flask and using a funnel, pour 3.0 ml of CH_3I down the condenser onto the silver salt. After a few seconds, an exothermic reaction starts and the contents of the flask warms up to the boiling point of CH_3I. (*Note:* the crystalline yellow-green salt reacts with CH_3I very slowly). After 15 minutes, distill off under vacuum the unreacted methyliodide and the water formed into a cold trap until dry (use a water vacuum pump). (Dispose of the distillate as instructed by the supervisor).

Sublime $SO_3 \cdot N(CH_3)_3$ from the dry reaction product *in vacuo* at 160–170 °C (use an oil bath) using an oil vacuum pump and the apparatus depicted in the Figure **3.22-1**. The crystals of the complex thus obtained are up to several centimeters long.

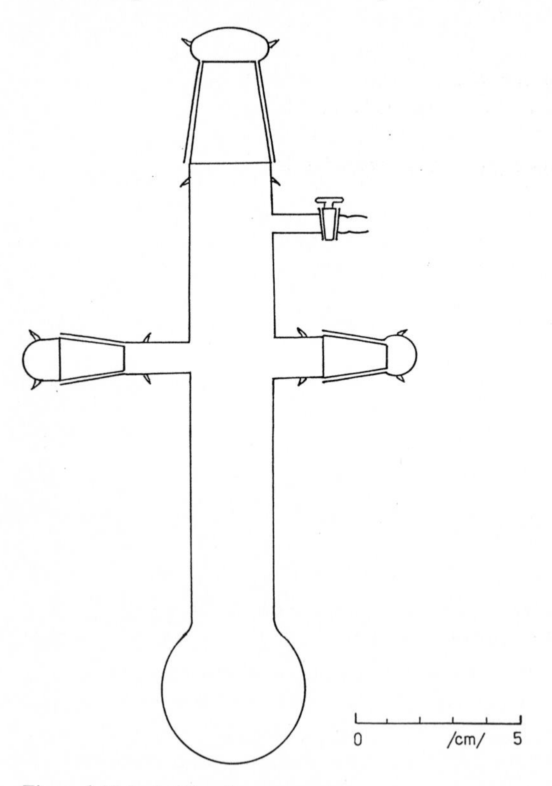

Figure 3.22-1. Sublimation apparatus.

Determine the melting point in a sealed glass capillary (240 °C). Record the IR spectrum and compare it (400–1600 cm^{-1}, KBr pellet) with that in the literature.

d) Recovery of silver

$$2\,AgNO_3 + NaOH \xrightarrow{H_2O} Ag_2O \cdot xH_2O + 2\,NaNO_3$$

$$Ag_2O + NaOH + HCOH \rightarrow 2\,Ag + HCOONa + H_2O$$

Transfer the mother liquid from the filtration of $AgSO_3NAg_2 \cdot H_2O$ into a 1000 ml beaker and keep adding a 10% solution of NaOH until all the silver is precipitated as hydrated silver oxide. Then add the same amount of 10% hydroxide solution as was required for the silver oxide precipitation and bring the contents to boil. Add *dropwise* under continuous stirring a 37% water solution of formaldehyde until all the Ag_2O is reduced to elemental silver – the colour of the suspension will change from dark brown to grayish and the supernatant will clear rapidly when stirring is interrupted. Allow the contents to cool, decant the supernatant, fill the beaker with distilled water, stir, leave to settle and decant. Repeat this procedure several times until the neutral reaction of the liquid is reached. Vacuum filter, wash the silver with water and dry in an oven at 100 °C.

References

K. A. Hofmann, E. Bielsacki, E. Söderlund, *Ber. Dtsch. Chem. Ges.* **1912,** *45,* 1731.
R. Paetzold, K. Dostál, A. Růžička, *Z. Anorg. Allg. Chem.* **1966,** *348,* 1.
A. B. Burg, *J. Amer. Chem. Soc.* **1943,** *65,* 322.
F. Belaj, Ch. Kratky, E. Nachbaur, A. Popitsch, *Mh. Chem.* **1987,** *118,* 947.
F. Watari, *Z. Anorg. Allg. Chem.* **1964,** *332,* 322

3.23 Four Methods for Analysing First Order Kinetics

Véronique Pimienta, Dominique Lavabre, Jean-Claude Micheau and Gaston Levy

Special Safety Precautions

1. Mercury acetate is toxic. Its toxicity stems from the ability of mercury to combine with sulphhydryl groups (–SH) that are essential for enzymatic activity.
2. Diphenylthiocarbazone ($C_{13}H_{12}N_4S$) is known as dithizone and has long been used as a reagent for the colorimetric analysis of metal traces. Exercise caution as it is potentially dangerous.
3. Concentrated ammonia solutions must always be used in a fume cupboard.
4. Xylene is flammable and toxic. Significant neuro and cardiac toxicity is described for this compound.
5. Limit exposure values: mercury vapour (0.05 mg m^{-3}); ammonia: (18 mg m^{-3}); xylene (130 mg m^{-3}).

A chemical reaction always corresponds to the relaxation of a nonequilibrium situation ($\Delta G \rightarrow 0$). The concentrations of the substrates, intermediates and products change as a function of time. The simplest method to induce a chemical reaction is to mix two compounds with different chemical potentials. An alternative and convenient way is to exploit the property of photochromism (a single compound + light). A photochromic compound is one that undergoes a reversible colour change under UV or visible irradiation. Photochromic compounds are used in such diverse applications as optical information storage, imaging and in the construction of sunglasses.

The purposes of this study is to familiarise yourself with four general methods for analysing first-order kinetics and investigating reaction mechanisms. In this experiment, you will be examining the thermally reversible orange/blue photochromism of the mercury dithizonate bis(1,5-diphenylthiocarbazonato-*NS*)mercury(II)). The kinetic study is based on the first-order thermal return reaction blue → orange. The colour change is easily monitored using a UV/visible spectrophotometer or a colorimeter.

A typical pattern of the absorbance at 606 nm (blue isomer) prior to, during and after irradiation of mercury dithizonate in xylene solution is shown in Figure **3.23-2.**

As the process can be repeated many times without appreciable fatigue, a single sealed or stoppered cell can be used for several kinetic runs.

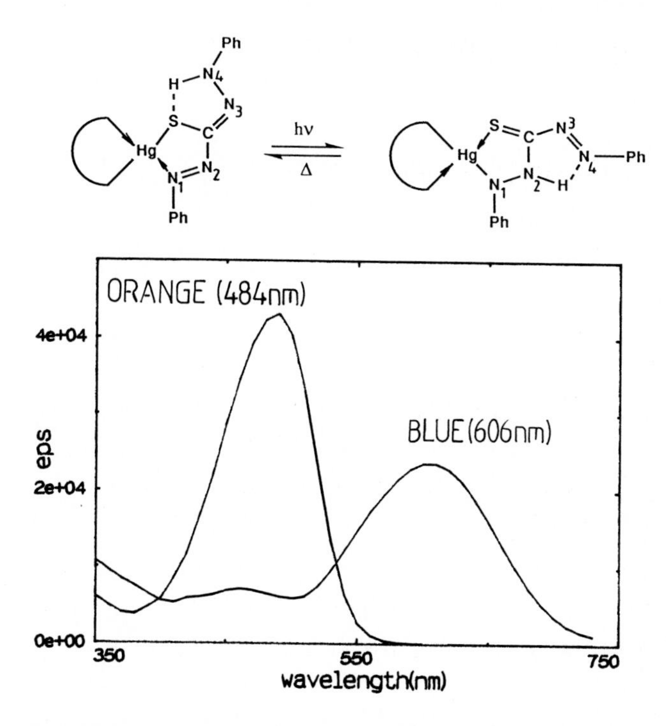

Figure 3.23-1. The photochromic mechanism of mercury dithizonate involves a N_4 to N_2 H-transfer and a *cis-trans* isomerization about the $C{=}N_3$ bond. There is an $S-H$ hydrogen bond and a $PhN_1N_2CN_3$ conjugated system in the stable orange isomer and an N_4-H bond and a Ph_4N_3CS conjugated system in the blue form.

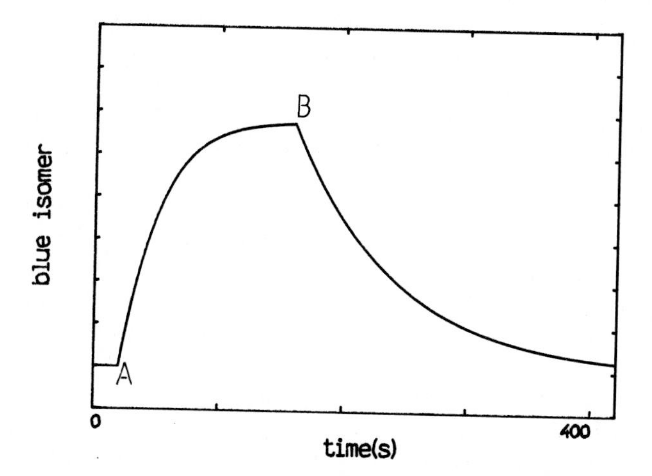

Figure 3.23-2. Photochemical formation and thermal decay of the blue isomer of mercury dithizonate. A) light on; B) light off. Note that under irradiation, the system reaches a steady state during which the thermal return rate exactly counterblances the photochemical rate. If is not necessary to start from the steady state for kinetic measurements.

Experimental

a) Mercury(II) dithizonate

3.2 g (0.01 mole) of mercury(II) acetate are dissolved in 6 M aqueous ammonia and added to 5.1 g (0.02 mole) of dithizone in hot 6 M ammonia. The solution is then poured onto ground ice and the mercury dithizonate precipitate filtered. The product is purified by recrystallisation from chloroform and is dried under vacuum. Red crystals are obtained (m. p.: 225–226 °C dec.) Analysis for $C_{26}H_{22}N_8S_2Hg$, FW = 710.6 (theoretical): C, 43.91; H, 3.10; N, 15.76; S, 9.01; Hg, 28.22. (Found): C, 43.8; H, 3.2; N, 15.6; S, 9.5; Hg, 27.2.

b) Kinetic and spectrophotometric measurements

Dry xylene is a good solvent for kinetic and spectrophotometric measurements. A 2×10^{-5} M mercury dithizonate solution exhibits an absorbance of about 0.86 at 484 nm ($\varepsilon(484) = 43000\ M^{-1}\ cm^{-1}$). Under irridiation (sunlight, overhead projector or flashlamp), the orange isomer is converted to the blue isomer and the solution turns deep blue ($\lambda_{max} = 606$ nm). After vigorous shaking, the cuvette (1 cm × 1 cm) is placed in the compartment of a spectrophotometer and the evolution of the absorbance at 606 nm (decreasing) or at 484 nm (increasing) is recorded continuously. In the dark, the blue complex reverts spontaneously to the stable orange isomer, following first-order kinetics. Be careful to screen from stray light which could induce partial rephotolysis.

c) Data processing

All studies of first-order kinetics are based on Eqn. (1).

$$y_t - y_\infty = (y_0 - y_\infty)\ \exp(-k_{obs}t) \quad (1)$$

in which y_t, y_0 and y_∞ are the magnitudes of the observed signal (here the absorbance) at time t, 0 and ∞; k_{obs} is the apparent first-order rate constant.

There are several methods for determining an accurate value of k_{obs}. Four main methods are presented here. If an estimation of y_∞ is easily available, the half-time and semi-log methods can be used.

Half-time method

The half-time equation (2) is derived directly from Eqn. (1) since at $t = t_{1/2}$, $(y_t - y_\infty)/(y_0 - y_\infty) = 1/2$. For pure first-order kinetics, the successive values of $t_{1/2}$ must be independent of the choice of y_0 (Fig. **3.23-3**).

$$k_{obs} = \ln 2/t_{1/2} \quad (2)$$

Semi-log method

The semi-log equation (3) is also easily derived from Eqn. (1).

$$\ln|y_t - y_\infty| = -k_{obs}t + \ln|y_0 - y_\infty| \quad (3)$$

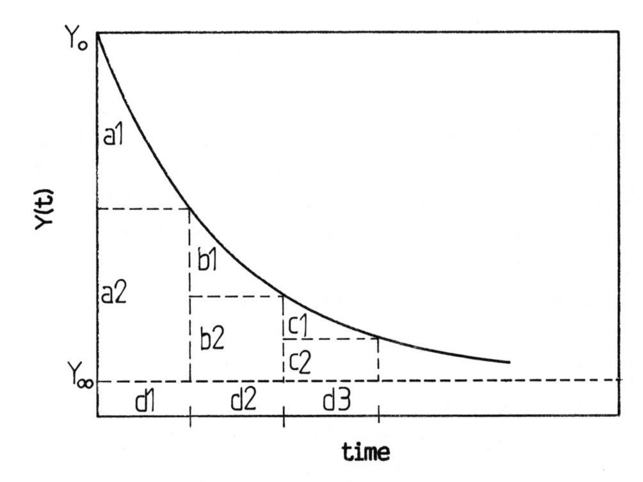

Figure 3.23-3. First-order decay leading to a good estimation of y_∞. Half-time method ($a_1 = a_2$; $b_1 = b_2$; $c_1 = c_2$). Note the regularity of the successive half-time ($d_1 = d_2 = d_3$).

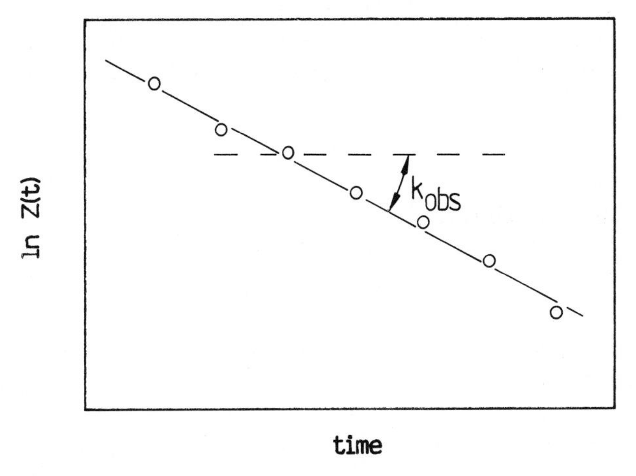

Figure 3.23-4. Linearization of a first-order kinetic curve. Semi-log plot: $Z(t) = |y_t - y_\infty|$. Guggenheim plot: $Z(t) = |y_{t+\Delta} - y_t|$. In both cases, k_{obs} is calculated from the slope of the straight line.

In the semi-log plot (Fig. **3.23-4**), since all the experimental points are (in principle) included in the linear regression analysis, it is more accurate than the simple half-time method.

If an estimation of y_∞ is not possible, one must use a time-lag method whereby the reaction time is divided into regular intervals (Δ), the value of which being chosen so as to obtain around six or more regular intervals over the whole reaction curve. Guggenheim's and the R/R_Δ methods are time-lag methods.

Guggenheim's method

It is quite similar to the semi-log plot, although the value of y_∞ is not required. Guggenheim's formula (Eqn. 6) derives from the relationship in Eqn. (1).

Thus, at any t:

$$y_t = (y_0 - y_\infty)\exp(-k_{obs}t) + y_\infty \quad (1)$$

and at $(t + \Delta)$:

$$y_{t+\Delta} = (y_0 - y_\infty)(\exp(-k_{obs}t))(\exp(-k_{obs}\Delta)) + y_\infty \quad (4)$$

and

$$y_{t+\Delta} - y_t = (y_0 - y_\infty)(\exp(-k_{obs}t))(\exp(-k_{obs}\Delta) - 1) \quad (5)$$

Taking the ln leads to Gugenheim's formula (6). See Figure **3.23-4**.

$$\ln|y_{t+\Delta t} - y_t| = -k_{obs}t + (\exp(k_{obs}\Delta) - 1)\ln|y_0 - y_\infty| \quad (6)$$

R/R_Δ method (ratios of rates)

This is a convenient alternative to Guggenheim's method which can be used directly on the recorded chart. The R/R_Δ relationship in Eqn. (10) derives from the rates at t and $(t + \Delta)$. At t:

$$R_t = (dy/dt)_t = -(y_0 - y_\infty)(\exp(-k_{obs}t))k_{obs} \quad (7)$$

and at $(t + \Delta)$:

$$R_{t+\Delta} = (dy/dt)_{t+\Delta} = -(y_0 - y_\infty)\exp(-k_{obs}t)\exp(-k_{obs}\Delta)k_{obs} \quad (8)$$

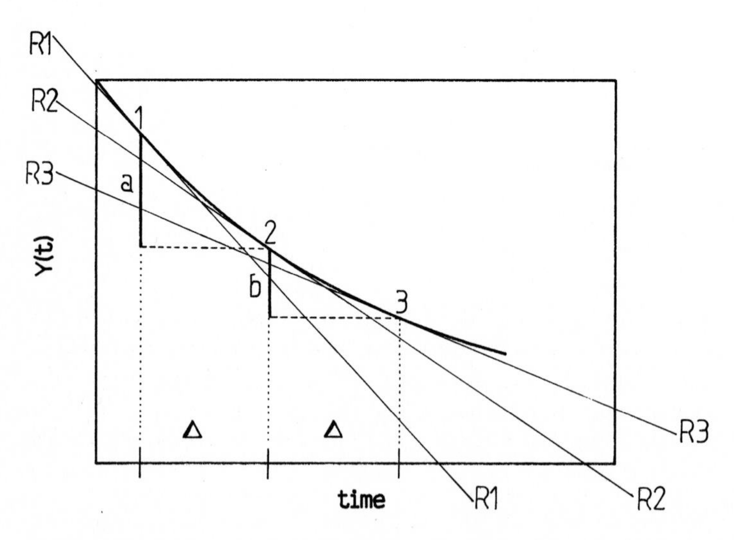

Figure 3.23-5. First-order decay leading to a poor estimation of y_∞. Δ is the time-lag. a), b) Guggenheim calculations of the successive steps $y_t - y_{t+\Delta}$. R_1, R_2, R_3: successive rates $(dy/dt)_t$, $(dy/dt)_{t+\Delta}$, $(dy/dt)_{t+2\Delta}$ for the R/R_Δ method. For the sake of clarity, only 2Δ intervals and 3 tangents are shown.

Taking the ratio $R/R_{t+\Delta}$ and the ln reduces to Eqn. (9).

$$k_{obs} = \ln[R_t/R_{t+\Delta}]/\Delta \tag{9}$$

With pure first-order decays, the successive ratio must remain constant.

$$k_{obs} = (\ln r_m)/\Delta \tag{10}$$

where r_m is the mean value of the successive ratios $R_t/R_{t+\Delta}$ along the whole kinetic curve.

Using a continuous recorded kinetic curve, it is quite easy to draw the tangents carefully by hand and estimate their respective slopes (dy/dt). As only the successive ratios of rates are needed, convenient arbitrary units can be used. This is shown in Figure **3.23-5**.

When used in a comparative fashion, these four methods give sufficient accuracy to demonstrate first-order decay and detect the effects of the purity of the solvent (traces or acid, bases, free dithizone, photolysis products, water, etc.), which accelerate the thermal return rate constant. A spontaneous monomolecular isomerization and/or a catalyzed pseudo first-order bimolecular process with a large excess of a catalytic species ([cat]) could be deduced from the data obtained. Thus $k_{obs} = k_1 + k_2[cat]$, the usual range lies between $5 \times 10^{-3} < k_{obs} < 5 \times 10^{-2}\ s^{-1}$.

References

R. L. Petersen, G. L. Harris, *J. Chem. Educ.* **1985,** *62,* 802.
A. T. Hutton, *J. Chem. Educ.* **1986,** *63,* 888.
B. Borderie, D. Lavabre, G. Levy, J. C. Micheau, *J. Chem. Educ.* **1990,** *67,* 459.

3.24 Symmetries of Inorganic Molecules and Applications of Group Theory and Graph Theory

D. Michael P. Mingos and Anthony F. Orchard

Symmetries of Inorganic Molecules and Applications of Group Theory and Graph Theory

Part 1: Symmetry and Group Theory

In everyday language, we use the word symmetry in one of two ways and, correspondingly, the Oxford English Dictionary gives the following two definitions:

1. Mutual relation of the parts of something in respect of magnitude and position; relative measurement and arrangement of parts; proportion.
2. Due or just proportion, harmony of parts with each other and the whole; fitting, regular or balanced arrangement and relation of parts of elements; the condition or quality of being well proportioned or balanced.

The first definition of symmetry clearly has a more scientific ring to it than the second, which is related more to the aesthetic concepts of harmony and beauty, but is still insufficient to define in an exact sense the features of a molecule which lead us to define it as symmetric.

The involvement of symmetry in chemistry has a long history; in 540 BC the society of Pythagoras held that the earth had been produced from the cube, fire from the tetrahedron, air from the octahedron, water from the icosahedron, and the heavenly sphere from the regular dodecahedron.

The first resolution of an optically active compound by Pasteur was achieved by the careful manual separation, with the aid of a magnifying glass, of crystals of tartaric acid, which has enantiomorphic crystal modifications for the two optical antipodes of the acid. This manifestation of optical activity led to the realisation that the carbon atom in organic compounds is tetrahedral, and to the development of modern stereochemistry. During the last fifty years, the symmetry properties of crystals and molecular lattices have simplified the solution of thousands of X-ray crystal structure determinations. The results of these determinations have made chemists appreciate more fully the symmetric arrangements adopted by atoms in crystals.

Today, a chemist intuitively uses symmetry every time he recognises which atoms in a molecule are equivalent; for example, in pyrene (Fig. **3.24-1**), it is easy to see that there are three sets of equivalent hydrogen atoms. The appreciation of the number of equivalent atoms in a molecule leads to the possibility of determining the number of substituted molecules that can exist: for example, there are only three possible monosubstituted pyrenes.

Symmetry also plays an important part in the determination of the structure of molecules using spectroscopic techniques. Here, a great deal of evidence comes from the measurement of IR spectra, UV spectra, dipole moments and optical activities. All concern properties which depend upon molecular symmetry.

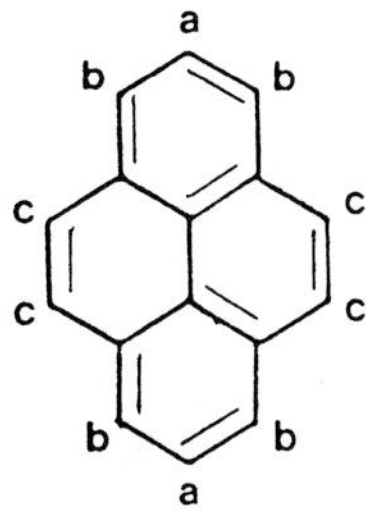

Figure 3.24-1. Pyrene.

The basis of our understanding of molecular structure (rather than its determination) lies in quantum mechanics and, therefore, any consideration of the role of symmetry in chemistry is basically a consideration of its role in quantum mechanics. The link between symmetry and quantum mechanics is provided by that branch of mathematics known as ***group theory.***

Symmetry elements and operations

We have noted that symmetry is related to equivalence, but a precise definition of symmetry must establish clearly the actual manipulations or mental operations necessary to establish equivalents. Also, it is necessary to make a clear distinction between symmetry operations and symmetry elements.

A **symmetry operation** is defined as ***a movement** of a molecule to a new orientation in which every point in the molecule is coincident with an equivalent point (or the same point) of the molecule in its original orientation.*

A **symmetry element** is ***a geometrical entity** such as a **plane**, a **point** or a **line**, with respect to which one or more symmetry operations may be carried out.*

The symmetry elements and operations necessary for defining molecular symmetry are summarised below:

Symmetry Element	**Symmetry Operation**
1. Plane (σ)	Reflection in the plane (σ)
2. Centre of symmetry (i)	Inversion of all points through the centre of symmetry (i)
3. Proper axis (C_n)	Rotation by an angle of $(360/n)°$ about the axis (C_n)
4. Improper axis (S_n)	Rotation by $(360/n)°$ followed by reflection in a plane perpendicular to the rotation axis (S_n)

Symmetry operations are perhaps best explained with reference to specific examples, and Figure **3.24-2 a** shows the effect of successive symmetry operations on a general point **a**. Rotation by 120° in a clockwise direction about the z-axis, which is directed into and perpendicular to the plane of the page, generates a new point **b**, and a 240° rotation yields a point **c**. By definition, these three points must be related by the symmetry operations C_3 (rotation by

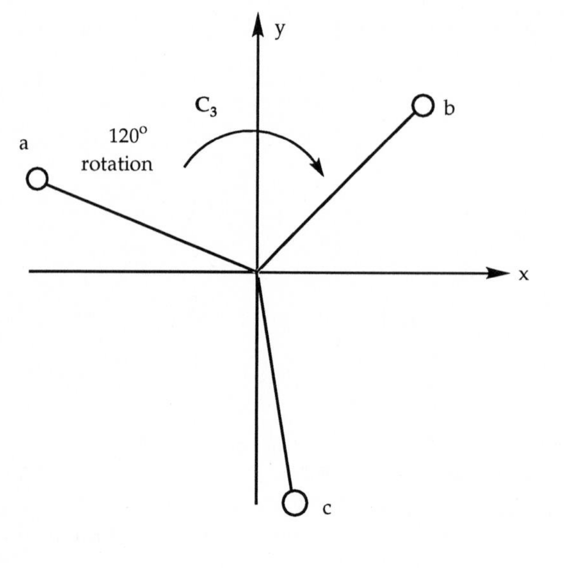

Figure 3.24-2 a

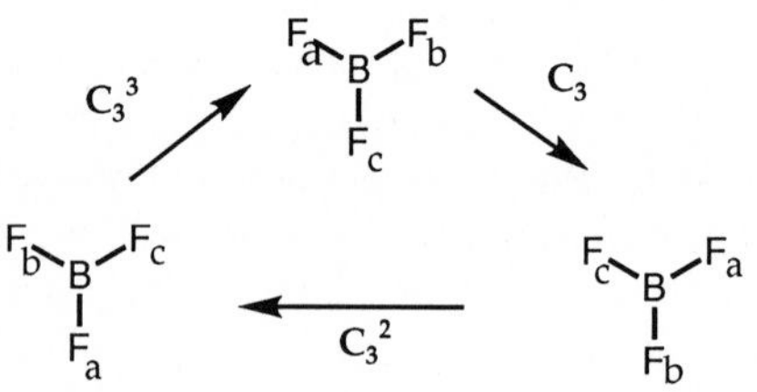

Figure 3.24-2 b

120°) and C_3^2 (rotation by 2 × 120° = 240°) about the symmetry axis z. Clearly the three fluorine atoms in the planar BF_3 molecule (Fig. **3.24-2 b**) must be related by such symmetry operations. Rotation of **2 b** about the symmetry axis by 120° (C_3) interchanges the fluorine atoms and gives an entirely equivalent configuration of BF_3 (Fig. **3.24-2**). A rotation by 240° (C_3^2) similarly gives an equivalent configuration. However, rotation by 360° (which could be designated as (C_3^3) gives a permutation of fluorine atoms identical with the original configuration, and therefore amounts to doing nothing with the molecule. Mathematically, the process of doing nothing is itself a symmetry operation, and is called the ***identity operation, E*** (from the German word *Einheit,* meaning unity). The effect of a 360° rotation can therefore be expressed in terms of the following equivalence, $C_3^3 \equiv E$. Quite generally, $C_n^n = E$.

A molecule may have more than one proper rotation axis. In such situations, the axis associated with the rotation operation of highest order is described as the ***principal axis.***

Figure **3.24-3 a** shows that when the general point **a** is reflected in the yz plane, the point **b** is generated. The hydrogen atoms of the water molecule (Fig. **3.24-3 b**) are clearly related by such a ***reflection operation,*** σ (after the German word *Spiegel,* meaning mirror). If the mirror plane is perpendicular to the principal axis, it is labelled σ_h (h = horizontal), and if it contains the principal axis, σ_v (v = vertical). If the mirror plane contains the principal axis and *bisects* the angle between two 2-fold axes that are perpendicular to the principle axis, it is labelled σ_d (d = diagonal or dihedral). Notice that $\sigma^2 = E$ always.

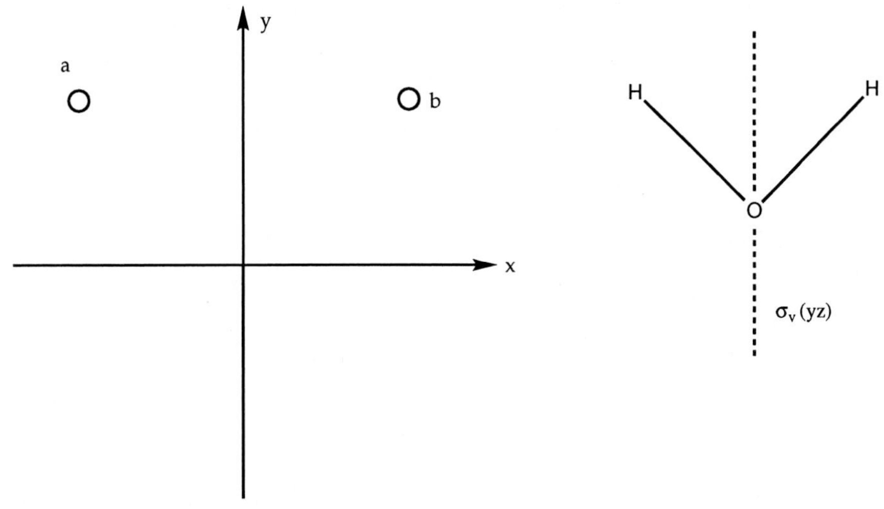

Figure 3.24-3 a **Figure 3.24-3 b**

Figure **3.24-4 a** illustrates the effect of an ***inversion operation i*** on the general point **a**, situated at (x, y, z). The new point that is generated, **b**, has coordinates $(-x, -y, -z)$. In the figure, empty and filled circles have been used to distinguish the points with coordinates z and $-z$ (*i. e.* above and below the plane of the paper) respectively. The chlorine and hydrogen atoms in the dichlorocyclobutane molecule are clearly related by such a symmetry operation (see Fig. **3.24-4 b**). Observe that $i^2 = E$ always.

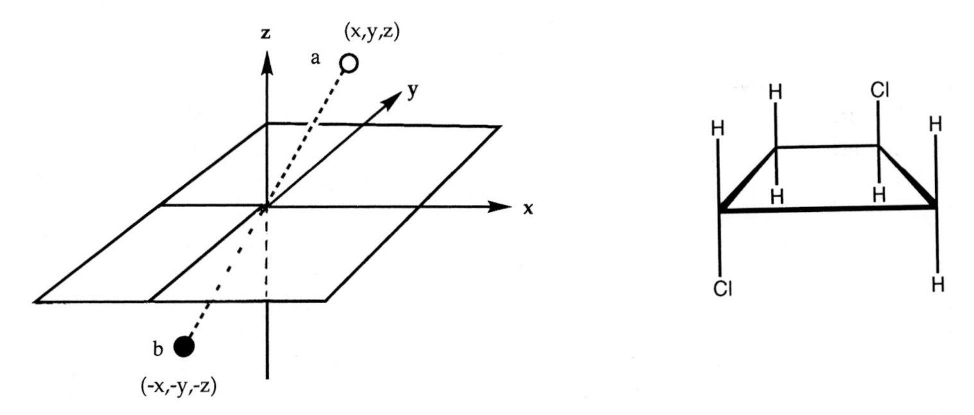

Figure 3.24-4 a **Figure 3.24-4 b**

The ***improper rotation operation*** S_n involves a rotation by $360°/n$ followed by a reflection in a mirror plane perpendicular to the rotation axis. Successive improper rotation operations regenerate the original point after a total rotation of 360° *if n is even,* but *if n is odd,* two complete revolutions are necessary (*i. e.* 720°). Figure **3.24-5 a** illustrates the set of operations S_6, S_6^2, S_6^3, S_6^4, and S_6^5, and Figure **3.24-6 a** the set S_3, S_3^2, S_3^3, S_3^4 and S_3^5. The first set generates a staggered arrangement of six points, which would correspond, for example, to the ligand positions in a regular octahedral complex (Fig. **3.24-5b**), and the second set to an eclipsed arrangement of points, such as the ligands of a trigonal prismatic complex (Fig. **3.24-6b**).

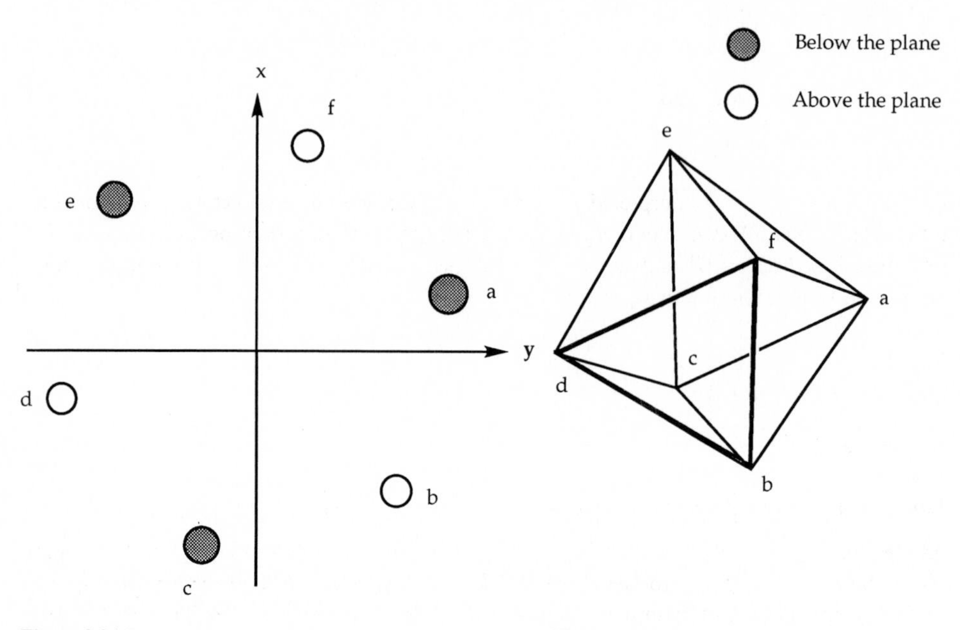

Figure 3.24-5a **Figure 3.24-5b**

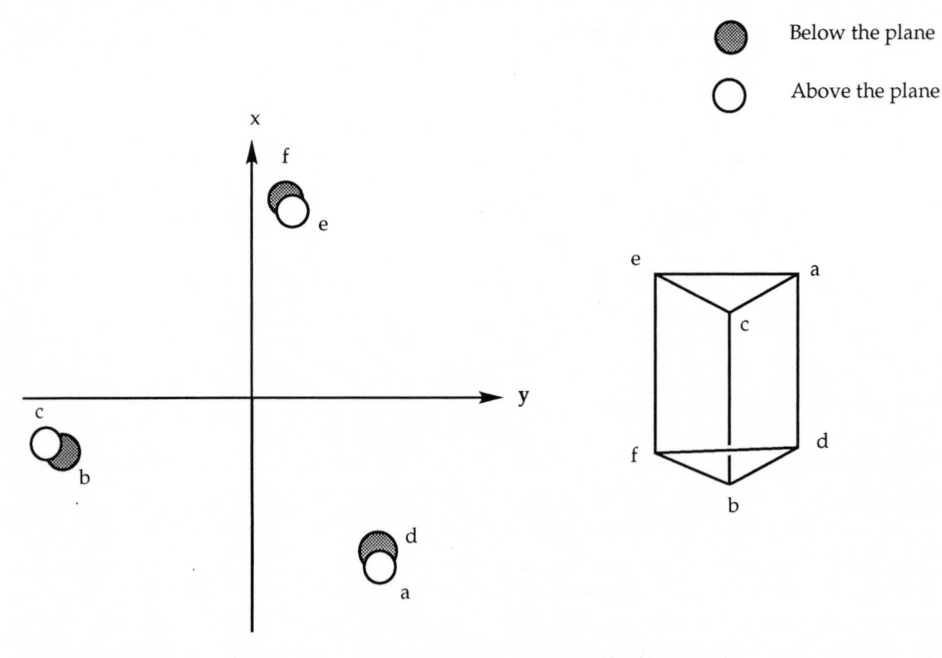

Figure 3.24-6a **Figure 3.24-6b**

It is clear that if a molecule has a C_n axis and a mirror plane of symmetry perpendicular to it, the C_n axis is also an S_n axis. It is easily seen that the application of C_n twice is the same as the application of S_n twice, because the effect of the reflection part of S_n is simply cancelled out, *i. e.* $S_n^2 = C_n^2$. In general, k applications of S_n will give

$$S_n^k = \sigma_h C_n^k \quad \text{if } k \text{ is odd, and}$$
$$S_n^k = C_n^k \quad \text{if } k \text{ is even.}$$

Consequently, S_n^k can be interpreted as a rotation C_n^k followed by a reflection in the horizontal mirror plane only if k is odd; the opposite is also true, *i. e.* a rotation by $2 \times 2\pi/3$ *plus* reflection is written as S_3^5 and not as S_3^2, which would be simply C_3^2. Furthermore, simple arguments lead to the following identities:

$$S_1 = \sigma$$
$$S_n^n = \sigma_h \quad \text{if } n \text{ is odd}$$
$$= E \quad \text{if } n \text{ is even}$$

Multiplication of Symmetry Operations

As was implicit above, the consecutive application of two symmetry operations may be represented algebraically by the product of the individual operations. Let us explore the multiplication of symmetry operations further, taking as a practical example the symmetry of the water molecule (Fig. **3.24-7**). A proper rotation by 180° about the *z*-axis (C_2) results in the interchange of the two hydrogen atoms; successive rotation by 180° returns the molecule to its original configuration and so we can write

$$C_2 \times C_2 = C_2^2 = E$$

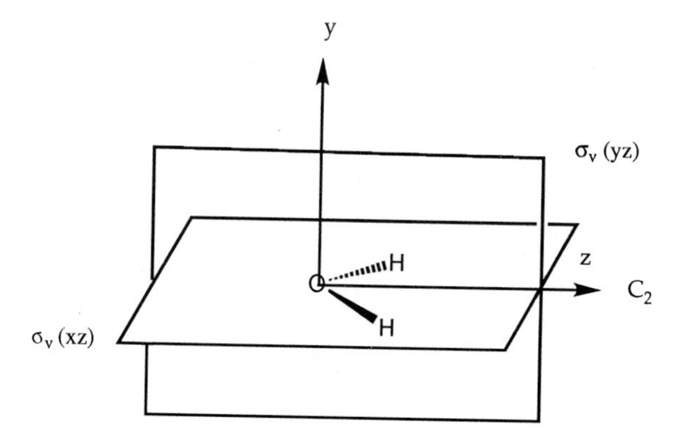

Figure 3.24-7.

The operation of reflection across the mirror plane *yz* (*i. e.* σ_v (*yz*)) also interchanges the hydrogen atoms, while reflection in the mirror plane *xz* (*i. e.* σ_v (*xz*)) leaves the hydrogen atom positions unchanged. Of course, the application of either reflection operation twice can also be written:

$$\sigma_v(xz) \times \sigma_v(xz) = E$$

$$\sigma_v(yz) \times \sigma_v(yz) = E$$

The consecutive application of $\sigma_v(xz)$ and C_2 has the overall effect of interchanging the hydrogen atoms, a result that could alternatively have been achieved by the single operation $\sigma_v(yz)$:

$$\sigma_v(xz) \times C_2 = \sigma_v(yz).$$

For this particular molecule, it would have made no difference if the two operations had been applied in the reverse order (the operations are said, therefore, to *commute*) but this is by no means always true. In general, it is necessary to define precisely the order in which symmetry operations are carried out. The multiplication of symmetry operations can be conveniently represented by a ***multiplication table*** as shown below:

	E	C_2	$\sigma_v(xz)$	$\sigma_v(yz)$
E	E	C_2	$\sigma_v(xz)$	$\sigma_v(yz)$
C_2	C_2	E	$\sigma_v(yz)$	$\sigma_v(xz)$
$\sigma_v(xz)$	$\sigma_v(xz)$	$\sigma_v(yz)$	E	C_2
$\sigma_v(yz)$	$\sigma_v(yz)$	$\sigma_v(xz)$	C_2	E

It is important to note that no new symmetry operations have been generated by these multiplications and, therefore, the four operations E, C_2, $\sigma_v(xz)$ and $\sigma_v(yz)$ constitute a self-contained or ***complete set.***

Mathematical Group

Mathematically, a **group** is defined as a collection of elements {*P, Q, R, etc.*}. with the following properties:

1. There is a rule for "combining" the elements of the group. When two elements of the group are combined, the result must also be a member of the group, *e. g.*

$$P \cdot Q = R$$

2. There is an element *E*, such that

$$E \cdot P = P \cdot E = P$$

and

$$E \cdot Q = Q \cdot E = Q \text{ etc.}$$

E is called the *identity element.*

3. *Associative* (but not necessarily commutative) multiplication applies, *i. e.*

$$P \cdot (Q \cdot R) = (P \cdot Q) \cdot R$$

4. Every element must have and *inverse* that is also an element of the group. The inverse $S = Q^{-1}$ of an element Q is defined in the following way:

$$Q \cdot S = S \cdot Q = E$$

Symmetry Point Groups

It can readily be seen that the set of operations defined for the water molecule satisfy the requirements of a mathematical group.

Rule 1 The combination (multiplication) of two symmetry operations does indeed result in a symmetry operation that is also a member of the group.

Rule 2 Clearly, the identity element is equivalent to the symmetry operation that leaves the molecule unchanged, namely *E*.

Rule 3 Associative multiplication can be shown to hold by the following examples:

$$C_2 \cdot (\sigma_v(xz) \cdot \sigma_v(yz)) = C_2 \cdot C_2 = E$$

$$(C_2 \cdot \sigma_v(xz)) \cdot \sigma_v(yz) = \sigma_v(yz) \cdot \sigma_v(yz) = E$$

Rule 4 For the water molecule, the symmetry operations C_2, $\sigma_v(xz)$ and $\sigma_v(yz)$ are their own inverses, *i. e.*

$$C_2 \cdot C_2 = E$$

$$\sigma_v(xz) \cdot \sigma_v(xz) = E$$

$$\sigma_v(yz) \cdot \sigma_v(yz) = E$$

It should be noted, however, that it is not generally true that symmetry operations are their own inverses.

The complete set of symmetry operations of a *molecule,* always form a mathematical group, termed in fact a ***point group*** because all the symmetry elements intersect at a point within the molecule which is not shifted by any of the symmetry operations. The point group of the water molecule is denoted C_{2v} according to the ***Schoenflies notation*** which is commonly used.

Using similar reasoning, the symmetry elements and operations of other point groups can be derived, and the appropriate Schoenflies symbol assigned. In Table **3.24-1,** the *essential* symmetry elements for the various point groups are listed. The word "essential" is used since some of the symmetry elements listed in this table for a given point necessarily imply the existence of others which are not listed. In Table **3.24-2** an exhaustive list of symmetry elements for each point group is given.

Table 3.24-1. The point groups and their essential symmetry elements.

point group	*essential symmetry elements* *
C_s	one symmetry plane
C_i	centre of symmetry
C_n	one n-fold (proper) axis of symmetry
D_n	one C_n axis plus n C_2 axes perpendicular to it
C_{nv}	one C_n axis plus n vertical planes σ_v
C_{nh}	one C_n axis plus a horizontal plane σ_h
D_{nh}	those of D_n plus a horizontal plane σ_h
D_{nd}	those of D_n plus n dihedral planes
S_n (n even)	one n-fold improper axis of symmetry
T_d	those of regular tetrahedron
O_h	those of regular octahedron
I_h	those of a regular icosahedron
K_h	those of the sphere

* These elements are all in addition to the identity element E which is possessed by all point groups.

Table 3.24-2. The point groups and all their symmetry elements [a]

point group	*symmetry elements*
C_s	E, σ
C_i	E, i
C_1	E
C_2	E, C_2
C_3	$E, C_3-C_3^2$
C_4	$E, C_4-C_2-C_4^3$
C_5	$E, C_5-C_5^2-C_5^3-C_5^4$
C_6	$E, C_6-C_3-C_2-C_3^2-C_6^5$
C_7	$E, C_7-C_7^2-C_7^3-C_7^4-C_7^5-C_7^6$
C_8	$E, C_8-C_4-C_8^3-C_2-C_8^5-C_4^3-C_8^7$
D_2	$E, 3\times C_2$ (mutually $\perp$)
D_3	$E, C_3-C_3^2, 3\times C_2$ ($\perp$ to C_3)
D_4	$E, C_4-C_2-C_4^3, 2\times C_2', 2\times C_2''$ [b]
D_5	$E, C_5-C_5^2-C_5^3-C_5^4, 5\times C_2$ ($\perp$ to C_5)
D_6	$E, C_6-C_3-C_2-C_3^2-C_6^5, 3\times C_2', 3\times C_2''$ [b]
$C_{1v}=C_s$	
C_{2v}	$E, C_2, \sigma_v, \sigma_v'$
C_{3v}	$E, C_3-C_3^2, 3\times\sigma_v$
C_{4v}	$E, C_4-C_2-C_4^3, 2\times\sigma_v, 2\times\sigma_d$ [c]
C_{5v}	$E, C_5-C_5^2-C_5^3-C_5^4, 5\times\sigma_v$
C_{6v}	$E, C_6-C_3-C_2-C_3^2-C_6^5, 3\times\sigma_v, 3\times\sigma_d$ [c]
$C_{\infty v}$	E, $\infty\times$ coincidental C, $\infty\times\sigma_v$
$C_{1h}=C_s$	
C_{2h}	E, C_2, i, σ_h
C_{3h}	$E, C_3-C_3^2-S_3-S_3^5, \sigma_h$ [d]

Table 3.24-2. (Cont.)

point group	*symmetry elements*
C_{4h}	E, $C_4-C_2-C_4^3-S_4-S_4^3$, i, σ_h
C_{5h}	E, $C_5-C_5^2-C_5^3-C_5^4-S_5-S_5^7-S_5^3-S_5^9$, σ_h [d]
C_{6h}	E, $C_6-C_3-C_2-C_3^2-C_6^5-S_6-S_3-S_3^5-S_6^5$, i, σ_h [d]
D_{2h}	E, $3\times C_2$ (mutually ⊥), i, $3\times\sigma$ (mutually ⊥)
D_{3h}	E, $C_3-C_3^2-S_3-S_3^5$, $3\times C_2$ (⊥ to C_3), σ_h, $3\times\sigma_v$ [d]
D_{4h}	E, $C_4-C_2-C_4^3-S_4-S_4^3$, $2\times C_2'$, $2\times C_2''$, i, σ_h, $2\times\sigma_v$, $2\times\sigma_d$ [b, c]
D_{5h}	E, $C_5-C_5^2-C_5^3-C_5^4-S_5-S_5^7-S_5^3-S_5^9$, $5\times C_2$ (⊥ to C_5), σ_h, $5\times\sigma_v$ [d]
D_{6h}	E, $C_6-C_3-C_2-C_3^2-C_6^5-S_6-S_3-S_3^5-S_6^5$, $3\times C_2'$, $3\times C_2''$, i, σ_h, $3\times\sigma_v$, $3\times\sigma_d$
$D_{\infty h}$	E, $\infty\times C$ and S (all coincidental), σ_h (=S_1), i, $\infty\times\sigma_v$, $\infty\times C_2$
D_{2d}	E, $C_2-S_4-S_4^3$, $2\times C_2'$ (⊥ to S_4), $2\times\sigma_d$ (through S_4)
D_{3d}	E, $C_3-C_3^2-S_6-S_6^5$, $3\times C_2$ (⊥ to C_3), i, $3\times\sigma_d$
D_{4d}	E, $C_4-C_2-C_4^3-S_8-S_8^3-S_8^5-S_8^7$, $4\times C_2'$ (⊥ to C_4), $4\times\sigma_d$
D_{5d}	E, $C_5-C_5^2-C_5^3-C_5^4-S_{10}-S_{10}^3-S_{10}^7-S_{10}^9$, $5\times C_2$ (⊥ to C_5), i, $5\times\sigma_d$
D_{6d}	E, $C_6-C_3-C_2-C_3^2-S_{12}-S_4-S_{12}^5-S_{12}^7-S_4^3-S_{12}^{11}$, $6\times C_2$ (⊥ to C_6), $6\times\sigma_d$
$S_1=C_s$	
$S_2=C_i$	
$S_3=C_{3h}$	E, $C_2-S_4-S_4^3$
S_4	
$S_5=C_{5h}$	E, $C_3-C_3^2-S_6-S_6^5$, i
S_6	
$S_7=C_{7h}$	E, $C_4-C_2-C_4^3-S_8-S_8^3-S_8^5-S_8^7$
S_8	
T_d	E, $4\times C_3-C_3^2$, $3\times C_2-S_4-S_4^3$ (mutually ⊥), $6\times\sigma_d$
O_h	E, $4\times C_3-C_3^2-S_6-S_6^5$, $3\times C_4-C_2-C_4^3-S_4-S_4^3$ (mutually ⊥), $6\times C_2$, i, $3\times\sigma_h$, $6\times\sigma_d$

[a] Axes which coincide are linked, *e.g.* $C_3-C_3^2-S_6-S_6^5$.
[b] C_2' and C_2'' denote two-fold axes ⊥ to the principal axis.
[c] The use of the label σ_d in connection with, in particular, the groups C_{4v} and C_{6v} is anomalous, but has become the convention. [R. S. Mulliken, *J. Chem. Phys.* **1955**, *23*, 1997.]
[d] S_3^5 is the element corresponding to rotation about an axis by $5\times 2\pi/3$ (or $2\times 2\pi/3$) followed by reflection in a plane ⊥ to that axis. We cannot use the symbol S_3^2 since this is identical with C_3^2. A similar argument holds for S_5^7 and S_5^9.

A systematic procedure for symmetry classification of molecules

In this section we describe a systematic procedure for determining the point group to which a given molecule belongs. This will be done in a practical "how to do it" manner, but the close relationship of this procedure to the arguments used in deriving the various groups should be evident. The following steps will lead systematically to a correct classification:

1. We determine whether the molecule belongs to one of the "special" groups (Table **3.24-1** above), that is $C_{\infty v}$ or $D_{\infty h}$, or to one of those with multiple high-order axes. Only *linear molecules* can belong to $C_{\infty v}$ or $D_{\infty h}$, so these occasion no uncertainty. The especially high symmetry of the others is usually obvious.

All of the *cubic groups* – T, T_h, T_d, O and O_h – require four C_3 axes, while the *icosahedral groups*, I and I_h, require 10 C_3's and six C_5's. These multiple C_3's and C_5's are the key things to look for. In practice, only molecules built on a central tetrahedron, octahedron, cuboctahedron, cube or icosahedron will qualify, and these polyhedra are usually very conspicuous.

2. If the molecule belongs to none of the special point groups, we search for proper or improper axes of rotation. If no axes of either type can be found, we look for a plane or centre of symmetry. If a plane only is found, the group is C_s. If a centre only is found (this is a very rare) the group is C_i. If no symmetry elements at all are evident, the group is the trivial one containing only the identity operation and is designated C_1.

3. If an *even*-order improper axis (in practice only S_4, S_6 and S_8 are common) is found but no planes of symmetry or any proper axis except a collinear one (or more) whose presence is automatically required by the improper axis, the group is S_4, S_6, S_8 ... An S_4-axis requires a C_2-axis; an S_6-axis requires a C_3-axis; and an S_8-axis requires both C_4 and C_2 axes. The important point here is the S_n (*n even*) groups consists exclusively of the operations generated by the S_n-axis. If any additional operation is possible, we are dealing with a D_n, D_{nd} or D_{nh} type of group. Molecules belonging to the S_n groups are relatively rare, and the conclusion that a molecule belongs to such a group should be checked thoroughly before it is accepted.

4. Once it is certain that the molecule belongs to none of the groups so far considered, we look for the highest-order proper axis. It is possible that there will be no one axis of uniquely high order but instead three C_2 axes. In such a case, we look to see whether one of them is geometrically unique in some sense – for example, collinear with a unique molecular axis. This occurs with the molecule allene, which is one of the examples to be worked through later. If all of the axes appear quite similar to one another, then any one may be selected at random as the axis to which the vertical or horizontal character of the planes is referred. Suppose that C_n is our reference or principal axis. The crucial question now is whether there exists a set of n C_2 axes perpendicular to the C_n axis. If so, we proceed to step **5**. If not, the molecule belongs to one of the groups C_n, C_{nv} or C_{nh}. If there are no symmetry elements except the C_n axis, the group is C_n. If there are n vertical planes, the group is $\boldsymbol{C_{nv}}$. If there is a horizontal plane, the group is $\boldsymbol{C_{nh}}$.

5. If, in addition to the principal C_n axis, there are n C_2 axes lying in a plane perpendicular to the C_n axis, the molecule belongs to one of the groups D_n, D_{nh} or D_{nd}. If there are no symmetry elements besides C_n and the n C_2 axes, the group is D_n. Should there also be horizontal plane of symmetry, the group is D_{nh}. A D_{nh} group will also contain, necessarily, n vertical planes; these planes contain the C_2 axes. If there is no σ_h but there is a set of n vertical planes which pass between the C_2 axes, the group is D_{nd}.

This five-step procedure is summarised in the flow-sheet of Figure **3.24-8.**

Illustrative Examples

The scheme just outlined for allocating molecules to their point groups will now be illustrated. We shall deal throughout with molecules which do not belong to any of the special groups, and we shall also omit molecules belonging to C_1, C_s and C_i. Thus, each illustration will begin at step **3**, the search for an even-order S_n axis.

Example 1: H_2O

3. H_2O possesses no improper axis.

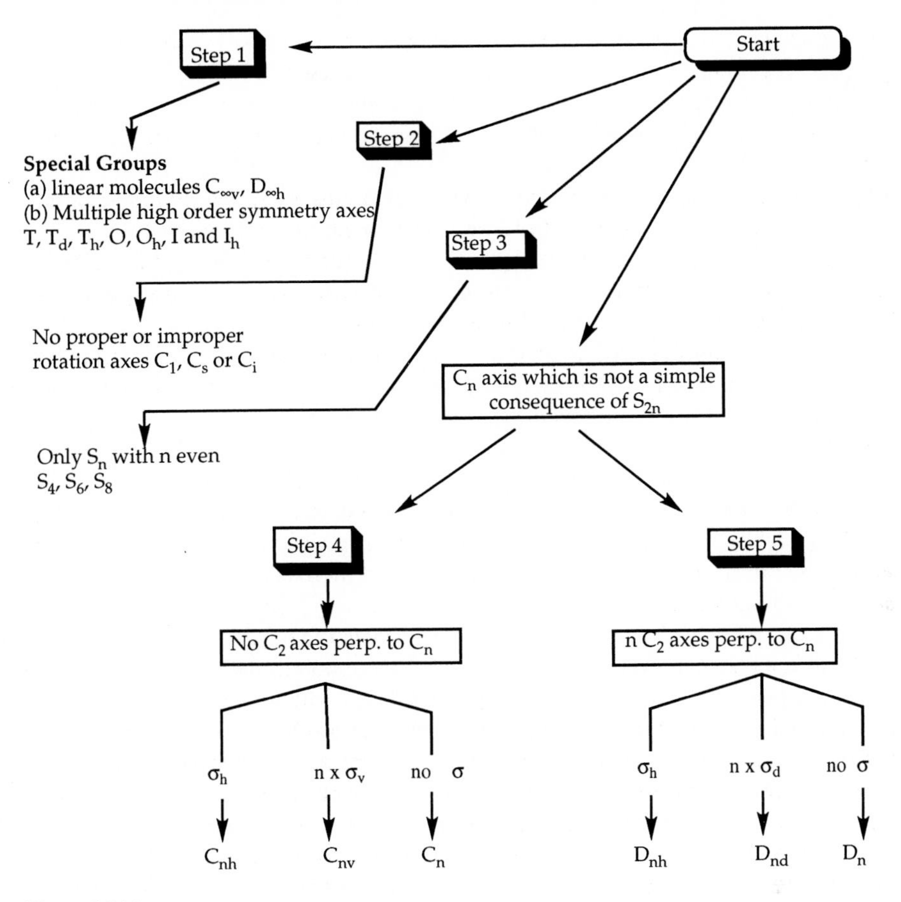

Figure 3.24-8.

4. The highest-order proper axis is a C_2 axis passing through the oxygen atom and bisecting a line between the hydrogen atoms. There are no other C_2 axes. Therefore, H_2O must belong to C_2, C_{2v} or C_{2h}. Since it has two vertical planes, one of which is the molecular plane, it belongs to the group C_{2v}.

Example 2: NH_3

3. There is no improper axis.

4. The only proper axis is a C_3 axis; there are no C_2 axes at all. Hence, the point group must be C_3, C_{3v} or C_{3h}. There are three vertical planes, one passing through each H atom. The group is thus C_{3v}.

Example 3: Allene, $H_2C{=}C{=}CH_2$

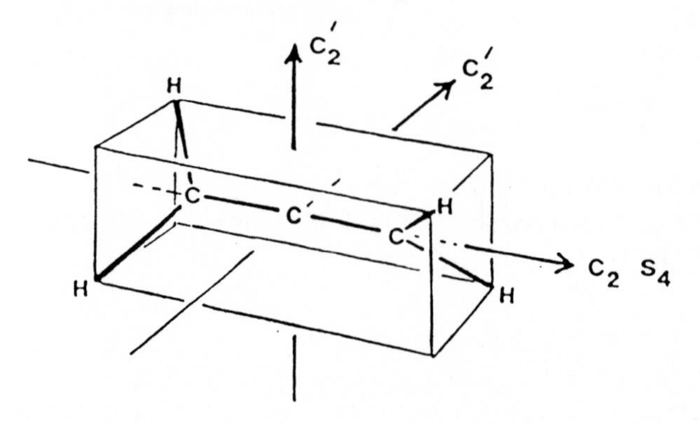

3. There is an S_4 axis coinciding with the main, molecular (C=C=C) axis. However, there are also other symmetry elements besides that C_2 axis, which is a necessary consequence of the S_4. Most obvious, perhaps, are the planes of symmetry passing through the $H_2C{=}C{=}C$ and $C{=}C{=}CH_2$ sets of atoms. Thus, although an S_4 axis is present, the additional symmetry rules out the point group S_4.

4. As noted, there is a C_2 axis lying along the C=C=C axis. There is no higher-order proper axis. There are two more C_2 axes perpendicular to this one, as shown in the sketch. Thus, the group must be of *D*-type, and we proceed to step **5**.

5. Taking the C_2 axis lying along the C=C=C axis of the molecule as the reference axis, we look for a σ_h. There is none, so the group D_{2h} is eliminated. There are, however, two vertical planes (which lie between the C_2' axes), so the group is D_{2d}.

Example 4: H_2O_2

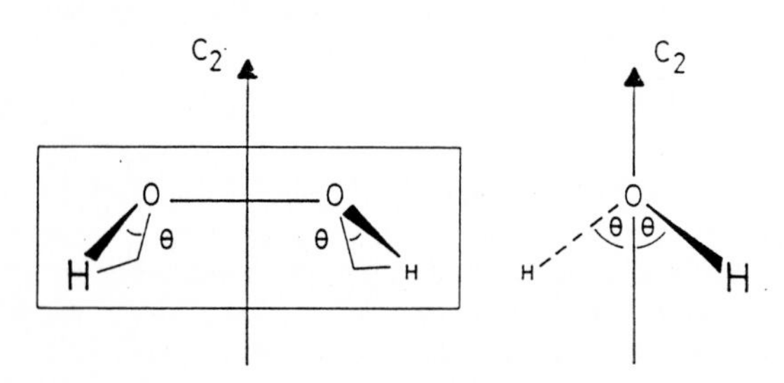

a) The nonplanar equilibrium configuration

3. There is no improper axis.

4. As indicated in the sketch, there is a C_2 axis and no other proper axis. There are no planes of symmetry. The group is therefore C_2. Note that the C_2 symmetry is in no way related to the value of the angle θ except when θ equals 0° or 90°, in which case the symmetry is higher. We shall next examine these two non-equilibrium configurations of the molecule.

b) The cis-planar configuration ($\theta = 0°$)

3. Again, there is no even-order S_n axis.

4. The C_2 axis, of course, remains. There are still no other proper axes. The molecule now lies in a plane, which is a plane of symmetry, and there is another plane of symmetry intersecting the molecular plane along the C_2 axis. The group is C_{2v}.

c) The trans-planar configuration ($\theta = 90°$)

3. Again, there is no even-order S_n axis, (except $S_2 = i$)
4. The C_2 axis is still present, and there are no other proper axes. There is now a σ_h, which is the molecular plane. The group is C_{2h}.

Symmetry and Optical Activity

Many substances can rotate the plane of polarisation of a ray of polarised light. These substances are said to be optically active. In molecular terms, the one necessary and sufficient condition for a substance to exhibit optical activity is that its molecular structure be such that it cannot be superimposed on its image obtained by reflection in a mirror. When this condition is satisfied the molecule exists in two forms, showing equal but opposite optical properties, and the two forms are called *enantiomers.*

Whether a molecule is or is not superimposable on its mirror image is a question of symmetry. A molecule that contains an n-fold improper rotation axis of symmetry S_n is always superimposable on its mirror image. This is true because the operation S_n consists of two parts: a rotation C_n and a reflection σ. Since a reflection creates the mirror image, the operation S_n is equivalent to rotating in space the mirror image. By definition, a molecule containing an S_n axis is brought into coincidence with itself by the operation S_n and hence its mirror image, after rotation, is superimposable. As $S_1 = \sigma$ and $S_2 = i$, a molecule with either a plane or a centre of symmetry is also optically inactive. However, the most general and economical rule is: *a molecule with an S_n axis is optically inactive.*

Exercises

Using molecular models answer **all** the following questions.

1) List all the ***symmetry elements*** and ***symmetry operations*** of the molecules SF_4 and NH_3, and indicate those symmetry operations which commute. In the case of NH_3 draw up a multiplication table for the symmetry operations and demonstrate that the operations which you have listed constitute a mathematical ***group***.

2) Which of the following molecules cannot be optically active?
H_2O_2, $[Co(en)_3]^{3+}$, *cis-* and *trans-*$[CoCl_2(en)_2]^+$ and the two molecules illustrated below:

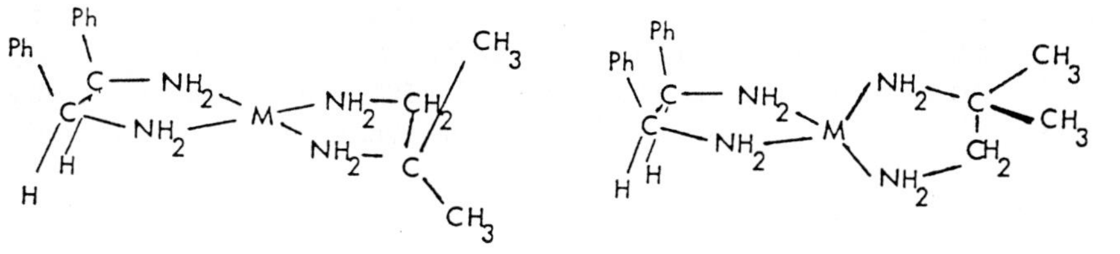

3) Assign ***point groups*** to the following molecules:
   a) SF_5Cl
   b) $[PtCl_4]^{2-}$
   c) a tetrahedral molecule AB_4 squashed along one of its S_4 axes, but short of planar – *e.g.* $[CuCl_4]^{2-}$ (*dihedral* symmetry)
   d) B_2H_6
   e) $[Co(en)_3]^{3+}$
   f) $(C_6H_6)_2Cr$ (both *staggered* and *eclipsed* forms)
   g) SF_6
   h) SiF_4

Part II: Mathematical Properties of Convex Polyhedra

The mathematical study of polyhedral solids started over 2000 years ago in ancient Greece. The five regular solids (tetrahedron, octahedron, cube, dodecahedron and icosahedron) were studied by Theotus, Euclid, Plato, Hypiscles and Pappus. Unfortunately for chemists, the Greeks chose a system of nomenclature based on the number of *faces* (Gk. *εδρα*-seat) rather than vertices – a convention that has persisted and is unlikely to be changed. The Greeks also discovered some of the semi-regular solids, but complete lists of these *isogonal* and *isohedral* solids were not made until the 17th and 19th centuries (by Kepler and Catalan respectively). The isogonal solids can generally be derived from the regular solids by symmetrically shaving off the corners of the latter, and therefore they are usually described as *truncated* or *snub* polyhedra (*e.g.* truncated tetrahedron, snub cube, *etc.*) The isohedral solids are reciprocal to the isogonal solids and would be described by chemists as *symmetrically capped* regular polyhedra, but mathematicians have developed an alternative nomenclature; *e.g.* tetrahedron is described as a *triakistetrahedron.*

Convex Polyhedra

A convex polyhedron can be economically defined as the convex hull in three-dimensional Euclidean space of a finite set of points which do not all lie in a common plane. This definition therefore excludes the convex polytopes discovered by the Swiss mathematician Schafli, which are regular figures in four or more dimensional space. These figures are fascinating but have little relevance to current chemical problems.

The type of projection or perspective drawing used to illustrate a polyhedron depends on the particular features that need to be emphasised. The *Schlegel diagram* is particularly useful for illustrating the mathematical relationships between the edges, faces and vertices of the polyhedron. In this type of diagram, the faces of the polyhedron are projected onto one face of the polyhedron that lies in the plane of the paper. It is assumed that the polyhedron is resting on one of its faces and is viewed through that face. The perspective is greatly exaggerated to give the impression that the other faces of the polyhedron lie within the chosen face. This method of projection is illustrated in Figure **3.24-9** for the tetrahedron, cube, dodecahedron, octahedron and icosahedron. The trigonal prism, which has triangular and rectangular faces, offers a choice of two unique Schlegel diagrams; these are shown in Figure **3.24-10.**

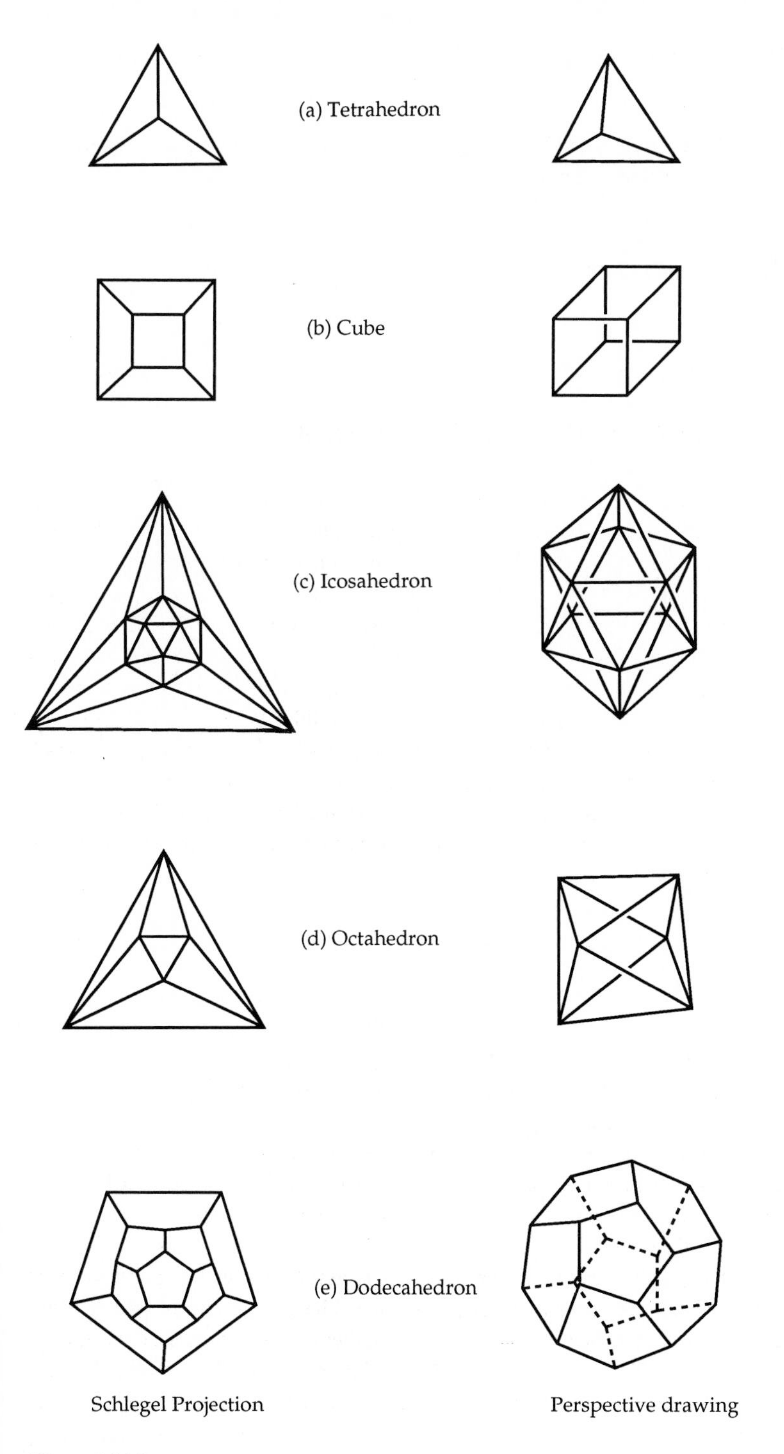
(a) Tetrahedron
(b) Cube
(c) Icosahedron
(d) Octahedron
(e) Dodecahedron
Schlegel Projection
Perspective drawing

Figure 3.24-9.

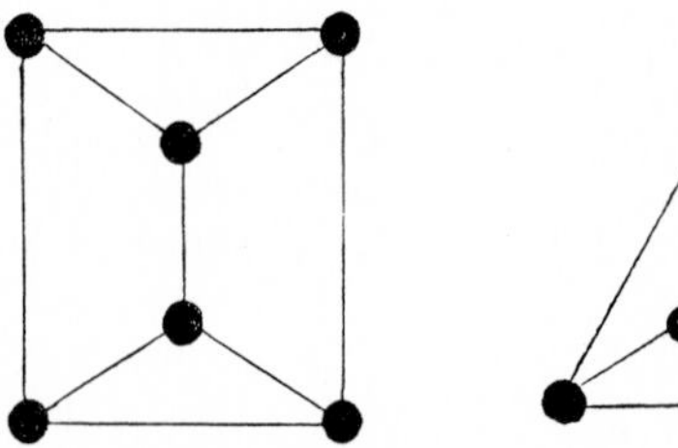

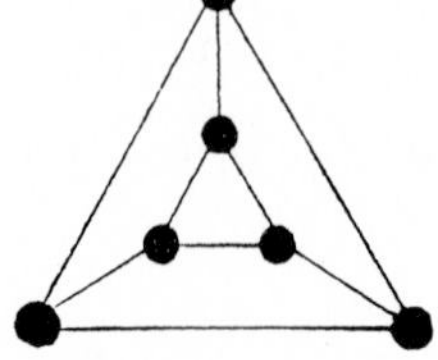

Figure 3.24-10.

Schlegel Diagrams and Graph Theory

It will be appreciated that a Schlegel diagram represents the three-dimensional polyhedron as a two-dimensional graph consisting of points (nodes or vertices) joined by straight lines. Convex polyhedra and planar graphs are related by a fundamental theorem discovered earlier this century by the German mathematician E. Steinitz. This theorem is based on the following definition of a *path* on a graph. A path is any sequence of edges such that each consecutive pair of edges meets at a node and no other pair of edges intersects. A graph is said to be n-connected if *every* pair of nodes can be joined by at least n disjoint (non-intersecting) paths. For example, the graph shown in Figure **3.24-11 a** is 1-connected since every pair of nodes can be connected by at least one path, but it is not 2-connected since a pair of nodes such as **a** and **b** must contain the edge **c-d** and therefore it is impossible to accommodate two paths along this edge. The graph in Figure **3.24-11 b** is 2-connected but not 3-connected and the graph in Figure **3.24-11 c** is 3-connected but not 4-connected.

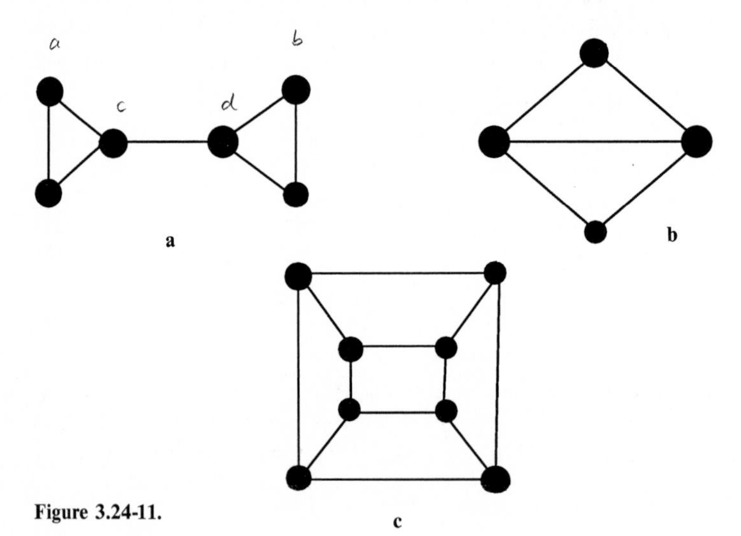

Figure 3.24-11.

Steinitz Theorem

The *Steinitz Theorem* states that a planar graph represents a three dimensional polyhedron *if and only if* it is 3-connected. Therefore, of the examples given above, only Figure **3.24-11 c** represents a three dimensional convex polyhedron. Actually, Figure **3.24-11 c** is identical to the Schlegel diagram for the cube (Fig. **3.24-9**). Steinitz's theorem therefore provides a test for deciding whether the open bonding networks found in boranes or metal clusters correspond to three-dimensional polyhedral solids. Clearly, the boron–boron bond networks in B_4H_{10} and B_5H_{11} shown in Fig. **3.24-12** are only 2-connected and therefore do not correspond to complete three-dimensional polyhedral solids.

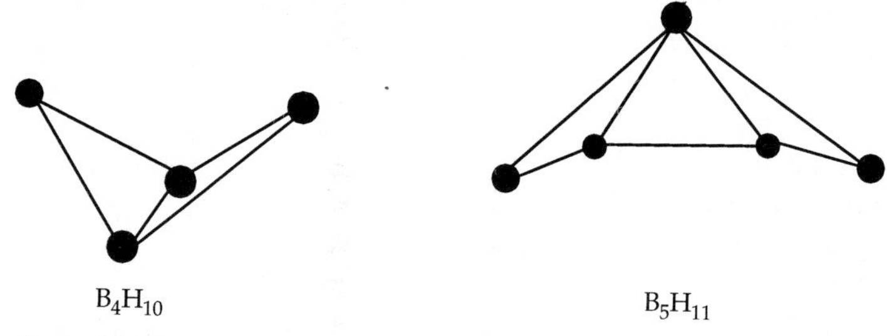

Figure 3.24-12.

Euler's Law

The law which relates the numbers of edges, faces and vertices of a polyhedron was first presented in 1738 by Euler in his *Elements Doctrinae Solidorium.* This law states that the number of vertices (V) plus the number of faces (F) minus the number of edges (E) of a polyhedron is equal to 2 (Eqn. 1).

$$V + F - E = 2 \qquad (1)$$

Table **3.24-3** illustrates Euler's law for some common polyhedra.

Table 3.24-3.

	V	F	E	$(V + F - E)$
Tetrahedron	4	4	6	2
Cube	8	6	12	2
Octahedron	6	8	12	2
Trigonal Prism	6	5	9	2

This law has a clear implication for chemical problems, namely that only two of three possible variables (vertices, faces and edges) need be specified to define the polyhedron. Polyhedra can have triangular, square, pentagonal, ... k-polygonal faces and the total number of faces may be expressed as a sum of f_k, where f_k is the number of k-polygonal faces. Similarly, the total number of vertices may be expressed as the sum of v_r, where v_r is the number of vertices that have r edges meeting there. Since every edge is common to two faces and two vertices, the total number of edges (E) can be counted in two ways:

$$E = \sum r v_r/2 = \sum k f_k/2$$

These relationships can be substituted into Euler's Eqn. (1) to give:

$$F - \sum r v_r/2 + V = 2$$

and

$$F - \sum k f_k/2 + V = 2$$

In addition,

$F = \sum f_k$ and $V = \sum v_r$ and, therefore:

$$\sum f_k - \sum r v_r/2 + \sum v_r = 2$$

and

$$\sum f_k - \sum k f_k/2 + \sum v_r = 2$$

Rearranging these gives the fundamental equations (2) and (3).

$$\sum f_k - \sum v_r(r-2)/2 = 2 \quad (2)$$

$$\sum v_r - \sum f_k(k-2)/2 = 2 \quad (3)$$

Equations (2) and (3) define the relationships between the vertices and faces for all convex polyhedra.

The total number of possible polyhedra which satisfy Euler's law increases as the number of vertices is increased. To explore all these possibilities thoroughly would require a computer. Fortunately, for chemical purposes, it is not generally necessary to investigate them all, and most of the important cases are generated by imposing the following limitations:
a) the vertices are constrained to have the same number of edges radiating from them,
b) all vertices are constructed only from polygons with the same number of edges,
c) the polyhedra have equivalent vertices and faces.
We consider these in turn.

a) All vertices constrained to have the same number of edges radiating from them.

If the principal atoms of a cage compound have well defined valencies and all the atoms are isovalent, then the number of polyhedral arrangements of atoms is limited. In evaluating the total number of polyhedral arrangements of carbon atoms which can be adopted by the hydrocarbons C_nH_n, the trivalent nature of the CH radical imposes the limitation $v_3 \neq 0$, $v_4 = v_5 = v_6 = \ldots = 0$ on Eqns. (2) and (3). This leads to the following simplified equations.

$$\sum f_k - v_3/2 = 2$$

and

$$v_3 - \sum f_k(k-2)/2 = 2$$

Eliminating v_3 gives

$$2\sum f_k - \sum f_k(k-2)/2 = 6$$

$$\sum f_k(6-k) = 12 \quad (4)$$

i.e. $3f_3 + 2f_4 + f_5 + 0f_6 - f_7 \ldots = 12$

This equation leaves the number of hexagonal faces undefined, but this may be obtained from the relationship:

$$\sum r v_r = \sum k f_k$$

which is also derived from Eqns. (2) and (3). In this case, $r = 3$ and $\sum v_k$ is equal to the total number of vertices of the polyhedron $\equiv V$.

$$\sum k f_k = 3V$$

Example

For the hydrocarbon C_8H_8, each CH radical is three-valent, and we obtain the following solutions of Eqn. (4):

(1) $f_4 = 6$ $\quad\quad$ $\sum k f_k = 4 \times 6 = 3V = 24$

(2) $f_3 = 2;\ f_4 = 2;\ f_5 = 2$ $\quad\quad$ $\sum k f_k = 2 \times 3 + 4 \times 2 + 5 \times 2 = 24 = 3V$

which are illustrated below.

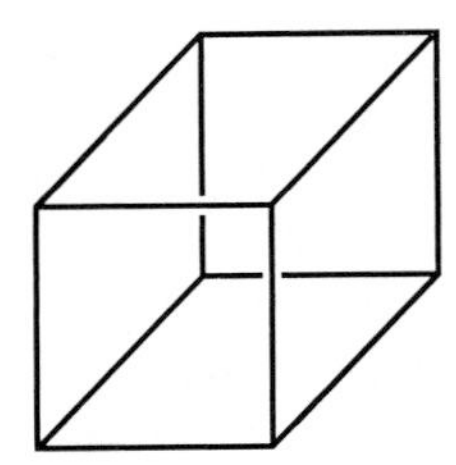

(1) cubane

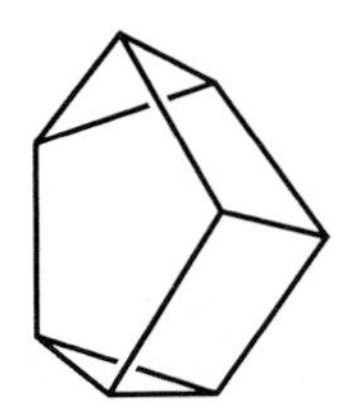

(2) cuneane

These two possibilities have identical Euler's law solutions, *i.e.* $V = 8$, $E = 12$, $F = 6$, and therefore can be distinguished only by specifying the polygonal faces.

The equations for four-valent and five-valent radicals corresponding to Eqn. (4) are given in Eqns. (5) and (6).

$$\sum f_k (4 - k) = 8 \tag{5}$$

$$\sum f_k (10 - 3k) = 20 \tag{6}$$

b) Polyhedra derived from polygonal faces with equal numbers of edges

An alternative simplification of Eqns. (2) and (3) results if the constraint that all faces are equal is applied. For example, many borane and transition-metal cluster polyhedra have triangular faces exclusively, and these polyhedra may be generated by substituting $f_4 = 0$, $f_5 = 0$, $f_6 = 0$, *etc.* into Eqns. (2) and (3). This gives the formula in Eqn. (7), which is very similar to Eqn. (4) above.

$$\sum v_r (6 - r) = 12 \tag{7}$$

These polyhedra have been described as *deltahedra, triangulated polyhedra* or *simplicial polyhedra.*

As a simple example, we will consider the case with six vertices (*i. e.* $\sum v_r = V = 6$). Clearly, the solutions for this problem will be similar to those given above for the polyhedron with six faces and three-valent vertices (*i. e.* C_8H_8). These are

(3) $v_4 = 6 \qquad 6(6-4)) = 12$

and

(4) $v_3 = 2$, $v_4 = 2$ and $v_5 = 2 \qquad (2(5-3) + 2(6-4) + 2(6-5)) = 12$

Solution (3) corresponds to the *octahedron* found in a number of octahedral metal carbonyl cluster compounds, borides and boron hydrides. Solution (4), which corresponds to a *bicapped tetrahedron,* has been established for $Os_6(CO)_{18}$. These two polyhedral geometries are illustrated below.

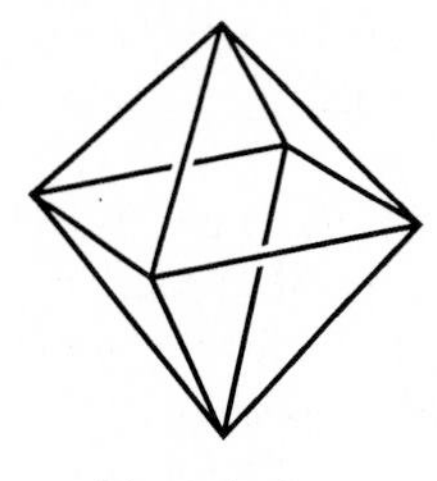

(1) octahedron

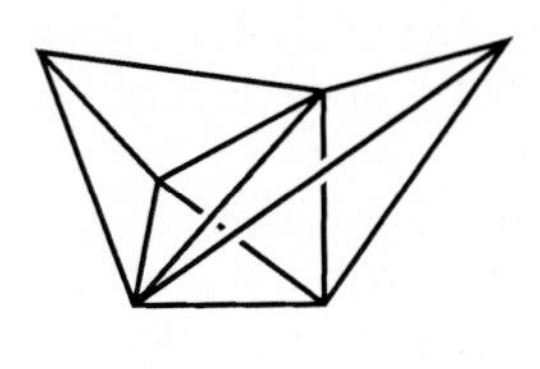

(2) bicapped tetrahedron

Both polyhedra have 8 faces and 12 edges and therefore can be distinguished only by the connectivities of the vertices.

The equations for rectangular and pentagonal-faced polyhedra corresponding to Eqn. (7) are given in Eqns. (8) and (9).

$$\sum v_r(r-4) = 8 \tag{8}$$

$$\sum v_r(10-3r) = 20 \tag{9}$$

c) Polyhedra with equivalent vertices and faces

A set of polyhedra which conform simultaneously to conditions a) and b) can be generated by adding additional constraints to Eqns. (4), (5) and (6). The following solutions result from these limitations:

$f_3 = 4$	$v_3 = 4$	tetrahedron
$f_3 = 8$	$v_4 = 6$	octahedron
$f_4 = 6$	$v_3 = 8$	cube
$f_5 = 12$	$v_3 = 20$	dodecahedron
$f_3 = 20$	$v_5 = 12$	icosahedron

If these polyhedra are constructed from regular polygons, the figures correspond to the regular Platonic solids (shown in Fig. **3.23-9**).

Chemical Implications of these Topological Aspects

The *Polyhedral Skeletal Electron Pair Theory* has related the closed-shell electronic requirements of main group and transition-metal polyhedral molecules to the topological properties of polyhedra. For example, three-connected main group polyhedral molecules (see above) are characterised by a total of $5V$ valence electrons, because the bonding can be described in terms of localised two-centre, two-electron bonds. Cubane has, for example, a total of 40 valence electrons (32 from the carbon atoms and 8 from the hydrogen atoms). Of these electrons, 16 occupy 8 C–H bonding molecular orbitals and 24 occupy 12 edge-localised C–C molecular orbitals.

What localised orbitals are occupied in tetrahedral P_4?

Transition-metal three-connected polyhedral molecules are characterised by $15V$ valence electrons, because each metal contributes an additional $5d$ valence orbitals. $[Ir_4(CO)_{12}]$ has a tetrahedral geometry, and a total of 60 valence electrons. How many of these electrons can be assigned to metal–metal and metal–carbonyl orbitals? How many non-bonding orbitals remain localised on the iridium atoms?

In four-connected and deltahedral molecules, the skeletal bonding can no longer be related directly to the number of edges of the polyhedron, but it can be shown using Molecular Orbital Theory that such molecules are characterised by $4V + 2$ valence electrons (if the vertices are occupied by main-group atoms) or $14V + 2$ electrons (if the vertices are occupied by transition-metal atoms). For example, $[B_6H_6]^{2-}$ has a total of 26 valence electrons – 12 occupying localised B–H bonding molecular orbitals and 14 occupying delocalised B–B skeletal molecular orbitals, $[Ru_6(CO)_{18}]^{2-}$ also has an octahedral geometry and a total of 86 valence electrons.

These structure-valence electron correlations can be extended to a wide range of complex polyhedral structures derived from the condensation of the primary polyhedral fragments described above.

References

K. Wade, *Advances in Inorganic and Radiochemistry* **1976,** *18,* 1
D. M. P. Mingos, *Accounts of Chemical Research* **1984,** *17,* 311
A. J. Stone, *Inorganic Chemistry* **1981,** *20,* 563
D. M. P. Mingos, *Chemical Society Reviews* **1986,** *15,* 31

Exercises

1. Using models, establish the point groups of the regular *Platonic solids, i.e.* the dodecahedron, icosahedron, octahedron, tetrahedron and cube.
2. Structural and NMR studies have established that the polyhedral frameworks in $[B_7H_7]^{2-}$ and $Os_7(CO)_{21}$ have triangular faces exclusively. Use the equations given to generate possible polyhedral structures for these molecules and use the additional information given below to identify the actual structures and the point groups of the molecules.

a) The ^{11}B NMR spectrum of $[B_7H_7]^{2-}$ has two types of resonance with intensities in the ratio 5 : 2.
b) The ^{13}C NMR spectrum of $Os_7(CO)_{21}$ has three resonances with intensities in the ratio 3 : 3 : 1.
What does the latter evidence indicate with regard to the behaviour of the carbonyl ligands at the temperature of the measurement?

3. The structures of the borane anions $[B_nH_n]^{2-}$ in solution have been studied using variable temperature ^{11}B NMR spectroscopy. Some of the borane anions, *e.g.* $[B_8H_8]^{2-}$ and $[B_{11}H_{11}]^{2-}$, are "stereochemically non-rigid" and the boron atoms appear to be equivalent in the NMR experiment, although their solid state structures indicate several distinct boron environments. The nonrigidity has been attributed to low energy rearrangement pathways. Other boranes, *e.g.* $[B_7H_7]^{2-}$ and $[B_9H_9]^{2-}$, are "stereochemically rigid" and their NMR spectra reflect the different boron chemical environments observed in the solid state.

   It has been proposed by Lipscomb that these polyhedral molecules rearrange by a *diamond* → *square* → *diamond* mechanism involving four adjacent vertices as shown:

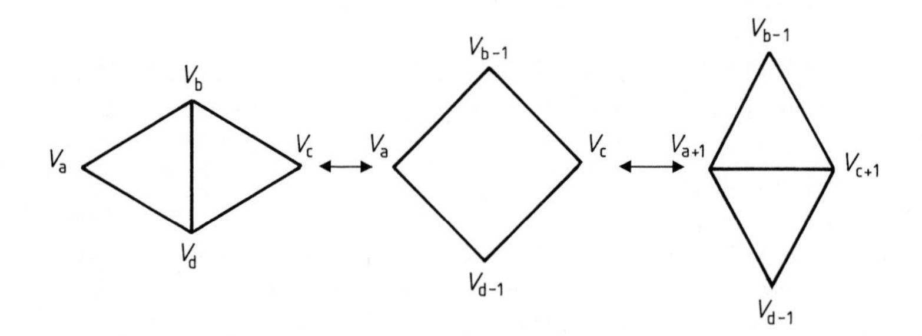

Also shown in the diagram are the connectivities of the four vertices involved in the Diamond-square-diamond (DSD) process *a,b,c,d.* The breaking of the diagonal bond to form the square has no effect on the connectivities of vertices V_a and V_c, but reduces the connectivities of the vertices V_b and V_d by one. In the final step forming the new bond between V_a and V_c increases their connectivities by one. Therefore the connectivities of a − d ($V_a - V_d$) have undergone the following mapping process.

$$\begin{bmatrix} V_a \\ V_b \\ V_c \\ V_d \end{bmatrix} \longrightarrow \begin{bmatrix} V_{a+1} \\ V_{b-1} \\ V_{c+1} \\ V_{d-1} \end{bmatrix}$$

If the rearranged polyhedron is to be identical to the initial polyhedron, then the DSD process must result in four vertices with connectivities and geometric distribution identical to the original. This can only be achieved if

$$b + d = a + c + 2$$

It follows that if all the connectivities of the vertices of the polyhedron are equal, then it is impossible for the DSD process to generate an identical polyhedron. If the vertices V_a, V_b,

V_c and V_d are identical in the transition state of the DSD process, the face has a four-fold symmetry axis. This can only be achieved for deltahedra with $4n + 1$ ($n = 1, 2 \ldots$) atoms. Can you explain why? For deltahedra with $4n + 1$ atoms, the DSD process leads to pseudo-rotation of the polyhedron by 90°, *i.e,* V_{a+1} replaces V_d, V_{b-1} replaces V_a, etc.

If V_a, V_b, V_c and V_d are related only by a mirror plane in the transition state, as shown below,

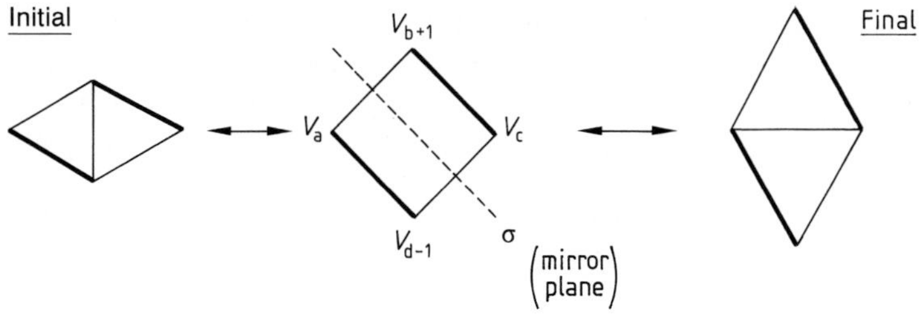

then the inital and final deltahedra are not related by a pseudo- 90° rotation but a pseudo-reflection operation. The symmetry constraints on the transition state require

$$a = b - 1 \text{ and } d - 1 = c$$

Using models, confirm that only $[B_5H_5]^{2-}$, $[B_8H_8]^{2-}$ and $[B_9H_9]^{2-}$ can rearrange by a single *diamond* → *square* → *diamond* transformation process to generate an identical polyhedron. Does the rearrangement lead to a *pseudoreflection* or *pseudorotation* of the polyhedron? What happens when you perform a single diamond → square → diamond rearrangement on $[B_6H_6]^{2-}$ and $[B_7H_7]^{2-}$? Can you identify the new polyhedron? (R. B. King, *Inorg. Chimica Acta* **1981,** *49,* 237. For a discussion of orbital symmetry effects see B. M. Gimarc, *Inorg. Chem.* **1986,** *25,* 83 and 2708).

Show that the polyhedra $[B_8H_8]^{2-}$, $[B_9H_9]^{2-}$ and $[B_{10}H_{10}]^{2-}$ conform to the equations given above p. 179 for triangulated polyhedra. What are the point groups of these three molecules?

4 Advanced Experiments

4.1 Chlorotris(*tert*-Butylimido)Manganese(VII)

Andreas A. Danopoulos and Geoffrey Wilkinson

Special Safety Precautions

1. Trimethylchlorosilane and *tert*-butylsilylamine hydrolyse and the products (HCl, and $tBuNH_2$) are harmful. A well ventilated fume cupboard should be used.
2. Acetonitrile and $tBuNH_2$ are toxic and should be handled using gloves in a fume cupboard.
3. The toxicity of $MnCl(NtBu)_3$, which has an appreciable vapour pressure at room temperature, is unknown. It is best kept in a sealed flask in a refrigerator.
4. Aqueous manganese residues should be safely disposed of.

Although the manganese (VII) and (VI) compounds, Mn_2O_7, MnO_3X (X = F, Cl) salts of MnO_4^{-1}, and MnO_4^{-2} have been known for over 140 years [1], no compounds without Mn=O bonds existed until the synthesis of $MnCl(NtBu)_3$ [2], which is the analogue of the explosive MnO_3Cl. Imido groups, $R\ddot{N}=$ (R = alkyl or aryl), are isoelectronic with $\ddot{O}=$ groups and can give similar compounds of transition metals in high oxidation states that are poorer oxidants than the corresponding oxides, *cf.* OsO_4 and $Os(NtBu)_4$.

$MnCl(NtBu)_3$ can be reduced, *eg.* by Na/Hg to the manganese dimer $[(tBuN)_2Mn(\mu\text{-}NtBu)]_2$ and a variety of derivatives made by substitutions of Cl, *eg.* by CH_3CO_2 [2].

Experimental

All glassware should be carefully dried to eliminate possibility of hydrolysis. Although $MnCl(NtBu)_3$ can be handled in air, the following syntheses must be carried out under dry, oxygen-free N_2 or Ar.

a) Tert-butylsilylamine [3]

In a 3 l 3-neck round-bottomed flask fitted with mechanical stirrer, a 1000 cm^3 pressure equalising dropping funnel and coil condenser connected to a N_2 line are placed 1.5 l of Et_2O (Note 1) and 315 cm^3 (3 mol) of $tBuNH_2$ (Note 2). The flask is cooled in an ice bath.

From the dropping funnel is added a solution of Me_3SiCl (190 cm^3, 1.5 mol) in 200–300 cm^3 Et_2O (made up under dry N_2). The rate of addition is adjusted so that refluxing of Et_2O is avoided. A white precipitate of $tBuNH_3Cl$ is formed and the final thick white slurry is allowed to warm to ambient temperature and stirred for 3–4 h. The liquid is then sucked off using a ground-joint filter stick. The residue is washed with Et_2O (3×300 cm^3) and the combined liquids distilled under N_2 at atmospheric pressure using a short (*ca.* 30 cm) column. After removal of Et_2O, the $tBuNH(SiMe_3)$ is collected at 118–120 °C in yields *ca.* 50–60%.

b) *Chlorotris(tert-butylimido)manganese* (VII)

In a 1l 3-neck flask with N_2 inlet, mechanical stirrer and dropping funnel as above is placed $Mn_{12}O_{12}(CO_2Me)_{16} \cdot (H_2O)_4 \cdot CH_3CO_2H$ [4] (12.5 g, 6.5 mmol) and 450 cm^3 acetonitrile (Note 3) and the flask cooled to *ca.* −35 °C to −40 °C in a dry-ice/acetone or /isopropanol bath. At this temperature (Note 4), an excess of neat Me_3SiCl (40 cm^3, 315 mmol) is added dropwise (Note 5) with stirring. The initial dark brown solution (which may contain some unreacted acetate) becomes dark red and finally shows the dark red-purple color of $MnCl_3$. After stirring for 0.5 h, neat *t*BuNH($SiMe_3$) (50 cm^3, 370 mmol) is added dropwise from the dropping funnel. The reaction mixture becomes orange-brown in colour. After stirring for *ca.* 12 h whilst allowing to warm to ambient temperature, the colour changes to dark green. At this point, the mixture is very moisture sensitive and failure to work under N_2 results in low yields. The green solution is filtered under N_2 using a G3 glass frit covered with a Celite 521 pad (3–4 cm) and the filtrate evaporated in a rotary evaporator (water bath temperature should not exceed 50 °C) connected to a conventional vacuum/N_2 manifold. A liquid N_2 trap is used to collect volatiles and prevent them entering the vacuum line. After removal of volatiles, the green residue (which can now be handled in air if necessary) is shaken thoroughly with petroleum (40 °C–60 °C, 200 cm^3) and water (200 cm^3). The green organic phase is separated, the aqueous phase (also yellow-green due to formation of the $MnCl_4^{2-}$ ion) is extracted with petroleum (3 × 100 cm^3) and the combined solvent dried over anhydrous Na_2SO_4 for 2 h. After filtration and removal of petroleum on a rotating evaporator, the green crystalline MnCl(N*t*Bu)$_3$ is collected and recrystallised from petroleum. Yield (2 crops): 1.6–1.9 g; m.p.: 92–93 °C. For spectroscopic data, refer to the literature [2]. Since the complex is air and moisture stable, no special handling precautions are necessary.

Notes

1. Et_2O dried over Na wire can be used without further treatment.
2. *t*BuNH$_2$ is dried with solid KOH or with Na wire and preferably distilled under N_2.
3. Acetonitrile is dried over CaH_2 and distilled under N_2.
4. The temperature is not critical for the formation of the puple solution of $MnCl_3$ in MeCN, but it is important to keep the temperature *ca.* −40 °C during addition of *t*BuNH($SiMe_3$).
5. Excess Me_3SiCl is used to ensure removal of H_2O and CH_3CO_2H, which will solvate the oxoacetate.
6. Excess *t*BuNH($SiMe_3$) is used to scavenge any acidic compounds formed by hydrolysis in solution.

References

[1] J. W. Mellor, *A Comprehensive Treatise on Inorganic and Theorectical Chemistry,* Vol XII, Longmans Green, London, **1932.**

[2] A. A. Danopoulos, G. Wilkinson, T. K. N. Sweet, M. B. Hursthouse, *J. Chem. Soc. Dalton Trans.* **1994**, 1037.

[3] R. M. Pike, *J. Org. Chem.* **1961,** *26,* 232.

[4] F. Lis, *Acta Crystallogr., Sect. B.* **1980,** *36,* 2042; S. P. Perlepes, A. G. Blackman, J. C. Huffman, G. Christou, *Inorg. Chem.* **1991,** *30,* 1965.

4.2 2,2′ : 6′,2″-Terpyridine Complexes and Metal Directed Reactivity

Edwin C. Constable

Special Safety Precautions

1. 2-Acetylpyridine and 4-pyridinecarboxaldehyde are toxic and irritants. They also smell very unpleasant. *Always* use these reagents in a fume cupboard, and perform part a) *entirely* in the fume cupboard. The residues from this preparation should be quenched in hydrochloric acid.
2. Tetrafluoroborate salts can hydrolyse in contact with water to give hydrofluoric acid. Do not allow either solid ammonium tetrafluoroborate or its solutions to come in contact with your skin. Do not keep solutions of ammonium tetrafluoroborate made up for long periods of time.
3. Methyl iodide is toxic and may be carcinogenic. Always use it in a fume cupboard. Remember that it is very volatile.
4. The health hazards of the 4′-(4-pyridyl)-2,2′ : 6′,2″-terpyridine and its complexes are unknown. Like all chemicals, they should be treated with extreme caution.

Ligands such as 2,2′-bipyridine and 2,2′ : 6′,2″-terpyridine have been of great importance to coordination chemists over the past one hundred years. The photochemical and photophysical properties of their transition-metal complexes have been the subject of intense interest. Recently, derivatives of these ligands have been developed in which the photochemical and redox properties of the metal centre may be tuned by the nature of the peripheral substituents on the ligands. In modern coordination chemistry, it is usually necessary to prepare the organic ligands that you require. In this experiment, you will prepare 4′-(4-pyridyl)-2,2′ : 6′,2″-terpyridine, and then make an iron complex and investigate its reaction with methyl iodide.

Experimental

a) 1,5-bis(2-pyridyl)-3-(4-pyridyl)-1,5-pentanedione

2-Acetylpyridine (8.4 ml) and 4-pyridinecarboxaldehyde (3.0 ml) are dissolved in ethanol (35 ml) in a 100 ml round-bottomed flask and a solution of sodium hydroxide (2.0 g) in water (25 ml) added. The mixture is stirred for one hour at room temperature, and then 30 ml of water added. This should give an off-white precipitate of 1,5-bis(2-pyridyl)-3-(4-pyridyl)-1,5-pentanedione. Collect this by filtration, wash well with water and a little *cold* ethanol, and dry in a dessicator.

Calculate the yield of your product and measure its IR spectrum.

b) 4'-(4-Pyridyl)-2,2':6',2''-terpyridine (pytpy)

Heat a solution of 1,5-bis(2-pyridyl)-3-(4-pyridyl)-1,5-pentanedione (0.40 g) and ammonium acetate (5.0 g) in 50 ml of ethanol to reflux for two hours. Allow the solution to cool and add 50 ml of water to precipitate the product. Recrystallise the product from a small amount of ethanol.

Calculate the yield of your product and measure its IR spectrum as a Nujol mull and, if possible, its ^{1}H NMR spectrum in $CDCl_3$ or CCl_4.

c) $[Fe(pytpy)_2][BF_4]_2$

Add 0.366 g of pytpy to 50 ml of methanol in a 100 ml twin-necked round-bottomed flask, attach a water condenser and a small dropping funnel containing a solution of $[Fe(H_2O)_6][BF_4]_2$ (0.1 g) in 10 ml of methanol to the flask, and then heat the pytpy solution to boiling. Add the solution of the iron salt dropwise over 10 minutes, and then continue heating for another ten minutes. Cool the dark purple solution and add to it a solution of $[NH_4][BF_4]$ (0.5 g) in 5 ml of methanol. Collect the purple solid and dry it.

Calculate the yield of your product and measure its IR spectrum as a Nujol mull and, if possible, its ^{1}H NMR spectrum in CD_3COCD_3. Record the electronic spectrum (250–700 nm) of your product in acetonitrile solution. You should accurately weigh your sample so that you can report extinction coefficients. As a guide to the concentration required, the bands have ε values between 10000 and 40000 $dm^3\ mol^{-1}\ cm^{-1}$.

d) $[Fe(Mepytpy)_2]_2[BF_4]_2$

Fit a 100 ml round-bottomed flask with a double surface reflux condenser. Dissolve $[Fe(pytpy)_2][BF_4]_2$ (0.1 g) in 50 ml of acetonitrile and add the solution to the flask. Add 2 ml of methyl iodide and heat to reflux for one hour. Add a solution of $[NH_4][BF_4]$ (0.5 g) in 5 ml of methanol and concentrate the dark blue solution on a rotary evaporator to about 10 ml volume and allow to cool. Collect and dry the blue-black solid.

Calculate the yield of your product and measure its IR spectrum as a Nujol mull and, if possible, its ^{1}H NMR spectrum in CD_3COCD_3. Record the electronic spectrum (250–700 nm) of your product in acetonitrile solution. You should accurately weigh your sample so that you can report extinction coefficients. As a guide to the concentration required, the bands have ε values between 20000 and 65000 $dm^3\ mol^{-1}\ cm^{-1}$.

You will find additional details about these reactions in the references.

References

E. C. Constable, *Adv. Inorg. Chem. Radiochem.* **1987,** *30,* 69; **1989,** *34,* 1.

R. J. Watts, *J. Chem. Educ.* **1983,** *60,* 6536.

L. De Cola, F. Barigelletti, V. Balzani, P. Belser, A. von Zelewsky, C. Seel, M. Frank, F. Vögtle, *Coord. Chem. Rev.* **1991,** *111,* 255.

N. Armaroli, V. Balzani, E. C. Constable, M. Maestri, A. M. W. Cargill Thompson, *Polyhedron* **1992,** *11,* 2707.

E. C. Constable, A. M. W. Cargill Thompson, *J. Chem. Soc. Dalton Trans.* **1992,** 2947.

E. C. Constable, Metals and Ligand Reactivity, Ellis Horwood (Chichester), **1990.**

4.3 Electronic Characterisation of a Transition-Metal Complex Using Electrochemical, UV/Vis and EPR Techniques

Lesley J. Yellowlees

Many students will have encountered electroanalytical techniques where redox active species are detected both *qualitatively* and *quantitatively*. This experiment is designed to illustrate other uses of the technique – how electrochemistry can be used to probe the electronic characteristics of a redox active transition-metal complex and in the preparation and study of an air unstable species. In particular, you will investigate the frontier orbitals (HOMO and LUMO) of $[Fe(bpy)_3]^{2+}$ (bpy = 2,2′-bipyridine). UV/vis and EPR spectroscopies will also be employed in this study.

Experimental

$[Fe(bpy)_3][BF_4]_2$

$FeCl_2$ (0.1 g) is dissolved in a minimum amount of H_2O and bpy (0.4 g) is dissolved in a minimum amount of ethanol. The two solutions are mixed and a solution of $NaBF_4$ (0.2 g) in H_2O is added. The resulting deep red precipitate is filtered and washed with cold water and finally cold ethanol.

Record the UV/vis spectrum of the $[Fe(bpy)_3]^{2+}$ in CH_3CN from 850 to 250 nm. Calculate the maximum molar absorption coefficient, ε_{max}, for each of the absorption bands ($\varepsilon_{max} = Ac^{-1}l^{-1}$, where A = measured maximum absorption, c = concentration in mol dm^{-3} and l = pathlength of solution in cm). (*Note:* it will be necessary to record spectra of $[Fe(bpy)_3]^{2+}$ over a range of concentrations in order to observe all transitions).

Construct a qualitative molecular orbital diagram for $[Fe(bpy)_3]^{2+}$, remembering that bpy is a bidentate N-donor ligand and is a fairly good π-acceptor. Where are the HOMO and LUMOs *primarily* based? Which electronic transition is responsible for the deep red colour?

You should be able to probe the electronic character of the HOMO and LUMO electrochemically since an oxidation will involve removing an electron from the HOMO and a reduction involves adding an electron to the LUMO. Furthermore, if either of the redox processes is chemically reversible, *i. e.* the complex does *not* undergo a chemical raction following the electron transfer reaction, then the redox product may be studied in order to test the molecular orbital diagram derived qualitatively.

Electrochemical Studies

Initially, you will probe the redox behaviour of $[Fe(bpy)_3]^{2+}$. There are approximately 30 distinct perturbations of potential, current and charge which can be used in the study of electrode processes – however, only about 10 of these are widely used. In this experiment, you will consider three of the most common, namely *cyclic voltammetry* (CV), *stirred voltammetry* (SV) and *coulometry.*

Electrochemical experiments will be performed in a conventional three-electrode cell using a Pt micro-disc (diameter approximately 1 mm) electrode, a Ag/AgCl reference electrode and a Pt counter electrode. The solvent is acetonitrile and the supporting electrolyte is tetraethylammonium tetrafluoroborate ($TEABF_4$) (0.1 M). Coulometric experiments will use an H-type cell in which the counter electrode compartment is separated by a frit from the working electrode compartment so as to avoid product contamination.

Make up 50 ml CH_3CN/0.1 M $TEABF_4$, set up the electrochemical cell and bubble the solution with Ar or N_2 for approximately 20 minutes. Why is it necessary to bubble the gas through the solution prior to experimentation? Check that the background solvent is free of any redox processes by running positive and negative CV's. Add approximately 20 mg of compound. Set the starting potential to 0 V and scan rate to 100 mV s^{-1}, sweep to +1.5 V and −2.0 V. Do SV at 20 mV s^{-1} to confirm whether the observed redox processes are oxidations or reductions. Note that the energy gap between the first reduction process and the oxidation process is in fairly good agreement with the energy of the visible absorption maximum of $[Fe(bpy)_3]^{2+}$. Why should this be?

Concentrating on the oxidative process, do a reversibility study by varying the scan rate from 20 mV s^{-1} to 500 mV s^{-1}. A reversible redox process means that the electron transfer rate from/to the electrode to/from the complex is rapid and the complex is not undergoing a major geometric change on electron transfer. Furthermore, the complex is not undergoing a chemical reaction following the electron transfer step. If the redox couple is fully reversible, then the following criteria must hold:

a) 'E_p' is independent of v (scan rate) and $E_a - E_c = 59/n$ mV (n = no. of electrons involved in the electron transfer process) at 298K and $E_p = (E_a + E_c)/2$ where E_a = anodic peak potential and E_c = cathodic peak potential.

b) $i_a/i_c = 1.0$ and i_a vs. $v^{1/2}$ is a straight line through zero. i_a = anodic peak current, i_c = cathodic peak current.

Is the $[Fe(bpy)_3]^{2+}$ oxidation process reversible?

Next, set up the coulometric cell and bubble the solution with Ar or N_2 for 20 minutes. Run background CV's and then add an accurately weighed amount of complex (approximately 20 mg). Run CV's and SV's. Perform electrogeneration of oxidised species at relevant potential recording i vs. t curve throughout generation. After generation of oxidised product, calculate the number of moles of electrons passed. Repeat CV's and SV's. Note that CV's are identical before and after electrogeneration, but SV's are not. Why is this so?

Provided it is kept in an inert atmosphere, the oxidised product is relatively stable and may be studied using a variety of spectroscopic techniques. Record the UV/vis and EPR (Fig. **4.3-1**) spectra of the oxidised species. Discuss them and decide whether they are in agreement with your qualitative molecular orbital diagram for $[Fe(bpy)_3]^{2+}$.

Comment on the electronic nature of the reductive processes and what you might expect to observe in the UV/vis and EPR spectra of the first reduction product.

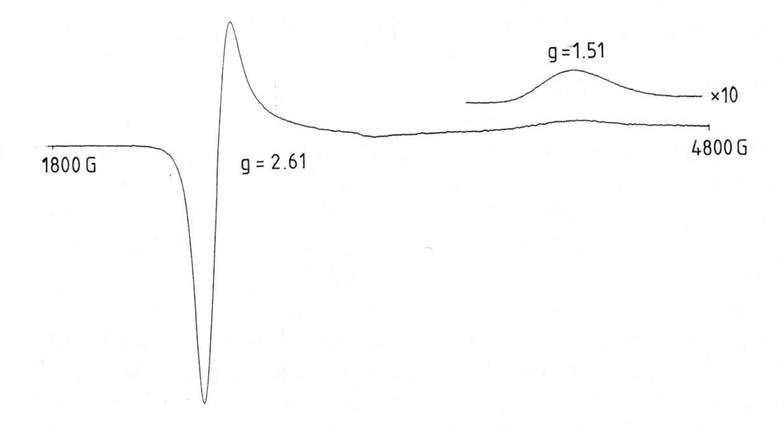

Figure 4.3-1. EPR spectrum of $[Fe(bpy)_3]^{3+}$ in CH_3CN/0.1 M $TEABF_4$ at 77 K.

4.4 The Synthesis of η^6-Arenetricarbonyl Chromium(0) Complexes

David A. Widdowson

Special Safety Precautions

Chromium hexacarbonyl is a colourless crystalline solid with a high vapour pressure. Highly toxic by ingestion or inhalation. All manipulations should be carried out in a fume cupboard.

The exploitation of the modification of reactivity of an organic molecule by complexation to a metal is one of the major areas of development in organic synthesis in recent times. One particular area is the study of the enhanced reactivity of arenes upon complexation by Group VI metal and manganese carbonyls. By far the most studied among these are the arenechromiumtricarbonyl complexes. The effect of the metal moiety on the arene ring is an apparent electron withdrawal from the π-system and this manifests itself in a variety of ways as shown below.

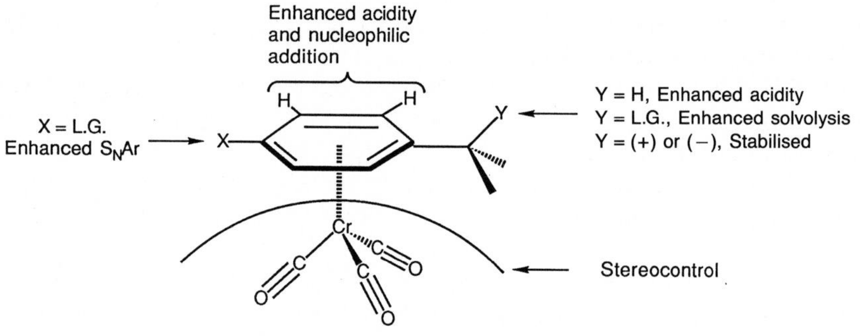

Enhanced acidity and nucleophilic addition

X = L.G. Enhanced S_NAr

Y = H, Enhanced acidity
Y = L.G., Enhanced solvolysis
Y = (+) or (−), Stabilised

Stereocontrol

It is important for the ease of use of such complexes that they should be readily prepared and efficiently decomplexed. This experiment demonstrates the most convenient method of preparation *via* the use of the Strohmeier apparatus and a simple vacuum line/nitrogen manifold (Schlenk line). The complexes are synthesised by the direct reaction between the arene and chromium hexacarbonyl in an ether solvent mixture (Bu_2O : THF 10 : 1).*

* This solvent mixture has been determined empirically to give the optimum reaction temperature and a good backwash of the volatilised chromium hexacarbonyl (see C. A. L. Mahaffy, P. L. Pauson, *Inorg. Syn.* **1979,** *19,* 154).

X, Y + $Cr(CO)_6$ $\xrightarrow{\Delta}$ X, Y, $(OC)_3Cr$

The need for the Schlenk line arises from the fact that the intermediates generated during the synthesis, the coordinatively unsaturated chromium carbonyl species, are very oxygen sensitive and so rigorously anaerobic conditions are essential. Once formed, the arene complexes are air stable in the solid state and can be handled without difficulty by conventional techniques.

The Strohmeier apparatus is, in effect, an inverted condenser. This is a convenient way of dealing with the problem of the volatility of the metal carbonyl which tends to condense above the level of the solvent in a normal condenser and block it. In the Strohmeier apparatus, any hexacarbonyl condensing ahead of the solvent is washed back into the reaction vessel *via* the syphon.

Experimental

A mixture of di-*n*butyl ether (purified by distillation from sodium-benzophenone* (60 ml), THF (purified by distillation from sodium-benzophenone (6 ml), the arene (*e.g.* fluorobenzene, chlorobenzene, anisole, 1-(*tri-iso*propylsilyl)indole, 2-methylthiophene; quantity: 1 equivalent – for valuable substrates – or 10 ml – for readily available substrates) and chromium hexacarbonyl (1.0 g) are placed, together with a magnetic stirrer bar, in a 100 ml round-bottomed flask and attached to the Strohmeier apparatus. The assembly is connected to the Schlenk line using semi-pressure tubing. A magnetic stirrer and a heating bath on a lab jack are placed beneath the round-bottomed flask.

Check that all joints are sealed, then evacuate the apparatus by carefully turning the 3-way tap on the line. As soon as the solvent begins to boil, let in the dry, oxygen-free nitrogen *via* the 3-way tap. Repeat this cycle nine more times to ensure that the system is completely anaerobic. With the system maintained under a slight positive pressure of nitrogen, commence stirring and heating the contents of the flask. Maintain the solution at a steady reflux for $\approx$24 h.

At the end of the reaction, the flask is cooled then detached from the Strohmeier apparatus. The solution is chromatographed over a short ($\sim$5 cm) column of silica gel 60 using ether to effect the elution of the yellow-orange complex. Evaporation of the solvents yields the product.

Record the m.p. and record and interpret the IR and NMR spectra. Comment on the synthetic uses of these complexes.

---

* A special communal still for the purification of this solvent should be set up in a fume cupboard. It is essential that you consult a demonstrator or a technician before you use it.

References

For broad surveys of organo metallic chemistry in organic syntheses see: *Comprehensive Organometallic Chemistry,* Vols. 7 and 8 (Eds.: G. Wilkinson, F. G. A. Stone, E. W. Abel), Pergamon Press, Oxford, **1982;** S. G. Davies, *Organotransitionmetal Chemistry: Applications to Organic Synthesis,* Pergamon Press, Oxford, **1982;** E.-I. Negishi, *Organometallics in Organic Synthesis,* Wiley-Interscience, New York, **1980.**

M. F. Semmelhack, G. F. Clark, J. L. Garcia, J. J. Harrison, Y. Thebtaranonth, W. Wulff, A. Yamashita, *Tetrahedron* **1981,** *37,* 3957.

G. Jaouen, *Ann. N. Y. Acad. Sci.* **1977,** *295,* 59. M. F. Semmelhack, *Ann. N. Y. Acad. Sci.* **1977,** *295,* 36.

D. A. Widdowson, *Phil. Trans. Roy. Soc. Lond.* **1988,** *A 326,* 595.

W. Strohmeier, *Chem. Ber.* **1961,** *94,* 2490.

4.5 Migratory Insertion Reactions in Organo Transition-Metal Chemistry

Mark J. Winter

Special Safety Precautions

The reactions described should all be carried out in a fume cupboard. Mercury gives off a dangerous vapour; sodium chips react violently with water to give flammable gas; tetrahydrofuran is extremely flammable, may form peroxides and is irritating to the eyes and the respiratory system. MeI is toxic in contact with the skin and gives off a poisonous vapour.

Many transition metal anions are *nucleophilic.* They can therefore replace the halides of alkyl halides (Eqns. 1 and 2).

$$[Mn(CO)_5]^- \xrightarrow{PhCH_2Br} Mn(CH_2Ph)(CO)_5 \qquad (1)$$

$$[(\eta^5\text{-}C_5H_5)Fe(CO)_2]^- \xrightarrow{PhC(O)Cl} (\eta^5\text{-}C_5H_5)Fe(CO)_2C(O)Ph \qquad (2)$$

Migratory insertion reactions (or simply "insertion reactions") are those in which an atom, or group of atoms, is inserted between two mutually bonded atoms (Eqn. 3).

$$M-L \xrightarrow{XY} M-XY-L \qquad (3)$$

The term "insertion reaction" is a little unfortunate since many intramolecular rearrangements and intermolecular additions are classified under the same heading. Migratory insertion reactions are very important in industrial homogeneous catalysis as well as synthetic organotransition metal chemistry in university research laboratories.

Of particular interest is the transformation of alkyl carbonyl complexes into acyl species. The incoming ligand L′ may be an entity such as phosphine, amine, halide ion or carbon monoxide itself. Examples of such transformations are known for most transition metals. One reaction that has been studied particularly extensively is that of alkyl complexes $MnR(CO)_5$ with ligands such as PPh_3 or CO (Eqn. 4).

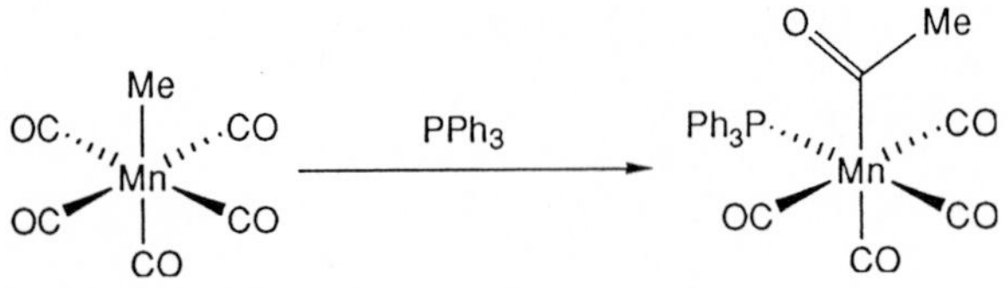

These studies, and those on some other systems, produced a set of observations consistent with a general reaction sequence for the alkyl to carbonyl insertion reaction. It is generally assumed that all alkyl to carbonyl reactions proceed in a similar fashion. You must understand that this is a tacit assumption and it does not necessary follow that all alkyl to carbonyl insertions proceed by the same mechanism (there are a few examples of those that do not).

The following five points refer to Scheme **4.5-1.**

1. The R group migrates to an *adjacent* (that is, *cis*) carbonyl group.
2. In a second step, the coordinatively unsaturated (16 e^-) intermediate is attacked by the incoming (nucleophilic) ligand.

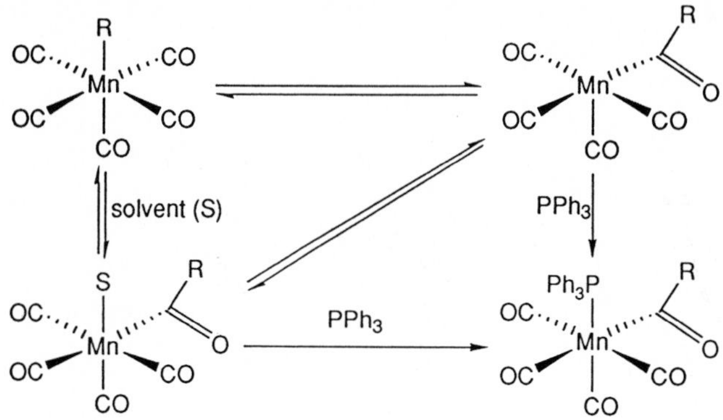

Scheme 4.5-1.

3. The 16 e^- intermediate may be "transiently stabilised" by a solvent molecule. This is particularly likely when the solvent has coordinating ability of its own, good examples are ethers and acetonitrile. Donation of a lone pair from an ether or acetonitrile transforms the 16 e^- intermediate into an 18 e^- intermediate, but one with a very loose ligand (the solvent). Although this intermediate has a stable 18 e^- configuration, it behaves essentially as an entity with a vacant coordiantion site.
4. If the alkyl group possesses chirality, retention of configuration is observed.
5. First row transition-metal complexes react more quickly than either second or third row compounds. the M-alkyl bond is stronger for second and third row transition metals and therefore more difficult to break (a necessary step to achieve the migratory insertion step).

The *reverse* of the alkyl to carbonyl insertion process is also known, in which case it is sometimes referred to as an *extrusion.* This process is frequently achieved by thermal or photochemical activation. An elegant series of labelling studies have demonstrated that the mechanism is the microscopic reverse of the migratory insertion process.

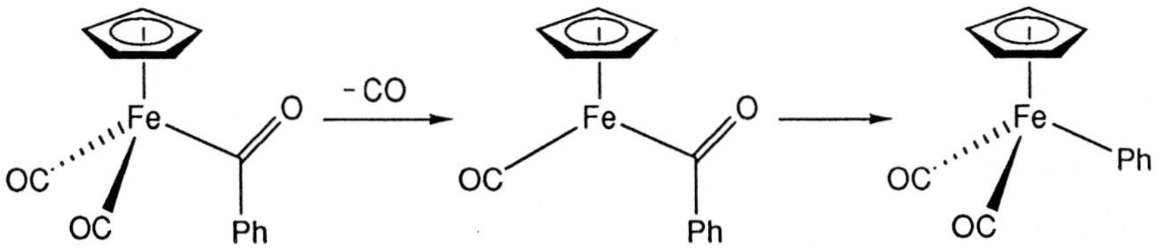

In the extrusion reaction a carbonyl ligand is lost and this is followed by phenyl migration to the resulting vacant coordination site.

It would be possible in a practical assignment of this type to study the manganese systems above (Scheme **4.5-1**), but we choose not to do so as the compounds are rather volatile and (perhaps more to the point!) very pricey. Instead, we shall examine a series of related reactions based upon the molybdenum complex $MoMe(CO)_3(\eta^5\text{-}C_5H_5)$ shown in Scheme **4.5-2.**

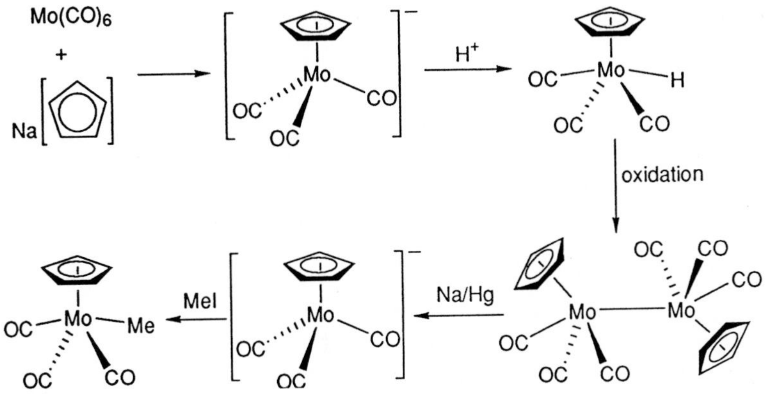

Scheme 4.5-2.

Experimental

a) $MoMe(CO)_3(\eta^5\text{-}C_5H_5)$

It is most important to exclude air from the apparatus until the reaction with MeI is complete, the anion $[Mo(CO)_3(\eta^5\text{-}C_5H_5)]^-$ is particularly air sensitive. Discuss how to do this with a demonstrator. Set up the apparatus *in a fume cupboard* as shown in Figure **4.5-1.**

Flush out the apparatus with N_2, add mercury (5 cm^3) and a magnetic stir bar. Add sodium chips (0.3 g, weighed under light petroleum, cut into 5–6 pieces) one at a time and stir rapidly until the sodium reacts with the mercury (*exothermic, vigorous reaction, fumes*). Amalgam formation is best achieved with fresh sodium surfaces. If there is a problem, try holding the sodium chip under the mercury surface with a spatula and press it firmly against the glass wall of the Schlenk tube to create a new surface.

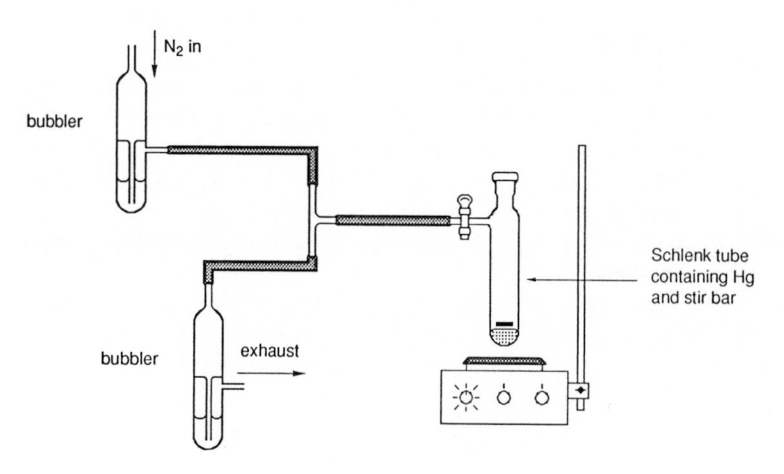

Figure 4.5-1. Apparatus for synthesis under N_2.

After the amalgam has formed, add tetrahydrofuran (THF) (50 cm^3, freshly distilled) followed by $[Mo(CO)_3(\eta^5\text{-}C_5H_5)]_2$ (1.0 g). Stir under N_2 until the maroon colour of the dimer is discharged (about 20 minutes to 2 hours). The idea is to stir the mixture as fast as possible in order to facilitate the reduction. The resulting murky yellow-green solutions contains $[Mo(CO)_3(\eta^5\text{-}C_5H_5)]^-$ and is very air sensitive. *The introduction of any air once the anion is generated will cause a pink colouration as the dimer is reformed.* If this happens, continue stirring over the amalgam until the pink colour is discharged.

Separate the anion solution from excess amalgam with a syringe. Stop stirring the reaction and rotate the Schlenk tube until it is nearly horizontal and allow any solids to settle. This may take a couple of minutes. Set up a *second* Schlenk tube and flush it out with N_2. Transfer the anion solution to the second tube *via* syringe. Use "subaseals" to prevent air getting into either Schlenk tube.

Add more THF (10 cm^3, freshly distilled) to the first tube and stir for a minute or so. If there is a pink colouration (a little oxidation of the anion), continue stirring until the colour is discharged. Allow the solution to settle and transfer the second batch of anion solution to the second Schlenk tube in the same way as before.

Add MeI (1.0 g) (weigh out and handle in fume cupboard) to the anion solution, replace the "subaseal" with a glass stopper, close the sidearm tap and leave to stir overnight.

Remove the solvent under reduced pressure, *at ambient temperature.* Extract the solid into light petroleum (*extremely flammable*) (4 × 20 cm^3 portions or until colourless), and filter the resulting solution through a short Al_2O_3 plug (about 2.5 cm in a filter stick). Allow the product to pass through the alumina under gravity. Wash through with extra light petroleum until the eluant is colourless.

Remove the solvent. Scrape all the solid into a sublimation unit and sublime out the product under vacuum ($<10^{-1}$ torr, about 40–60 °C, use a water bath for heating). It is most important to maintain a good vacuum while this is going on.

Record the IR spectrum of your product as a *solution* in light petroleum (CaF_2 cells) in the range 2200–1500 cm^{-1}, determine the yield.

b) [Mo(CO)$_2$(COMe)(η^5-C_5H_5)(PPh_3)]

Stir an acetonitrile (10 cm^3) solution containing MoMe(Co)$_3$(η^5-C_5H_5) (0.13 g, 0.5 mmol) and PPh_3 (0.2 g, 0.76 mmol) under nitrogen overnight.

This should give a bright yellow precipitate. Collect the yellow solid [CpMo(COMe)(CO)$_2$(PPh_3)] on a sinter and wash with a little light petroleum. Dry under vacuum. If the product looks a little grubby, recrystallise from CH_2Cl_2 and light petroleum.

Record the *solution* IR spectrum of the product in CH_2Cl_2(CaF_2 cells) in the range 2200–1500 cm^{-1}. Determine the yield.

4.6 Organometallic Molybdenum Compounds

Manfred Bochmann and Paul Hammerton

Special Safety Precautions

$Mo(CO)_6$ and cycloheptatriene are toxic. Perform the raction in a fume cupboard.

Transition metals in low oxidation states are able to form complexes with many unsaturated organic ligands, e. g. CO, olefins and acetylenes. This ability to coordinate such molecules and to activate them for subsequent reactions has led to an extremely rich chemistry and constitutes the basis of many important industrial catalytic processes. When CO or alkenes coordinate to a metal centre, electron density from their lone-pair or π-levels is donated to the metal, while at the same time the π^*-orbitals of the ligands receive electron density from the metal, thereby strengthening the metal–ligand bond. This is known as the "Dewar-Chatt-Duncanson model" of ligand bonding. CO is particularly powerful in withdrawing electron density from the metal in this way, i. e. it is a strong "π-acid". Since the C–O stretching mode is easily observed in the IR spectrum, the position of this band is a sensitive probe for the electron density of the metal centre, its oxidation state and the overall charge.

The following experiments illustrate several aspects of the chemistry of metal carbonyl complexes.

a) Some, but rarely all, CO ligands in the metal carbonyl complex can be displaced by other ligands such as alkenes. In our experiment, a cyclic olefin, cycloheptatriene, is used as the displacing ligand.
b) The cycloheptatriene ligand has a polar C–H bond. Hydride abstraction generates an η^7-C_7H_7 (tropylium) complex, a representative of the large family of 6π arene complexes. *Cationic and neutral Mo(0) and Mo(II) complexes are generated in a) and b) whose IR spectra show typical shifts in the CO band patterns.*
c) Cationic complexes are susceptible to nucleophilic attack, either on one of the ligands or on the metal centre.

Some of the following preparations work best if carried out under a protective atmosphere of nitrogen; use a nitrogen manifold in a fume cupboard. A rotary vacuum pump is recommended for the thorough drying of some of the products. Ask a demonstrator to instruct you in the use of these pieces of equipment.

Experimental

a) Cycloheptatriene molybdenum tricarbonyl

The experimental set-up is shown in Fig. 3.7-3. The compound is prepared by refluxing a mixture of $Mo(CO)_6$ and cycloheptatriene in a high-boiling solvent under nitrogen. Use thoroughly dry glassware. To 10 mmol $Mo(CO)_6$ in 25–30 ml of sodium-dried *n*-octane in a

250 ml flask add 15 mmol cycloheptatriene. Purge the flask with a stream of N_2 to expel the air and stopper it quickly. Assemble the straight condenser with a T-piece, connect the N_2 supply to the T-piece and purge the condenser with N_2. Maintain the N_2 stream while you quickly exchange the stopper on your flask for the condenser assembly. In this way, you should have avoided the entry of air as far as possible. Reflux the mixture for 5–6 h, or until no more $Mo(CO)_6$ is seen to sublime. The oil bath temperature should be 160 °C. Make sure the solvent level inside the flask is slightly higher than the oil bath level to avoid the decomposition of the product on overheated glass walls. During the refluxing, unreacted molybdenum hexacarbonyl sublimes from the reaction mixture onto the upper part of the flask or into the condenser and must be brought back into the reaction medium by swirling the flask or by poking it down with a rod at regular intervals (every 30 minutes or so; briefly remove T-piece to do this). *This reaction should not be left unattended for long periods.*

The resulting deep red solution is evaporated on a rotary evaporator (water bath temp. 70 °C) leaving a deep red solid. This is dissolved in 20–30 ml of diethyl ether. The solution is filtered quickly through a dry sintered glass filter funnel. Since any unreacted $Mo(CO)_6$ is less soluble in ether than the product, most of it is left on the filter. The filtrate is immediately taken to dryness on a rotary evaporator (water bath temperature *ca.* 20–30 °C). The bright red product may still contain small amounts of $Mo(CO)_6$ but should be pure enough for the subsequent reaction. Rinse the product quickly with *ca.* 5 ml light petroleum (b. p. 40–60 °C) to remove any residual cycloheptatriene and dry *in vacuo.* Pure samples are obtained by dissolving a fraction in light petroleum, concentrating the solution and placing it in the fridge to crystallise.
Yields of *ca.* 60% orange-red $(C_7H_8)Mo(CO)_3$ can be expected.

b) Cycloheptatrienyl molybdenum tricarbonyl tetrafluoroborate

Weigh out separately 4 mmol $(C_7H_8)Mo(CO)_3$ and 4 mmol "trityl" (= triphenylmethyl, Ph_3C^+) tetrafluoroborate, dissolve each in 15 ml *dry* dichloromethane separately, and combine the two solutions rapidly with stirring. Leave for 4–10 min, filter off through a G3 sintered glass frit, wash with dry diethyl ether and dry *in vacuo.* The product is obtained as an orange powder in almost quantitative yield.
Note: $[(C_6H_5)_3C]BF_4$ is the salt of a stable carbocation. Although it can be handled briefly in air, it is very susceptible to nucleophilic attack, *e. g.* by atmospheric moisture. It must be weighed out quickly and protected from moisture by dissolving it immediately in dry CH_2Cl_2. The container of $[(C_6H_5)_3C]BF_4$ must always be stored in a desiccator over a powerful drying agent (P_2O_5 with moisture indicator). Good quality "trityl"BF_4 is orange to orange-brown; brown or colourless material (hydrolysis!) gives poor results or is unsuitable.

c) Cycloheptatrienyl molybdenum dicarbonyl bromide

This experiment is best carried out in a fume cupboard. $[(C_7H_7)Mo(CO)_3]BF_4$ (1.5 mmol) is added in small portions to a stirred solution of 1.7 mmol anhydrous LiBr in 25 ml dry, degassed acetone under a moderate counter-current of nitrogen. (Acetone is dried over molecular sieves. Oxygen is removed by bubbling nitrogen through it for 2–3 minutes).

Stop stirring briefly and add a little $[(C_7H_7)Mo(CO)_3]BF_4$. What do you observe? Continue stirring for 20 min, remove the acetone *in vacuo,* extract the residue with 20 ml dichloromethane and filter in air through a G3 sintered glass frit. $LiBF_4$ and unreacted

$[(C_7H_7)Mo(CO)_3]BF_4$ are left behind. Wash with another 15 ml CH_2Cl_2, concentrate the combined filtrate to 3 ml (rotary vacuum pump), stopper the flask and leave on an ice bath to complete the crystallisation. If crystals fail to form, add a few mls diethyl ether with a Pasteur pipette. The green-black crystalline product is collected on a sintered glass filter. Yields of 60–70% can be expected.

d) Characterisation

Record the IR spectra of $Mo(CO)_6$ and of your products as Nujol mulls between KBr plates. Adjust the thickness of the mull so that the CO stretching bands are well resolved. Interpret your spectra, paying particular attention to the CO stretching region.

Record the ^1H NMR spectra of cycloheptatriene (in CCl_4) and of each of your products ($(C_7H_8)Mo(CO)_3$ in C_6D_6, the others in acetone-d_6). Use only freshly prepared solutions. Assign the spectra and show how the reaction sequence can be deduced from these data. Figure **4.6-1** shows the ^1H NMR spectrum and Figure **4.6-2** the COSY spectrum of $(C_7H_8)Mo(CO)_3$. Its mass spectrum is illustrated in Figure **4.6-3.**

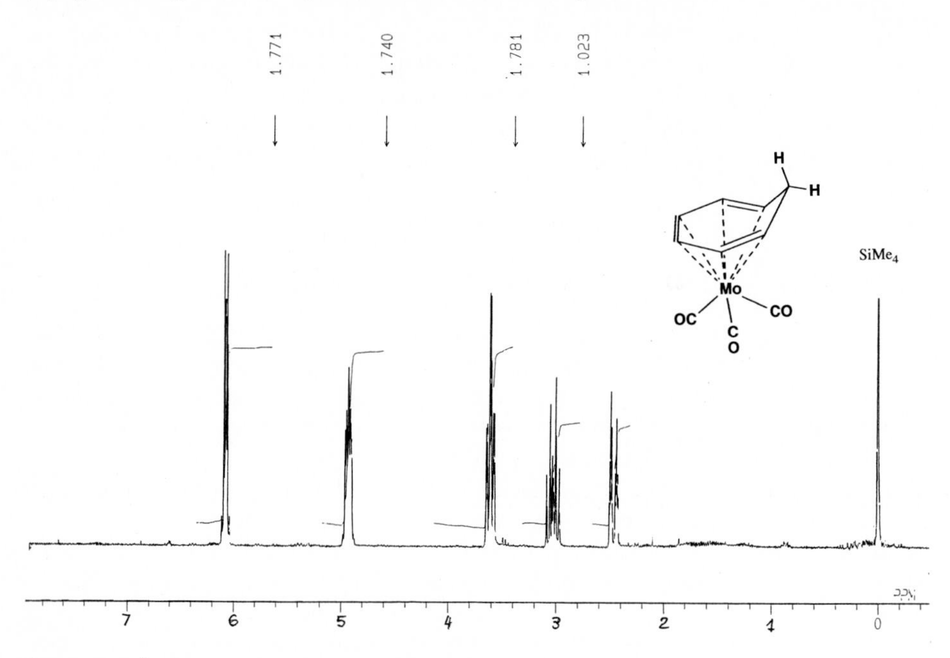

Figure 4.6-1. ^1H NMR spectrum of $(C_7H_8)Mo(CO)_3$.

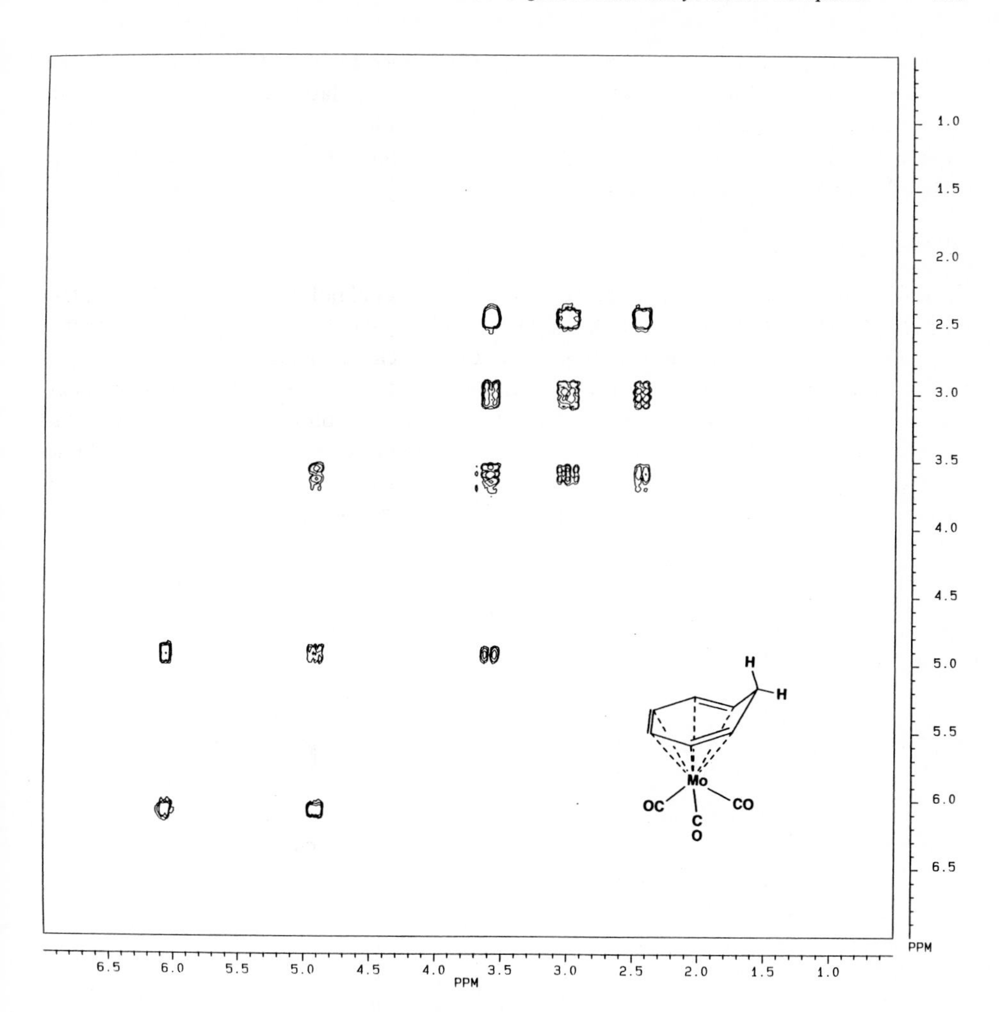

Figure 4.6-2. COSY spectrum of $(C_7H_8)Mo(CO)_3$.

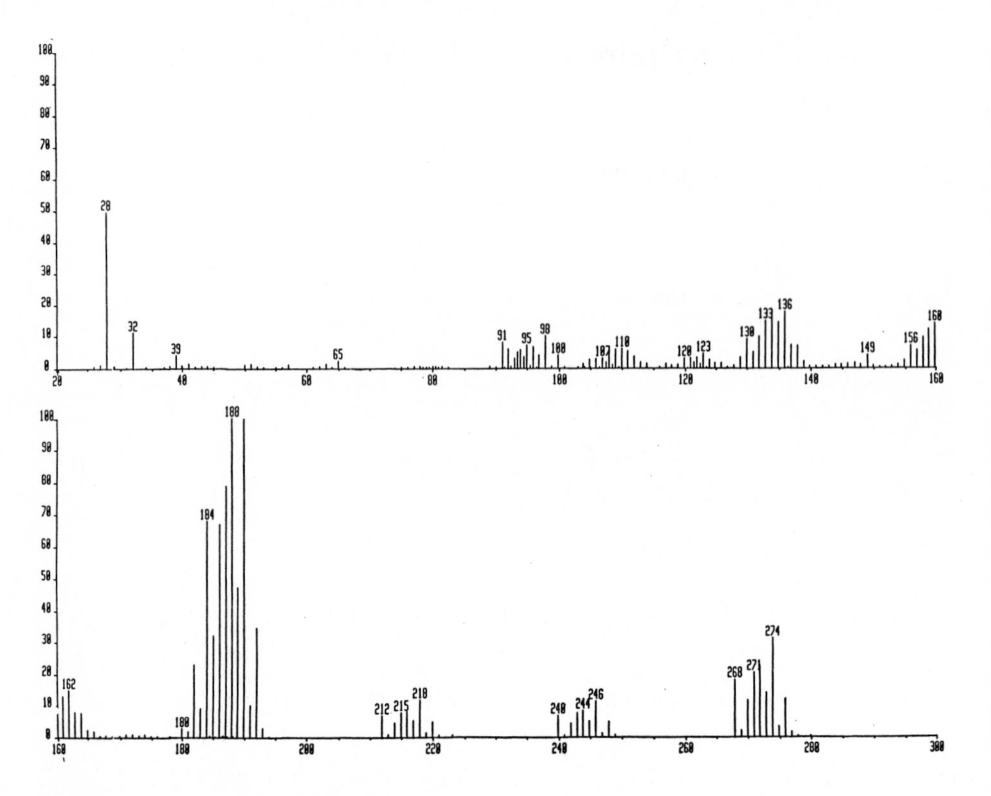

Figure 4.6-3. Mass spectrum of $(C_7H_8)Mo(CO)_3$.

Exercises

1. How is spectroscopic (and other) evidence used to arrive at chemical conclusions?
2. Write down the chemical equations, including possible reaction mechanisms, and suggest the structures of your products.

References

D. F. Shriver, P. W. Atkins, C. H. Langford, *Inorganic Chemistry,* Oxford University Press, **1990,** Chpt. 16.

C. Elschenbroich, A. Salzer, *Organometallics,* VCH, **1989,** Chpts. 12 and 14.5.

F. A. Cotton, G. Wilkinson, *Advanced Inorganic Chemistry,* 5th ed., J. Wiley, **1985,** Chpts. 22 and 28.

E. W. Abel, M. A. Bennett, R. Burton, G. Wilkinson, *J. Chem. Soc* **1958,** 4559.

R. B. King, *Organometallic Synthesis,* Vol. 1, Academic Press, **1965,** p. 141.

M. Bochmann, M. Green, H. P. Kirsch, F. G. A. Stone, *J. Chem. Soc., Dalton Trans.* **1977,** 714.

4.7 The preparation of *rac*-Ethylene-1,2-bis(1-indenyl)zirconium Dichloride and Dibenzyl Complexes

Manfred Bochmann and Simon J. Lancaster

Special Safety Precautions

1. All experimental procedures should be carried out on a vacuum/inert-gas manifold in a fume cupboard.
2. Petroleum ether, diethyl ether and indene are flammable. THF is both flammable and an irritant, toluene is flammable and toxic.
3. Zirconium(IV) chloride is corrosive and liberates HCl on hydrolysis; its inhalation must be avoided.
4. Dibromoethane is toxic and a cancer suspect agent. It should be handled in a fume cupboard.
5. Benzyl chloride is a lachrymator. It should be handled using rubber gloves in a fume cupboard.
6. Butyllithium is pyrophoric and reacts violently with water. It should be manipulated under inert gas using stainless-steel cannulae and gas-tight syringes. Rubber gloves should be worn!

Bis(cyclopentadienyl) metal dichloride complexes of titanium, zirconium and hafnium in the presence of aluminum alkyls as activators are well known catalyst precursors for the polymerisation of olefins. If the cyclopentadienyl ligands carry suitable substituents and are connected to each other by a bridge, stereorigid complexes result which are able to control the stereochemistry of the growing polymer chain during the polymerisation of 1-alkenes, notably propene. The best known example for complexes of this type is *rac*-ethylene-1,2-bis(1-indenyl)zirconium dichloride ($C_2H_4Ind_2ZrCl_2$), which possesses C_2 symmetry and leads to the formation of highly *isotactic* polypropene. The second possible isomer of the complex, with *meso* configuration, has C_s symmetry and is a minor by-product which is removed during the purification process; it is undesirable since it exhibits no stereocontrol.

Experimental

a) The ligand

Required are a 1l flask with central B24 neck and two outer B19 necks equipped with magnetic stir bar, a dropping funnel with 200 ml capacity, a B19 stopcock adaptor, two B19 stoppers, a B19 "subaseal", a large insulated dry-ice/acetone bath, two 100 ml glass syringes, a 2l separatory funnel, a large sintered glass frit and a 2l Büchner flask.

BuLi

Li⁺

2 Li⁺ + Br Br

Assemble the three-necked flask with dropping funnel and evacuate on a vacuum line, flame dry the flask and allow to cool before filling it with dry argon or nitrogen. Weigh 56.7 g (0.49 mol) indene (*Note:* to ensure purity, indene should be distilled on a long vigreux column and stored in a freezer) in the 100 ml syringe and transfer to the flask through the "subaseal". Add 300 ml dry, degassed THF through the "subaseal". Cool the solution to −78 °C on the dry-ice/acetone bath. Taking a clean 100 ml syringe, and with great care, charge the dropping funnel with 195 ml of a 2.5 M solution of *n*-BuLi. Add the *n*-BuLi dropwise over a period of one hour. The solution will turn red in colour. After complete addition, warm the solution to room temperature and stir for a further hour to ensure complete deprotonation.

Cool the solution again to −78 °C on a dry-ice/acetone bath and via the dropping funnel add 45 g 1,2-dibromoethane dissolved in a further 50 ml dry degassed THF dropwise with stirring over one hour. Then allow the reaction mixture to warm very slowly to room temperature over a period of two hours before stirring for a further two hours. Cool the reaction mixture to 0 °C in an ice bath, quench with 100 ml of a saturated solution of ammonium chloride and add a further 200 ml of distilled water to dissolve the LiCl. The procedure can pause overnight at this point.

Transfer the contents of the flask to a large separatory funnel and add a further 200 ml petroleum ether to dilute the organic phase. Discard the aqueous phase and wash the organic phase with a further two 200 ml portions of distilled water. Clean the original flask, remove the dropping funnel and transfer the organic phase to the flask. Add approximately 30 g anhydrous magnesium sulphate and stir for one hour to dry. Separate the magnesium sulphate by filtration through a G3 sintered glass frit. Reduce the solvent volume to about 100 ml or until crystals persist at 50 °C on a rotary evaporator. Allow the solution to cool slowly; the product will crystallise as off-white crystals which are filtered off and dried *in vacuo* for several hours. Yield: 40 g, 64.5%. The ligand is of sufficient purity for the preparation of the zirconocene dichloride. An analytically pure sample is obtained by recrystallisation from ethanol/acetone. ^{1}H NMR (400 MHz, $CDCl_3$): δ 2.95 (br s, 4H), 3.35 (br s, 4H), 6.29 (br s, 2H), 7.21 (d, 2H, $J = 6.9$ Hz), 7.31 (t, 2H, $J = 7.3$ Hz), 7.40 (d, 2H, $J = 7.3$ Hz), 7.48 (d, 2H, $J = 6.9$ Hz).

b) $C_2H_4Ind_2ZrCl_2$

The zirconocene dichloride complex was first prepared by Brintzinger *et al.* The following procedure is a modification of the original procedure and has proved more reliable, with less formation of the undesirable *meso* isomer, than variations published subsequently. The yield is not adversely affected by an increase in scale and gives typically and reliably 40–50% yields of the *rac* product. (*Note:* the synthesis of $ZrCl_4(THF)_2$ is described in the reference by Manzer.)

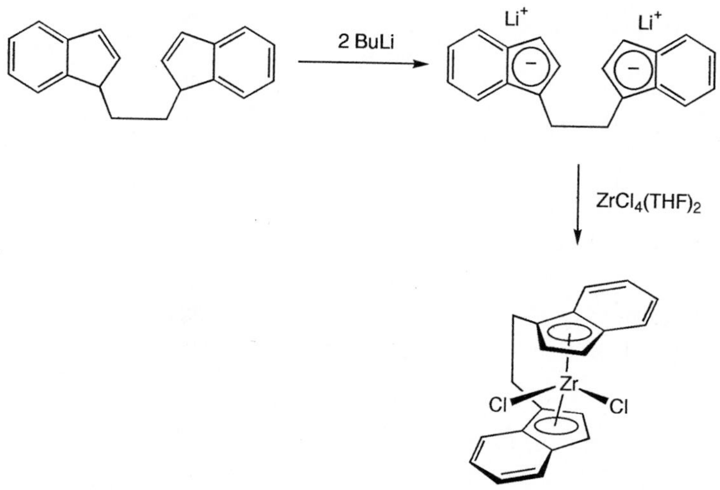

Required are two 250 ml flasks with central B24 and two outer B19 necks each equipped with magnetic stirrer-hotplates and bars, 100 ml dropping funnels, stopcock adaptors, two B19 stoppers, two bent fingers for solid handling, 20 ml glass syringe, transfer tubing, filtration cannula, two B19 "subaseals", one water bath and one dry-ice/acetone bath.

The first three-necked flask should be equipped with a dropping funnel and stopcock adaptor and evacuated before flame drying, then filled with inert gas. Using the bent-finger for weighing and transfer, place 6 g of the prepared ligand (23 mmol) in the flask and dissolve in 70 ml of dried, degassed THF. Cool to −78 °C using a dry-ice/acetone bath. Using the syringe, add 18.4 ml of a 2.5 M solution of *n*-BuLi (46 mmol) to the dropping funnel and add dropwise to the THF solution with stirring. The solution will become dark red in colour. In some cases, the dianion may be observed to precipitate as a cream coloured solid, causing difficulties with stirring. If this occurs, and after addition of all the *n*BuLi, the flask should be removed from the bath and allowed to warm to room temperature, at which point all the dianion will dissolve. To ensure complete deprotonation, the reaction should be stirred for a further 1 hour at room temperature.

The second flask should be prepared in the same manner as the first. Using a bent-finger for weighing and transfer, add 8.75 g $ZrCl_4(THF)_2$ (23 mmol) to the flask. Add 70 ml dried and degassed THF and place the flask in a water bath at 60 °C and stir to give a suspension.

The dianion solution should be transferred to the dropping funnel of the second flask using stainless-steel transfer tubing. Allow the dianion to add to the suspension of $ZrCl_4(THF)_2$ dropwise with vigorous stirring over a period of two hours, maintaining the temperature at 60 °C. The reaction mixture will become yellow and gradually darken while the suspended $ZrCl_4(THF)_2$ dissolves. Towards the end of the addition, some precipitation of a fine yellow solid may be observed. After complete addition of the dianion, stir for a further hour at 60 °C before stirring overnight at room temperature. A microcrystalline yellow solid separates from solution. This fraction should be filtered off using a filter cannula and washed with anhydrous, degassed diethyl ether. Further fractions can be obtained by concentration of the THF solution and/or addition of diethyl ether and cooling overnight at −16 °C. A yield of 4.2 g (44%) is typical. Over-concentration and addition of too much diethyl ether may lead to product contamination with the more soluble minor *meso* isomer, unreacted $ZrCl_4(THF)_2$ and other impurities. Although the material obtained is frequently very pure (by ^{1}H NMR), purification can be achieved by recrystallisation from hot anhydrous, degassed toluene. Elemental analysis: calcd. for $C_{20}H_{16}ZrCl_2$: C, 57.40; H, 3.86; Cl 16.94. found: C, 57.04; H, 3.78; Cl, 16.02. ^{1}H NMR (400 MHz, $C_2D_2Cl_4$, 25 °C): δ 3.66–3.82 (m, 4H, CH_2CH_2), 6.25 (d, 2H, C_5 of Ind, $J = 2.7$ Hz), 6.61 (d, 2H, C_5 of Ind, $J = 2.7$ Hz), 7.26 (m, 2H, Ar), 7.38 (m, 2H, Ar), 7.52 (m, 2H, Ar), 7.72 (m, 2H, Ar).

c) The dibenzyl complex, $C_2H_4Ind_2Zr(CH_2Ph)_2$

Required are two 250 ml flasks each with central B24 neck and two outer B19 necks, both equipped with magnetic stir bar, B19 stopcock adaptors, one with a 100 ml dropping funnel; a filter cannula, two B19 "subaseals" and a dry-ice/acetone bath.

Equip one 250 ml flask with the dropping funnel, flame dry and fill with inert gas. Add 1.1 g (3 mmol) of the dichloride complex to the flask using a bent-finger. Add 20 ml dry and degassed diethyl ether and cool to −78 °C. Fill the dropping funnel with 60 ml of a 0.1 M solution of $PhCH_2MgCl$ in diethyl ether (6 mmol). Add the Grignard solution dropwise to the suspension of the dichloride complex over a period of twenty minutes and then warm slowly to room temperature. The solution will take on a yellow colour which will increase in intensity to orange. The ensure complete reaction, stir for four hours at room temperature and then remove the diethyl ether under vacuum. Extract the residue with 50 ml of a 9 : 1 mixture of anhydrous degassed toluene/petroleum ether and filter by cannula into the second flame dried flask. Place this flask in the freezer overnight. Small orange cubic crystals of the product are formed. The product may take some time to crystallise and further concentration of the filtrate may be required. Yield: 0.6 g, 38%. If, for whatever reason, the product is impure, it should be recrystallised from toluene. Elemental analysis: calcd. for $C_{34}H_{30}Zr$: C, 77.07; H, 5.72. Found: C, 75.90; H, 5.68. ^{1}H NMR (90 MHz, C_6D_6); δ −0.34 (d, 2H, $J = 11.3$ Hz, CH_2Ph), 0.70 (d, 2H, $J = 11.3$ Hz, CH_2Ph), 2.57 (m, 4H, CH_2CH_2), 5.34 (d, 2H, Cp, $J = 3.4$ Hz), 5.69 (d, 2H, Cp, $J = 3.4$ Hz), 6.53–7.13 (m, 18H, Ar).

$C_2H_4Ind_2ZrCl_2$, activated by a large excess of methylaluminoxane, $[(-AlMe-O-)_n]$, is a highly active catalyst for the stereoselective synthesis of isotactic polypropene. The catalytically active species are cationic complexes of the type $[Cp_2ZrR]^+$. The reaction of $C_2H_4Ind_2Zr(CH_2Ph)_2$ with $[CPh_3)^+[B(C_6F_5)_4]^-$ gives such a cationic complex which has been shown to possess high activity for the isotactic polymerisation of propene in the absence of aluminum alkyl activators.

References

F. R. W. P. Wild, M. Wasiucionek, G. Huttner, H. H. Brintzinger, *J. Organomet. Chem.* **1985,** *288,* 63.
S. Collins, B. A. Kuntz, N. J. Taylor, D. G. Ward, *J. Organomet. Chem.* **1988,** *342,* 21.
R. Grossman, R. A. Doyle, S. L. Buchwald, *Oranometallics* **1991,** *10,* 501.
L. E. Manzer, *Inorganic Synthesis,* Vol. XXI, p. 136.
W. Kaminsky, K. Külper, H. H. Brintzinger, F. R. W. P. Wild, *Angew. Chem. Int. Ed. Engl.* **1985,** *24,* 507.
M. Bochmann, S. J. Lancaster, *Organometallics* **1993,** *12,* 633.

4.8 Halfsandwich Carbonyl Vanadium Complexes

Max Herberhold and Matthias Schrepfermann

Special Safety Precautions

1. Carbon monoxide, CO, is very toxic (MAK value 33 mg/m^3, 30 ppm). During the manipulations with the autoclave, a CO warning sensor (threshold value 30 ppm!!) should be worn.
2. Sodium sand is pyrophoric and must be handled under inert gas (N_2 or Ar). Excess sodium sand from the filtration of the autoclave reaction mixture should be destroyed by treating the residue with *i*PrOH, EtOH and water subsequently.
3. $V(CO)_6$ and $(\eta^7\text{-}C_7H_7)V(CO)_3$ are pyrophoric; $V(CO)_6$ loses CO slowly.
4. The residues of all filtration steps are pyrophoric and are allowed to decompose slowly in the fume cupboard after being poured into a large glass dish.

Following the synthesis of the first sandwich complex, di(cyclopentadienyl)iron ("ferrocene", $Fe(\eta^5\text{-}C_5H_5)_2$) in 1951/52, a series of sandwich and halfsandwich compounds of other transition metals has been described. In the case of vanadium, the halfsandwich complex η^5-cyclopentadienyl tetracarbonylvanadium, $(\eta^5\text{-}C_5H_5)V(CO)_4$ (**2**), can be prepared starting from either vanadocene, $V(\eta^5\text{-}C_5H_5)_2$ (**1**), or hexacarbonylvanadium, $V(CO)_6$ (**3**).

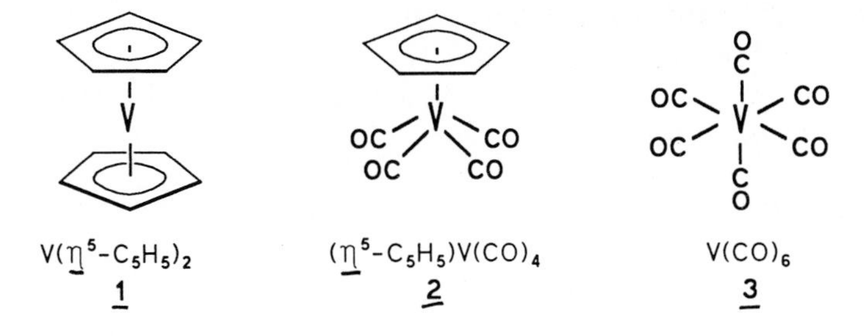

In contrast to the paramagnetic compounds **1** and **3**, the diamagnetic halfsandwich complex $(\eta^5\text{-}C_5H_5)V(CO)_4$ (**2**) conforms to the 18-electron rule (5 (V) + 5 (C_5H_5) + (4 × 2 (CO)) = 18 electrons). Halfsandwich carbonyl-metal complexes such as **2** are good precursors for oligonuclear $(\eta^5\text{-}C_5H_5)V$ compounds; they may also serve as models for either ring substitution or photo-induced displacement of CO ligands in organometallic complexes. The ring ligand $(\eta^5\text{-}C_5H_5)$ is able to screen one hemisphere of the coordination shell of the metal; the permethylated cyclopentadienyl ring ligand $(\eta^5\text{-}C_5Me_5)$ is an even better protecting group.

Although **2** is conveniently obtained by high pressure carbonylation of $V(\eta^5\text{-}C_5H_5)_2$ (**1**), the synthesis of halfsandwich carbonylvanadium complexes is generally carried out via hexacarbonylvanadium, $V(CO)_6$ (**3**), a blue-black 17-electron compound. The first two parts of this experiment describe the preparation of the homoleptic carbonyl-metal complex, $V(CO)_6$ (**3**), by reductive high pressure carbonylation of VCl_3 to give a [sodium(diglyme)$_2$]$^+$ salt of the hexacarbonylvanadate anion, $[V(CO)_6]^-$ (Eqn. 1). (A synthesis of the $[V(CO)_6]^-$ anion by reductive carbonylation of VCl_3 under atmospheric pressure in the presence of cyclooctatetraene (C_8H_8) in tetrahydrofuran (THF) solution has also been described, however, the salt $[Na(THF)_x][V(CO)_6]$ cannot be oxidised subsequently to $V(CO)_6$, which loses CO easily in the presence of THF).

$$VCl_3 \xrightarrow[\text{diglyme} \atop \text{200 bar CO}]{\text{4 Na (sand)}} [Na(C_6H_{14}O_3)_2{}^+][V(CO)_6]^- \xrightarrow[-1/2\ H_2]{H_3PO_4} \underset{\mathbf{3}}{VCl_3V(CO)_6} \quad (1)$$

Protonation of the $[V(CO)_6]^-$ anion by anhydrous phosphoric acid and thermal decomposition of the intermediate $\{HV(CO)_6\}$ leads to $V(CO)_6$ (**3**) in *ca.* 90% yield.

Three examples of the conversion of **3** into halfsandwich carbonylvanadium complexes such as **4–6** are presented, using the reactions with pentamethylcyclopentadiene, hexachlorocyclopentadiene and 1,3,5-cycloheptatriene.

$V(CO)_6$ **3**; C_5Me_5H, hexane, 69 °C → $(\eta^5\text{-}C_5Me_5)V(CO)_4$ **4**; C_5Cl_6 (25 °C) → $(\eta^5\text{-}C_5Cl_5)V(CO)_4$ **5**; C_7H_8, hexane, 69 °C → $(\eta^7\text{-}C_7H_7)V(CO)_3$ **6**

Experimental

a) Di[bis(2-methoxyethyl)ether]sodium-hexacarbonylvanadate(−1), [Na(diglyme)$_2$][V(CO)$_6$]

$$VCl_3 + 4Na + 2C_6H_{14}O_3 \xrightarrow[(Fe(CO)_5)]{160\,°C \atop 200\text{ bar CO}} [Na(C_6H_{14}O_3)_2][V(CO)_6] + 3NaCl \quad (2)$$

A 0.5 l autoclave (pressure capacity 300 bar) is charged with 15 g sodium sand (650 mmol), 17.2 g VCl_3 (109 mmol), 200 ml "diglyme" (bis(2-methoxyethyl)ether) and 1 ml $Fe(CO)_5$ under an inert gas. The autoclave is first flushed with nitrogen (50 bar, to check leaks) and then CO (50 bar). After expanding, 200 bar CO is pressed on the system which is then slowly heated to 160 °C under stirring.

After 48 h, the autoclave is allowed to cool down to room temperature and the reaction mixture is filtered over Na_2SO_4 (10 × 3 cm). The residue is washed with Et_2O until the filtrate has become colourless. The yellow solution is then extracted with 1 l hexane in a separatory funnel. The brown layer separating at the bottom of the funnel is allowed to drop slowly into cold (−78 °C) hexane (*ca.* 100 ml). This leads to precipitation of the hexacarbonylvanadate salt, $[Na(diglyme)_2][V(CO)_6]$.

The crude orange-red product is sticky and therefore reprecipitated several times by redissolving it in a small amount of Et_2O (30–40 ml), adding 100 ml hexane via syringe and decanting the pink supernatant solution as soon as possible. This procedure is repeated until the product is obtained as a yellow powder, which is then dried at the oil pump.

Yield: 27.8–33.4 g (50–60%).

The product should be stored in a refrigerator (−20 °C) in the dark.

b) Hexacarbonylvanadium, $V(CO)_6$ (3)

$$[Na(C_6H_{14}O_3)_2][V(CO)_6] + H_3PO_4 \xrightarrow{\text{via } [HV(CO)_6]} \underset{\mathbf{3}}{V(CO)_6} + 1/2\ H_2 + NaH_2PO_4 + 2\,C_6H_{14}O_3 \qquad (3)$$

All manipulations must be carried out under inert gas (N_2 or Ar).

In a special sublimation vessel (Fig. **4.8-1**), about 15 g P_4O_{10} are dissolved in *ca.* 65 ml commercial orthophosphoric acid (85% H_3PO_4) at 0 °C (charge the vessel without inert gas stream!). A 10 g portion of dry $[Na(diglyme)_2][V(CO)_6]$ (19.6 mmol) is added through a long funnel to the surface of the anhydrous acid. The lower end of the sublimation tube is loosely fitted with a lump of glass wool to protect the $V(CO)_6$ (**3**) product at the sublimation finger against the squirting acid during the sublimation process. Then the cooling finger is attached above the glass wool (2–3 cm) and fixed by a small clamp.

The cooling finger is cooled to −30 °C using a cryostat (isopropanol), and high vacuum (ca. 10^{-2} mbar) is applied. Whereas the water bath is heated to 40 °C, the blue-black $V(CO)_6$ (**3**) is subliming to the finger (the vacuum should be re-applied periodically). As soon as the phosphoric acid becomes slightly green, the sublimation is discontinued and the detached sublimation finger is placed on a frit. The product $V(CO)_6$ (**3**) is removed from the finger, washed with degassed air-free water (degassed for *ca.* 15 min at the aspirator vacuum) and finally dried at the aspirator vacuum until the product can be poured into an appropriate Schlenk tube. The blue-black crystals should be so dry that formation of ice is not observed when the tube is stored at −20 °C).

Yield: 3.8 g (90%), m.p. 70 °C (dec.).

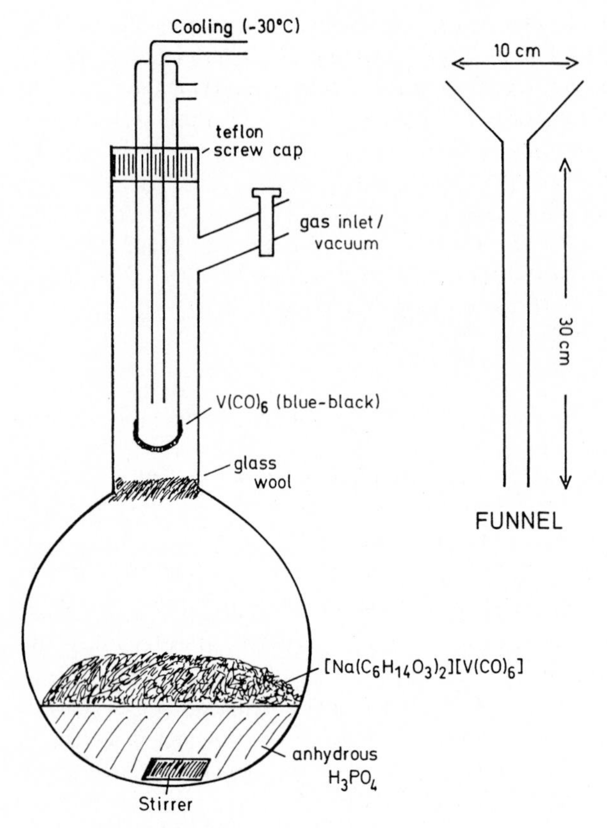

Figure 4.8-1. Sublimation Vessel.

c) (η^5-Pentamethylcyclopentadienyl)tetracarbonylvanadium, (η^5-C_5Me_5)$V(CO)_4$ ***(4)***

$$V(CO)_6 + C_5Me_5H \rightarrow (\eta^5\text{-}C_5Me_5)V(CO)_4 + 2CO + 1/2\ H_2 \quad (4)$$

3 **4**

All manipulations are carried out under inert gas (N_2 or Ar), and the direct influence of sun or UV light should be avoided.

In a Schlenk tube, 1.3 g $V(CO)_6$ (**3**) (5.94 mmol) are dissolved in 75 ml hexane (not completely soluble!) and treated with 1.15 ml (7.12 mmol) distilled pentamethylcyclopentadiene, C_5Me_5H. The reaction mixture is heated to boiling (paraffin valve at the reflux condenser) in a preaheated oil bath (*ca.* 100 °C) until all the $V(CO)_6$ (**3**) is consumed (*ca.* 1–2 h; monitor by IR spectroscopy!).

The cold mixture is chromatographed over silica* (5 × 3 cm) using hexane/toluene (5 : 1) as the eluant. The solvent is removed from the orange solution at the oil pump vacuum, the crude product is redissolved in 5 ml pentane and stored on dry ice overnight.

* The silica used for column chromatography should be periodically degassed under high vacuum and subsequently loaded with inert gas (N_2 or Ar) before use.

The supernatant liquid is discarded and the orange residue is dried at the aspirator vacuum. The crude (η^5-C_5Me_5)$V(CO)_4$ (**4**) product is then sublimed in a high vacuum using a water bath (90–100 °C). The sublimate can be scratched off the finger in the presence of air, but the pure product should be stored under inert gas. It can be kept at room temperature in the dark for unlimited periods.
Yield: 1.15 g (65%), orange needles, m.p. 144 °C.

d) (η^5-Pentachlorocyclopentadienyl)tetracarbonylvanadium, (η^5-C_5Cl_5)$V(CO)_4$ (5)

$$\underset{\mathbf{3}}{4\,V(CO)_6} + 3\,C_5Cl_6 \rightarrow \underset{\mathbf{5}}{3\,(\eta^5\text{-}C_5Cl_5)V(CO)_4} + VCl_3 + 12\,CO \qquad (5)$$

All manipulations are carried out under inert gas and the direct influence of sun or UV light should be avoided.

In a Schlenk tube, 0.25 g $V(CO)_6$ (**3**) (1.14 mmol) are dissolved in 30 ml hexane (not completely soluble!) and treated with 0.18 ml (1.14 mmol) distilled hexachlorocyclopentadiene, C_5Cl_6. After a few minutes, complex **5** is formed in the solution, while VCl_3 precipitates. The mixture is chromatographed over silica* (5 × 3 cm) using hexane as the eluant. The solvent is removed from the orange solution under reduced pressure and the product **5** is dried at the oil pump.
Yield: 0.2 g (45%), orange needles, m.p. 71 °C.
(η^5-C_5Cl_5)$V(CO)_4$ (**5**) should be stored under argon at −20 °C in the dark and prepared freshly for reactions.

e) (η^7-Cycloheptatrienyl)tricarbonylvanadium, (η^7-C_7H_7)$V(CO)_3$ (6)

$$\underset{\mathbf{3}}{V(CO)_6} + C_7H_8 \longrightarrow \underset{\mathbf{6}}{(\eta^7\text{-}C_7H_7)V(CO)_3} + 3\,CO + 1/2\,H_2$$

$$\xrightarrow{\text{by-product}} 1/2\ [(\eta^7\text{-}C_7H_7)V(\eta^6\text{-}C_7H_8)][V(CO)_6] + 3\,CO + 1/4\,H_2 \qquad (6)$$

All manipulations are carried out under inert gas.

In an appropriate Schlenk tube, 6.11 g $V(CO)_6$ (**3**) (27.9 mmol) are dissolved in 220 ml hexane (not completely soluble!) and treated with 6.7 ml (65.0 mmol) distilled 1,3,5-cycloheptatriene, C_7H_8. The mixture is refluxed in a preheated oil bath (*ca.* 100 °C) for 90 min (paraffin valve at the reflux condenser).

The cold reaction mixture is filtered over a fine glass frit to remove the brown by-products (such as [(η^7-C_7H_7)V(η^6-C_7H_8)][$V(CO)_6$]). The residue is repeatedly washed with 30 ml portions of hexane until the filtrate becomes colourless. The volume of the solution is reduced to *ca.* 60 ml. In case there should be unreacted $V(CO)_6$ in the cooling trap, the solution is brought to dryness and the residue is redissolved in 60 ml hexane. The hexane solution is kept on dry ice overnight. The halfsandwich complex **6** forms dark green needles. If the product is amorphous, it can be recrystallised easily from hexane.

* The silica used for column chromatography should be periodically degassed under high vacuum and subsequently loaded with inert gas (N_2 or Ar) before use.

The supernatant solution is decanted, and the product **6** is dried in a high vacuum. It should be stored under argon.
Yield: 2.14 g (34%), dark green crystals, m.p. 134–137 °C (dec.).

All complexes described can be characterised easily by ^{51}V NMR spectroscopy (with the exception of paramagnetic $V(CO)_6$ (**3**)) and IR spectroscopy. The spectroscopic data are summarised in Table **4.8-1.**

Table 4.8-1. Characteristic spectroscopic data.

Complex	$\delta(^{51}V)^a$	IR (ν(CO), [cm $^{-1}$])^b
$[Na(diglyme)_2][V(CO)_6]$	−1945	1861^c
$V(CO)_6$ (**3**)	paramagn.	1974
$(\eta^5\text{-}C_5Me_5)V(CO)_4$ (**4**)	−1466	2015 s, 1915 vs
$(\eta^5\text{-}C_5Cl_5)V(CO)_4$ (**5**)	−1008	2046 s, 1963 vs
$(\eta^7\text{-}C_7H_7)V(CO)_3$ (**6**)	−1518	1997 vs, 1901 vs

a) in C_6D_6 solution rel. neat $VOCl_3$ ($\delta(^{51}V) = 0$); b) in hexane solution; c) in Et_2O solution.

References

Ch. Elschenbroich, A. Salzer, *Organometallics – A Concise Introduction,* VCH, Weinheim (F.R.G.), **1989.**

E. O. Fischer, W. Hafner, *Z. Naturforsch. B* **1954,** *9b,* 503; E. O. Fischer, S. Vigoureux, *Chem. Ber.* **1958,** *91,* 2205.

M. Hoch, A. Duch, D. Rehder, *Inorg. Chem.* **1986,** *25,* 2907.

R. P. M. Werner, H. E. Podall, *Chem. Ind. (London)* **1961,** 144.

F. Calderazzo, G. Pampaloni, *J. Organomet. Chem.* **1983,** *250,* C33.

R. Ercoli, F. Calderazzo, A. Alberola, *J. Amer. Chem. Soc.* **1960,** *82,* 2966.

M. Herberhold, W. Kremnitz, M. Kuhnlein, M. L. Ziegler, K. Brunn, *Z. Naturforsch. B* **1987,** *42b,* 1520.

W. A. Herrmann, W. Kalcher, *Chem. Ber.* **1982,** *115,* 3886.

W. Priebsch, M. Hoch, D. Rehder, *Chem. Ber.* **1988,** *121,* 1971.

R. P. M. Werner, S. A. Manastyrskyj, *J. Amer. Chem. Soc.* **1961,** *83,* 2023.

cf. Ed.: G. Brauer, *Handbuch der Präparativen Anorganischen Chemie,* "Metallorganische Komplexverbindungen", W. P. Fehlhammer, W. A. Herrmann, K. Öfele, Vol. 3, Ferd. Enke Verlag, Stuttgart, **1981,** p. 1815, 1852 and 1856.

4.9 Covalent and Ionic Metallocene Hexafluoroarsenate Complexes

Inis C. Tornieporth-Oetting and Thomas M. Klapötke

Special Safety Precautions

1. Arsenic pentafluoride, AsF_5, and many non-metal fluorides are oxidisers, fluorinators and are toxic. Extensive care must be taken to avoid contact between the fluorides and oxidisable materials. The whole AsF_5 cylinder should be securely clamped inside a fume cupboard.
2. Hydrolysis of AsF_5 (and AsF_5/AsF_6^- containing compounds) can produce HF. Hydrogen fluoride (HF) can cause severe burns.
3. Protective clothing and face masks should be worn all times.
4. Sulfur dioxide is toxic and corrosive (bp: $-10\,°C$, $P(20\,°C): 3.30$ bar)
5. The residues from both preparations (also cold trap, etc.) can be decomposed by slow addition of aqueous sodium bicarbonate.
6. If in any doubt whatsoever about safe operation, consult a demonstrator.

The complex $Cp_2Ti(AsF_6)_2$ ($Cp = \eta^5\text{-}C_5H_5$), prepared either from Cp_2TiF_2 and AsF_5 or from Cp_2TiCl_2 and $AgAsF_6$, was the first metallocene hexafluoroarsenate complex containing a direct M···F···E interaction (E = P, As, Sb or Bi). It fertilized both the high oxidation-state organometallic chemistry of the early transition elements and the chemistry of cationic metallocene species.

The cationic metallocene dichloride salts $[Cp_2MCl_2]^{n+}[AsF_6]_n^-$ ($n = 1$: M = V, Nb, Ta; $n = 2$: M = Mo, W; $n = 3$: M = Re) were synthesised quantitatively by oxidation of the lower valent metallocene dichlorides with AsF_5. All group 5 ionic metallocene complexes possess pronounced antitumor properties against experimental as well as human tumors heterotransplanted to athymic mice tumors.

Experimental

a) $Cp_2Ti(AsF_6)_2$

$$Cp_2TiCl_2 + 2\,AgAsF_6 \rightarrow Cp_2Ti(AgF_6)_2 + 2\,AgCl \qquad (1)$$

In a dry box or glove bag, silver hexafluoroarsenate (1.00 g, 3.37 mmol) is placed in one 25 cm^3 bulb of a two-bulb Pyrex glass vessel (equipped with a J. Young Teflon-stemmed glass valve and a medium sintered glass frit; Fig. **4.9-1**). Titanocene dichloride (0.42 g, 1.69 mmol)

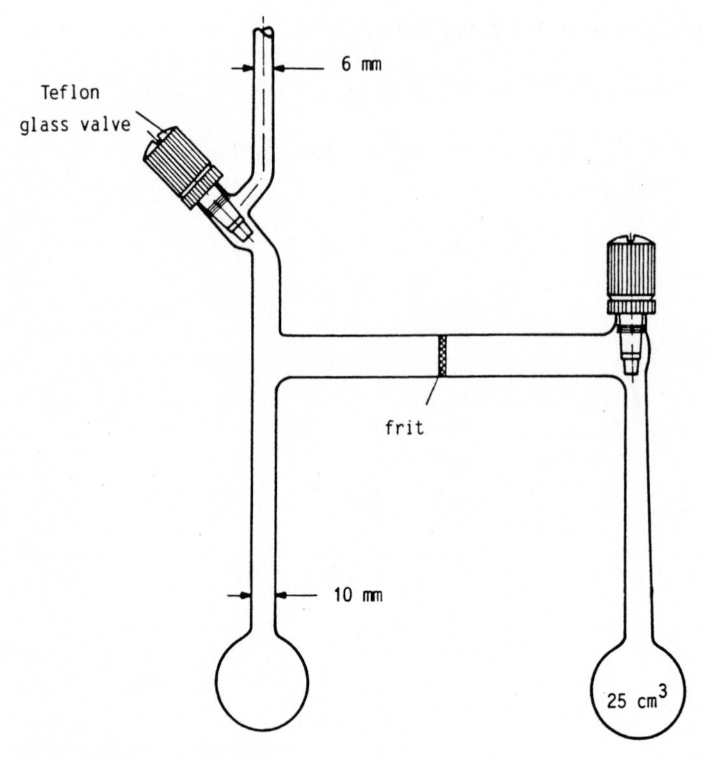

Figure 4.9-1. Two-bulb glass vessel.

is placed in the other 25 cm^3 bulb. The vessel is evacuated and at −78 °C (acetone/dry-ice) sulfur dioxide (dried over CaH_2) is now condensed onto both compounds (10 cm^3 each). The solution of $AgAsF_6$ is poured at room temperature onto the solution of Cp_2TiCl_2. The mixture is stirred for 2 hrs. The solution is filtered and the precipitate (AgCl) washed twice with recondensed solvent (10 cm^3). The solvent is removed under a dynamic vacuum leaving a dark red solid – $Cp_2Ti(AsF_6)_2$ – and AgCl. In the dry box or glove bag, the product is placed in a clean and dry two-bulb vessel (see above) and recrystallised from 10 cm^3 sulfur dioxide.

Calculate the yield of your product and measure its IR (glove bag, Nujol mull, KBr plates) and ^1H NMR ($CDCl_3$) spectra. Seal the product in a glass ampoule.

b) $[Cp_2MoCl_2]^{2+}[AsF_6]_2^-$

$$Cp_2MoCl_2 + 3\,AsF_5 \rightarrow [Cp_2MoCl_2]^{2+}[AsF_6]_2^- + AsF_3 \qquad (2)$$

Molybdenocene dichloride (1.01 g, 3.40 mmol) is placed in one 25 cm^3 bulb of a dry two-bulb Pyrex glass vessel (equipped with a J. Young Teflon-stemmed glass valve and a medium sintered glass frit, Fig. **4.9-1**). The vessel is evacuated and 15 cm^3 sulfur dioxide (dried over CaH_2) are condensed onto the Cp_2MoCl_2 at −78 °C (acetone/dry-ice). The SO_2 solution is frozen in liquid nitrogen and arsenic pentafluoride (1.73 g, 10.20 mmol) is condensed onto the frozen solution. The mixture is warmed to room temperature and stirred for 30 min. The solution is filtered (although there is nearly no precipitate) and the volatile materials (SO_2, AsF_3) are removed under a dynamic vacuum leaving a black solid. In a dry box or glove bag, the product is placed in a clean and dry two-bulb vessel (see above) and recrystallised from 15 cm^3 sulfur dioxide.

Calculate the yield of your product and measure its IR (glove bag, Nujol mull, KBr plates) and ^1H NMR spectra (sealed NMR tube, SO_2 solution). Seal up the product in a glass ampoule.

Instead of Cp_2MoCl_2 you can use Cp_2WCl_2 (1.20 g) and AsF_5 (1.58 g).

References

T. M. Klapötke, U. Thewalt, *J. Organomet. Chem.* **1988,** *356,* 173.
P. K. Gowik, T. M. Klapötke, T. S. Cameron, *J. Chem. Soc. Dalton Trans.* **1991,** 1433.
P. Gowik, T. M. Klapötke, P. S. White, *Chem. Ber.* **1989,** *122,* 1649.
P. Gowik, T. M. Klapötke, I. C. Tornieporth-Oetting, *Chem. Ber.* **1989,** *122,* 2273.
P. Gowik, T. M. Klapötke, K. Siems, U. Thewalt, *J. Organomet. Chem.* **1992,** *431,* 47.
P. Köpf-Maier, T. M. Klapötke, *J. Cancer. Res.* **1992,** *118,* 216.
P. Köpf-Maier, T. M. Klapötke, *Cancer Chemother. Pharmacol.* **1992,** *29,* 361.

4.10 Preparation of the Iron Tricarbonyl Complex of Methyl-2-Acetamidoacrylate

Simon R. Drake, M. Elena Lasterra-Sanchez and Susan E. Thomas

Special Safety Precautions

1. Methyl-2-acetamidoacrylate can act as a skin and nasal irritant. Wash off with alcohol followed by copious quantities of water.
2. Diethyl ether is highly flammable and should be handled in a fume cupboard.
3. Iodomethane is highly toxic and volatile. It must be handled with disposable gloves in a fume cupboard.
4. Di-iron nonacarbonyl is harmful and should also be handled in a fume cupboard.

Many organic compounds form stable iron carbonyl complexes. By forming these complexes, organic molecules which are too reactive to be isolated can be stabilised and studied, olefinic groups can be protected during the course of a multi-step organic synthesis, the chemical reactivity of organic compounds can be modified and the stereochemical course of organic reactions can be altered. In this experiment, you will form the iron tricarbonyl complex of the amino acid dehydroalanine using ultrasound (Eqn. 1). Formation of its iron carbonyl complex modifies the chemical reactivity of the dehydroalanine and this has been exploited in a synthetic route to unusual highly branched amino acids.

H N CO_2H O —(K_2CO_3, MeI)→ H N CO_2Me O —($Fe_2(CO)_9$, ultrasound)→ H N CO_2Me O Fe $(CO)_3$

Experimental

a) Methyl-2-acetamidoacrylate

2-Acetamidoacrylic acid (2.33 g, 18 mmol) is dissolved in dry acetone (300 ml) in a two-neck round-bottomed flask fitted with a condenser (carrying a $CaCl_2$ drying tube) and a stopper. To this solution is added K_2CO_3 (5.45 g, 39.5 mmol) and iodomethane (2.24 ml, 36 mmol). The mixture is then heated under reflux for 3 h, after which time additional iodomethane (1.12 ml, 18 mmol) is carefully added and the heating maintained for a further 3 h. The solution is then allowed to cool and the solid material removed by filtration and washed with dry acetone. The acetone solution and the acetone washings are evaporated to dryness (using a rotary evaporator) and the residue is redissolved in hot chloroform (700–800 ml) and filtered. Removal of the chloroform *in vacuo* affords the product either as white crystals or as an oil, which forms white crystals on standing overnight. Record the melting point, the IR spectrum (KBr) and the ^{1}H NMR spectrum ($CDCl_3$) of your white crystals.

b) (Methyl-2-acetamidoacrylate)tricarbonyliron(0)

The ester prepared above (1.5 g, 10.5 mmol) is dissolved in sodium-dried diethyl ether (75 ml) in a three-neck round-bottomed flask fitted with a condenser and a gas inlet. Nitrogen gas is bubbled through the stirred solution for 5 min to saturate the solution with nitrogen and then $Fe_2(CO)_9$ (7.65 g, 21 mmol) is added carefully. Taking care to maintain the nitrogen atmosphere, the apparatus is placed in an ultrasonic bath and sonicated at 35 °C for approximately 30 min. The product mixture is then filtered through deactivated alumina or celite using diethyl ether as eluent and the solvent is removed *in vacuo.*

If the product contains $Fe_3(CO)_{12}$ impurities (green in colour), it may be purified by column chromatography (SiO_2; 2:3 ethyl acetate: 40–60 °C petroleum ether) to give the iron complex as an air stable yellow-orange crystalline material. Record the melting point, the IR spectra (KBr and hexane solution) and the ^{1}H NMR spectrum ($CDCl_3$) of your product.

References

A. J. Pearson in *Comprehensive Organometallic Chemistry,* Vol. 8 (Eds.: G. Wilkinson, F. G. A. Stone, E. W. Abel), Pergamon Press, **1982,** p. 939.

R. Grée, *Synthesis* **1989,** 341.

A. de Cian, R. Weiss, J.-P. Haudegond, Y. Chauvin, D. Commereuc, *J. Organomet. Chem.* **1980,** *187,* 73.

M. P. Bueno, C. Cativiela, C. Finol, J. A. Mayoral, C. Jaime, *Can. J. Chem.* **1987,** *65,* 2182.

K. S. Suslick, *Adv. Organomet. Chem.* **1986,** *25,* 73.

S. V. Ley, C. M. R. Low in *Ultrasound in Synthesis,* Springer Verlag, **1989.**

J. Barker, S. L. Cook, M. E. Lasterra-Sanchez, S. E. Thomas, *J. Chem. Soc., Chem. Commun.* **1992,** 830.

4.11 A Transition-Metal Alkylidyne Complex

Stephen Anderson, Darren Cook and Anthony F. Hill

Special Safety Precautions

1. Tungsten hexacarbonyl and trifluoroacetic anhydride are extremely toxic. Deuterochloroform is highly toxic, mutagenic and an irritant. Accordingly, all manipulations should be carried out in an efficient fume cupboard. *N,N,N′,N′*-tetramethylethylene diamine (tmen) and diethyl ether are flammable liquids.
2. Lithium metal and organolithium reagents react violently with water so every effort must be made to exclude moisture. Unrequired excess lithium reagents may be destroyed by slow addition to propan-2-ol followed by ethanol and finally water.
3. Carbon monoxide is liberated in the decarbonylation step and so caution must be exercised.

Although alkylidyne complexes *i. e.* compounds with metal–carbon triple bonds, were once considered exotic laboratory curiosities, they have now found wide application in organic synthesis and as catalysts for alkyne polymerisation and metathesis. This experiment involves the synthesis of a thermally stable alkylidyne complex of tungsten, $[W(\equiv CC_6H_4Me\text{-}4)Br(CO)_2(tmen)]$, via the abstraction of oxide from an anionic tungsten acylate, following a method originally described by Mayr. The bromide ligand results from CF_3CO_2/Br metathesis by excess bromide present in the lithium reagent with an initially formed trifluoroacetato complex $[W(\equiv CC_6H_4Me\text{-}4)(O_2CCF_3)(CO)_2(tmen)]$.

Experimental

Equation (1) shows the reaction sequence to be performed.

$$[W(CO)_6] \xrightarrow[\text{(iii) tmen}]{\text{(i) } LiC_6H_4Me/LiBr \quad \text{(ii) } (CF_3CO)_2O,\ -78\,^\circ C} [W(\equiv CC_6H_4Me)Br(CO)_2(tmen)] \qquad (1)$$

All manipulations must be carried out under anaearobic conditions in a fume cupboard. It is essential that sodium-dried diethyl ether is used due to the moisture sensitive nature of the intermediates. All glassware should be oven dried and assembled whilst hot, flushing with nitrogen. High-sodium grade lithium (1% Na) gives the best results for the synthesis of the 4-lithiotoluene.

c) An ether solution of LiC_6H_4Me-4

A three-necked 250 ml round-bottomed flask equipped with a reflux condenser, a paraffin bubbler, a pressure-equalised dropping funnel and magnetic stir bar is flushed with nitrogen for 3–5 minutes. Lithium metal (1.2 g), cut into small pieces, is then added under a counterflow of nitrogen followed by petroleum ether (30 ml). The petroleum ether is then removed by syringe and discarded (to remove the protective coating of paraffin from the lithium). Diethyl ether (DRY, 50 ml) is then added and the flask immersed in an ice bath. A solution of 4-bromotoluene (10.0 g) in dry diethyl ether (25 ml) is then added via a dropping funnel over a period of 1 hour. The mixture is left to warm slowly to room temperature overnight by which time the bulk of the lithium metal should have dissolved.

b) [W≡(CC_6H_4Me-4)Br$(CO)_2$(tmen)]

A three-necked 250 ml round-bottomed flask equipped with a paraffin bubbler, a pressure-equalised dropping funnel and magnetic stir bar is flushed with nitrogen for 3–5 minutes. Tungsten hexacarbonyl (finely ground, **CAUTION,** 5.0 g) is then added against a counterflow of nitrogen followed by dry diethyl ether (100 ml). The solution of 4-lithiotoluene (50 ml) prepared in a) above is then transferred via syringe (or cannula) to the dropping funnel. The solution is then added dropwise to the tungsten hexacarbonyl solution with rapid stirring. After approximately 25 ml has been added, a solution infra-red spectrum should be measured to estimate the extent of reaction ($\nu(CO)_{max}$ $W(CO)_6$ = *ca.* 1970 cm^{-1}). Further aliquots of the 4-lithiotoluene solution are added whilst monitoring the effect by IR spectroscopy until only a trace of [$W(CO)_6$] remains (<2%). The dropping funnel is then removed and the remaining lithium reagent destroyed by adding slowly to propan-2-ol (50 ml). The flask is then immersed in a dry-ice/acetone bath and stirred for 5–10 minutes to ensure that the reaction mixture has cooled adequately. Trifluoroacetic anhydride (1.50 ml) is then added dropwise whereupon a red colour appears and then dissipates. The mixture is stirred for a further 10 minutes and then removed from the dry-ice bath. After stirring for a further 10 minutes, *N,N,N′,N′*-tetramethylethylene diamine (tmen) (3 ml) is then added and the mixture stirred for 10 minutes. Stirring is then stopped and the mixture allowed to warm up slowly overnight (**CAUTION:** CO evolves). The brown supernatant is then removed by decantation whilst a flow of nitrogen is maintained over the surface of the mixture. Dry diethyl ether (50 ml) is added and the mixture stirred and then left to settle. The ether is removed by decantation and the flask transferred to a rotary evaporator to remove the last traces of ether. The yellow solid should be transferred to a vial which has been flushed with nitrogen and stored in the dark.

Measure the yield, melting point, infra-red spectra (Nujol and dichloromethane) and ^1H NMR spectrum ($CDCl_3$, 60 MHz). NB: The ^1H NMR spectrum should be measured as soon as possible after preparing the solution in $CDCl_3$. Prolonged storage of such aerated solutions will lead to decomposition. After obtaining these data, you will be provided with a ^{13}C-{^1H} NMR spectrum, a full analysis of which should include identification of *J* ($^{183}W^{13}C$) for the carbonyl and alkylidyne carbon nuclei. Use the spectroscopic data obtained to determine the stereochemistry at the tungsten centre and explain why this is the most stable ligand arrangement. Briefly discuss the steps involved in the formation of the alkylidyne complex and suggest why [W(≡CC_6H_4Me-4)Br$(CO)_2$(tmen)] is a more stable compound than [W(≡CC_6H_4Me-4)Br$(CO)_4$].

References

H. P. Kim, R. J. Angelici, *Adv. Organomet. Chem.* **1987,** *27,* 51.
A. Mayr, H. Hoffmeister, *Adv. Organomet. Chem.* **1991,** *32,* 227.
A. Mayr, G. A. McDermott, *Organometallics* **1985,** *3,* 608.

Technical notes

(i) Diethyl ether should be dried over sodium wire.
(ii) Lithium wire containing 1% sodium (*e. g.* Aldrich: 27,832-7) should be used to ensure that there are no problems with the initiation of the lithiation step.
(iii) Because of the extreme moisture sensitivity of trifluoroacetic anhydride, for large classes this should be stored in a Schlenk or Youngs tube and dispensed by syringe. Ill treated and aged samples of the anhydride will contain considerable amounts of the acid which seriously diminishes the overall yield.
(iv) All other reagents may be used as received from commercial sources.
(v) The procedure also works well for a wide range of aryl bromides and also for $[Mo(CO)_6]$, however, that illustrated here is chosen for the simplicity of the NMR interpretation.

4.12 Preparation of $[Bu_4N][B_3H_8]$ and the Formation of $RuH[B_3H_8](CO)(PPh_3)_2$

Anthony F. Hill and J. Derek Woollins

Special Safety Precautions

This experiment involves the use of standard solvents which represent normal hazards as well as the *in situ* generation of diborane. You must adhere to the instructions closely to avoid the possibility of fire hazards. Carry out the first stage of the reaction in a fume cupboard behind a safety screen.

Boron-hydrides are often considered to be very reactive and explosive materials. In this experiment you will prepare a boron-hydride anion and its ruthenium complex. Both compounds are quite air stable.

Experimental

$[Bu_4N]B_3H_8]$

In a dry box (or a nitrogen atmosphere), 8.5 g (0.225 mole) of powdered sodium tetrahydroborate is slurried with 125 ml of anhydrous diglyme in a 500 ml three-necked flask. The flask is fitted with a 125 ml pressure-compensating dropping funnel whose tip is extended below the surface of the mixture. The third neck is connected to a bubbler containing a benzene-amine (4-picoline was used) mixture to scrub the gaseous boranes evolving from the reaction as minor products. During assembly, a gentle stream of dry nitrogen is maintained through the apparatus to minimize the possible entrance of air into the system. After rapidly transferring a solution of iodine into the funnel (11.3 g, 40.0 mmoles in 50 ml of dry diglyme), the entire reaction system is purged with nitrogen for about 10 minutes and the reaction flask is placed in a previously heated oil bath whose temperature is adjusted to 98–102 °C before the addition of iodine. After the nitrogen flow is stopped, the iodine solution is added dropwise during a period of 75–90 minutes to the hot, vigorously stirred reaction mixture. The hydrogen gas produced passes through the bubbler. The reaction mixture is stirred for two more hours while the temperature is maintained at about 95 °C. The volume is then reduced to 80 ml by passing dry nitrogen over the warmed reaction mixture at 50 °C, or alternatively by pumping. The cooled mixture, together with washings of 50 ml of water, is transferred to a 2 l beaker. About 500 ml of saturated aqueous tetra-*n*-butylammonium iodide is added slowly with vigorous stirring until no more precipitation takes place. The white precipitate is filtered on a Büchner funnel, washed with water (about 450 ml) and dried under vacuum. (Octahydrotriborate (−1) may be precipitated without removal of diglyme, a lower yield of the crude product being obtained).

A portion of the crude salt (*ca.* 2 g) is dissolved in 15–20 ml of dichloromethane, filtered and reprecipitated by adding 200 ml of diethyl ether. The precipitate is dried *in vacuo.* Record your yield and the IR spectrum of the crude and purified product. The ^{11}B NMR spectrum of the B_3H_8 anion is shown in Figure **4.12-1.**

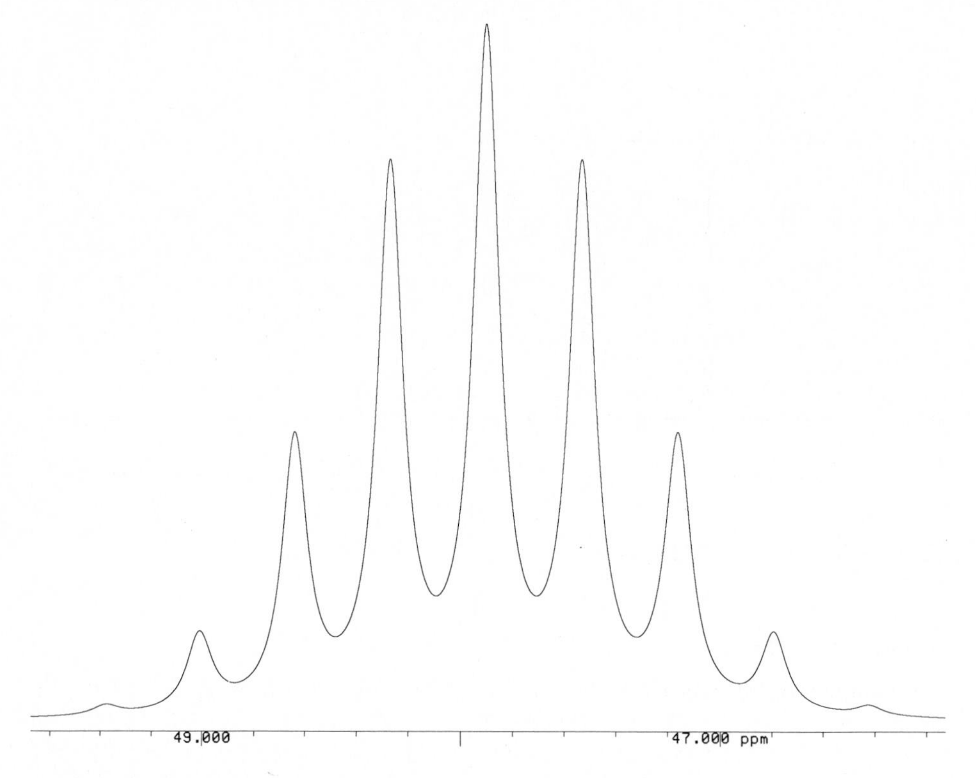

Figure 4.12-1. ^{11}B NMR spectrum (86.6 MHz, CD_2Cl_2) of the B_3H_8 anion.

b) [RuH(B_3H_8)(CO)(PPh_3)$_2$]

A mixture of [RuClH(CO)(PPh_3)$_3$] (1.00 g, 1.05 mmol, see c) and [nBu$_4$N][B_3H_8] (0.30 g, 1.05 mmol) in dichloromethane (30 cm^3) is stirred for 2 h. Ethanol (15 cm^3) is added and the suspension filtered through Kieselguhr to remove [nBu$_4$N]Cl. The filtrate is concentrated to *ca.* 50 cm^3 and the product, [RuH(B_3H_8)(CO)(PPh_3)$_2$], isolated by filtration, washed with ethanol (10 cm^3) and recrystallised from a mixture of dichloromethane and ethanol.

Your characterisation of the complex should include yield, melting point, infra-red (Nujol mull, 4000–600 cm^{-1}) and ^{1}H NMR ($CDCl_3$, 60 MHz). Figure **4.12-2** shows the ^{31}P NMR spectrum and you should use these data to identify the stereochemistry at the ruthenium centre and rationalise this with reference to the Skeletal Electron Pair theory.

N.B.: A stock of the ruthenium complex may be prepared as described below.

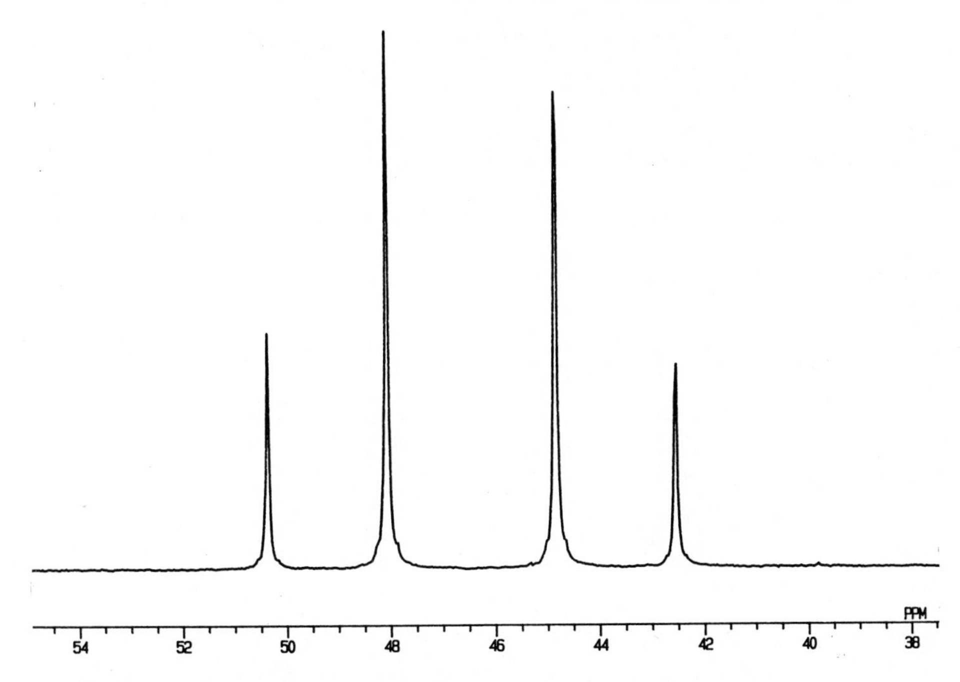

Figure 4.12-2. $^{31}P\text{-}\{^1H\}$ NMR (at 109.3 MHz) of $RuH(B_3H_8)CO(PPh_3)_2$.

c) $[RuClH(CO)PPh_3)_3]$

A suspension of commercial $RuCl_3 \cdot xH_2O$ (3.00 g) and triphenylphosphine (23.0 g) in 2-methoxyethanol (350 cm^3) is heated under reflux for 48 h. The suspension is then allowed to cool and stirred at room temperature for 3–5 hours to complete precipitation of the product. The product is then isolated by filtration and washed with ethanol (2 × 50 cm^3) and petrol (2 × 50 cm^3) and dried *in vacuo.* Yield *ca.* 9 g depending on quality of $RuCl_3 \cdot xH_2O$. The colour of the sample varies from pale yellow to pale pink, but this is not important.

References

G. E. Ryschkewitsch and K. C. Nainan, *Inorg. Synth.* **1974,** *15,* 110.

N. N. Greenwood, J. D. Kennedy, M. Thornton-Pett, J. D. Woollins, *J. Chem. Soc., Dalton Trans.* **1985,** 2397.

I. D. Burns, A. F. Hill, A. R. Thompsett, N. W. Alcock and K. S. Claire, *J. Organomet. Chem.* **1992,** *425,* C8.

For a review of M-B-H chemistry see J. D. Kennedy, *Prog. Inorg. Chem.* **1984,** *32,* 519.

4.13 Vacuum Line Techniques: Preparation of SiF_4 and $SiF_4(C_5H_5N)_2$

Dennis F. Evans and J. Derek Woollins

Special Safety Precautions

1. Read and note the introductions to this experiment.
2. Both pyridine and silicon tetrafluoride are toxic materials; avoid inhaling their vapours. If you do inhale either, inform a demonstrator. Rest and keep warm, breathe fresh air. Take milk of magnesia after SiF_4 inhalation.
3. Special hazards which may arise in this experiment include tap blockage, overheating the pyridine, condensation of liquid oxygen into the main or smaller traps, escape of SiF_4 or pyridine, a build up of the SiF_4 pressure (caused by the premature removal of the Dewar flasks while taps are closed).
4. The Dewar flasks which you use must all be protected with insulating tape or metal containers. Be *especially careful* when you *clamp* them; it is better to use a supporting ring at the base and a loose clamp as a top support.

Before starting this experiment, read these instructions carefully, then read them again in front of the vacuum line which you intend to use, identifying the parts referred to in Figure **4.13-1** and their function in the experiment.

Make sure that the vacuum line is clean (and remember that you are responsible for cleaning it when you have finished your experiment).

It is very important that the line should be free of leaks (sources of these are badly greased taps and joints, particularly the spherical joints). Check that all the taps and ground glass joints are free from streaks. Use Apiezon-M grease sparingly, there should be no grease in the tap-barrel bore. To test the line, evacuate it, isolate it from the pump and watch for an increase in pressure over a period of about 5 minutes (manometer).

Make sure that you keep the Dewar flask around trap B (Fig. **4.13-1**) topped up with liquid nitrogen throughout the experiment. Both the Dewar flask and the gasstorage flask must be taped.

At *no stage* in the experiment should you leave solid SiF_4 in a vessel with a closed tap unless it is surrounded by liquid nitrogen since it is a gas at room temperature.

Never let air into the vacuum line while the Dewar flask surrounds trap B, otherwise you will condense liquid oxygen with consequent dangers of explosion. When you have finished, isolate the line with tap H, turn off the pump and let air into it *via* tap A, let air into trap B *via* tap A and immetiately remove the Dewar surrounding trap B. Ease the trap B off at the joint and place it in a fume cupboard. Use a tisssue to hold the *top* of trap B as it will still be very cold.

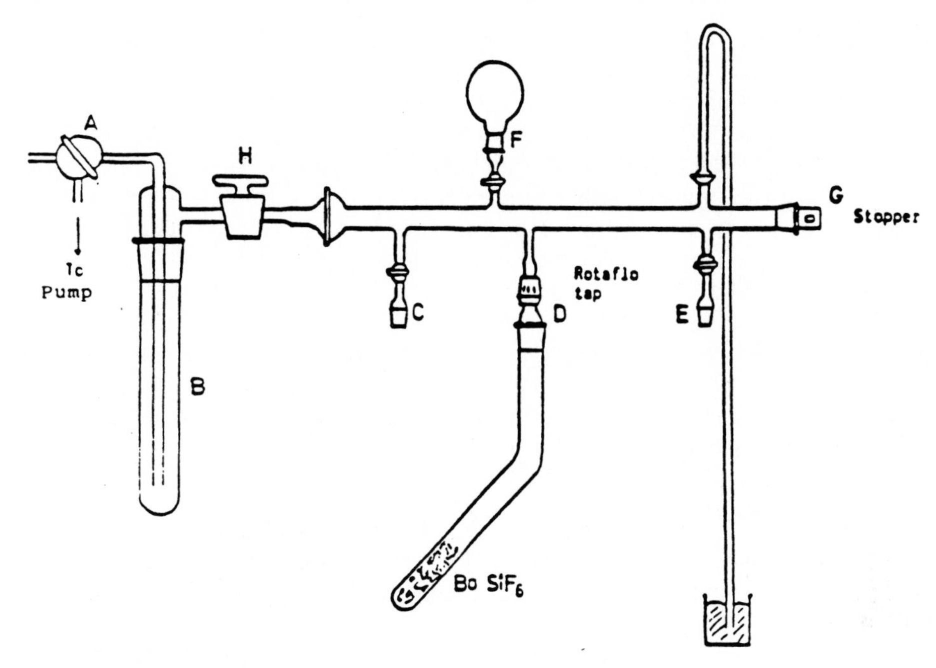

Figure 4.13-1. Vacuum line apparatus.

Carry out parts b) and, if possible, c) in one day. Keep the SiF_4 which has not been used in the storage bulb overnight (if necessary) to carry out part d) on the following day.

Experimental

a) Barium hexafluorosilicate, ($BaSiF_6$)

Barium chloride dihydrate (25 g, 0.1 mole) is dissolved in 70 ml of water. Aqueous H_2SiF_6 (30%) is added until precipitation is complete, this requires about 40 ml of the solution. The freshly precipitated white barium hexafluorosilicate is filtered and washed thoroughly until free from chloride. The product is dried in an oven at 110 °C and then in a desiccator over phosphorus(v) oxide. Yield is nearly 100%.

b) SiF_4

Place dry $Ba[SiF_6]$ (5.0 g) in the bent-tube provided at the PTFE "Rotaflo" tap equipped socket D. Attach two smaller tubes at sockets C and E and a 500 cm^3 gas storage flask at socket F (this *must* be protected with tape). Evacuate the whole assembly after ensuring that all connections, taps, etc. are properly greased.

Test for leaks with the manometer. Isolate the vacuum line from the pump and heat the $Ba[SiF_6]$ with a Bunsen burner (start heating at the top of the $Ba[SiF_6]$ to avoid this being blown into the line) until there is a pressure of about 10 mm (1 cm) of crude SiF_4 in the line. Stop heating, slowly pump this crude SiF_4 (contaminated with HF and fluorosilicic acid) into the main trap B (this is of course cooled with liquid nitrogen throughout). Isolate the line again (tap H) and cool the small tube at E, using a small Dewar flask almost full of liquid nitrogen, *cool only the bottom 3-4 cms.* Open tap E. This is to condense the SiF_4 in the next stage. Now heat the $Ba[SiF_6]$ to just *below* the softening point of the glass until no more SiF_4 is evolved. Cool the small tube at C with liquid nitrogen in the same way as that at E, then open tap C, remove the small Dewar surrounding the vessel E and allow some 4/5 of the SiF_4 to sublime from E into C. Isolate C from the vacuum line, keeping it cold with liquid nitrogen. Keeping tap E open, immediately pump the crude residue from E into the main trap. Isolate the line by closing tap H. Open tap C, remove the Dewar around the tube at C, and allow the pure SiF_4 to vapourise from C into the evacuated gas storage bulb F. Isolate F from the line. The gas can now be slowly drawn from this as required for the two experiments below. Open tap H and pump any remaining SiF_4 from the line into the trap. Slowly open taps C and E to also pump any remaining SiF_4 into the trap. Close taps C, D and E. After it has cooled, remove the bent tube.

c) Infra-red spectrum of SiF_4

Attach an infra-red gas cell (NaCl windows) to the vacuum line at E and evacuate it for five minutes. Allow SiF_4 to enter the cell from the storage bulb until a pressure of about 10 cm is attained in the cell. Isolate the cell *and* the storage bulb and pump out the line. Measure the infra-red spectrum over the range 2500–625 cm^{-1}, and then run another spectrum after pumping out most of the gas from the cell in order to measure ν_3 accurately (this will also give the cell "blank"). Using a 10 cm long cell with polythene windows, measure the infra-red spectrum of approximately 1 cm pressure of SiF_4 over the region 625–250 cm^{-1}. To compensate for the absorption of the polythene windows place two polythene discs cemented together with Apiezon-M grease in the reference beam. Pump all the SiF_4 out of the cells before letting air in. Give a complete assignment of all the bands observed from 2500–250 cm^{-1}.

d) $SiF_4 \cdot py_2$

Place *ca.* 4 cm^3 of dried pyridine, some 3A molecular sieves (check that these are new) and a very small PTFE stir bar in a small tube and attach it to the line at E. Attach a calcium chloride drying tube to D (leave tap of D closed). Freeze the pyridine in liquid nitrogen and evacuate E. Then close tap E and allow the pyridine to warm up to room temperature. With the tap still closed, refreeze the pyridine and re-evacuate by opening tap E. Isolate E from the line and evacuate the tube C. Condense SiF_4 into C by cooling the tube in liquid nitrogen (cover only the bottom 3–4 cm) and opening the storage bulb F. Quickly evacuate the line to remove any traces of uncondensed SiF_4 and again isolate the line. Now distill the dry pyridine onto the solid SiF_4 in C, moving the stir bar with an external magnet. If the pyridine does not distill under hand heat only from E, you have a leak in the line and must rectify this after turning off tap E. In this event, pump out the line again, rectify the leak, again evacuate the line and repeat the degassing of the pyridine. Never warm the pyridine other than with your hand. Allow the mixture in C to warm up *very slowly.* The tap C must be open to the manometer during this process. If the reaction becomes too violent, cool again with liquid nitrogen. When the mixture has attained room temperature, pump off al the excess pyridine into the main trap B, close H, close the tap at C, let *dry* air (*via* a calcium chloride drying tube at D) into the vacuum line. Remove the tube at C containing the adduct and immediately stopper it. Transfer it to a dry bag and run the IR spectrum. Turn the pump and line off as directed in (iv) above. Measure the spectrum of pyridine (POISON!) as a thin film between KBr plates over the same region. Make sure that you make this film up in the fume cupboard. Clean the plates afterwards in the fume cupbaord when you have finished. Comment on the spectrum of the adduct. Hand in a sealed sample of the adduct with your report.

At the end of the experiment (or the session), remove the Dewar, let air via A into the trap, allow the bent tube to cool and then remove the tubes form C, D and E. (There should only be slight resistance – you do not need to let the line up to air.) Remove the trap and place in a fume cupboard. Let air into the pump and switch off at mains. Let air into the line via tap H. Clean and return "quick-fit" tubes and magnetic stir bars to vacuum line. Keep NaCl cells in desiccator. Ba residues should be place in a fume cupboard in a marked container.

References

R. S. Armstrong, R. J. Clark, *J. Chem. Soc. Faraday II* **1976,** *72,* 11.

H. J. Campell-Ferguson, E. A. V. Ebsworth, *J. Chem. Soc. (A)* **1967,** 705.

J. Heicklen, C. Knight, *Spectrochim. Acta* **1964,** *20,* 295.

V. A. Bain, R. C. G. Killean, M. Webster, *Acta Cryst.* **1969,** *25B* 156.

$SiCl_4py_2$ X-ray structure: O. Bechstein, B. Ziemer, D. Hass, S. I. Troyanov, V. Rybakov, G. N. Man, *Z. Anorg. allg. Chem.* **1990,** *582,* 200.

4.14 Preparation and Structure of the Acetyl Chloride-Antimony Pentachloride Complex

Peter N. Gates

Special Safety Precautions

1. Antimony pentachloride is corrosive and toxic by inhalation and skin contact. Apart from the dry bag, always use in a fume cupboard.
2. Carbon tetrachloride is toxic by inhalation and skin contact.
3. Acetyl chloride is corrosive and toxic by inhalation and skin contact. Keep away from sources of ignition.
4. The product is moisture sensitive and will hydrolyse to HCl and ethanoic acid in moist air.

Reaction of a Lewis acid of the type MX_n (*e.g.* M=B, Al, Sb, etc.; X=Cl, Br) with acid halides, RCOX (R = alkyl, aryl, etc.), usually produces a 1:1 complex $RCOX \cdot MX_n$. The structure of such complexes depends on the particular acid halide and Lewis acid but is usually of one of two types.

$$\begin{matrix} CH_3 \\ X \end{matrix}\!\!>\!C{=}O \rightarrow MX_n \quad \text{or} \quad [CH_3{-}C{\equiv}O]^+[MX_{n+1}]^-$$

(I) **(II)**

An example of type **I** is the 1:1 complex formed between $TiCl_4$ and CH_3COCl which has the structure $\begin{matrix} CH_3 \\ Cl \end{matrix}\!\!>\!C{=}O \rightarrow TiCl_4$. Reaction between CH_3COF and BF_3 produces a complex of type **II**, $CH_3C{\equiv}O^+BF_4^-$.

A very quick and convenient distinction between such structural types can be made by infrared spectroscopy and relies on the characteristic group frequencies of carbon-oxygen bonds. In the parent acid halide the vibrational mode which corresponds mainly to the stretching of the CO bond lies in the region of 1800 cm^{-1}. In a complex of Type **I**, a polarisation of electron density in the carbonyl bond would be predicted with a consequent reduction in the stretching force constant and hence the vibrational frequency. Such reductions have been observed in many Lewis acid complexes of carbonyl containing molecules. In a type **II** complex, where the formal bond order increases to 3, an increase in the CO force constant would be predicted with a consequent increase in the vibrational frequency (CH_3CO^+ is isoelectronic with CH_3CN).

In this experiment, the object is to determine the structure of the complex formed between CH_3COCl and $SbCl_5$. The complex is very sensitive to hydrolysis and must be prepared and handled under anhydrous conditions.

Experimental

Both reactants and the product are sensitive to hydrolysis and all operations should be carried out in a dry bag. You are provided with a sealed ampoule containing a known amount (*ca.* 5 g) of $SbCl_5$. This should be opened in the dry bag by making a small nick with a glass knife in the constriction of the ampoule and carefully breaking the neck. Pour the $SbCl_5$ into CCl_4 (15 cm^3) contained in a dropping funnel fitted with a silica gel drying tube. Calculate the stoichiometric quantity of CH_3COCl required for a 1 : 1 molar ratio with the $SbCl_5$ you have used and place this in a 100 cm^3 B24 round-bottomed flask together with 20 cm^3 of CCl_4. Remove the dropping funnel from the dry bag and fit it to the round-bottomed flask *via* a connector with a side-arm. The apparatus should be protected from the atmosphere by attachement of a silica gel drying tube to the side-arm. Add the solution of $SbCl_5$ in CCl_4 dropwise, slowly with constant shaking (or stirring with a magnetic stir bar). When the addition is complete, the white solid may be filtered off as follows: connect a 100 cm^3 flask to the lower end of the closed sintered filter (Fig. **4.14-1**) and connect the filter (via tap B) to an oil pump protected by a liquid nitrogen trap. (Consult a demonstrator). With tap A on the filter closed, quickly remove the side-arm connector from the reaction vessel, replace it with the top

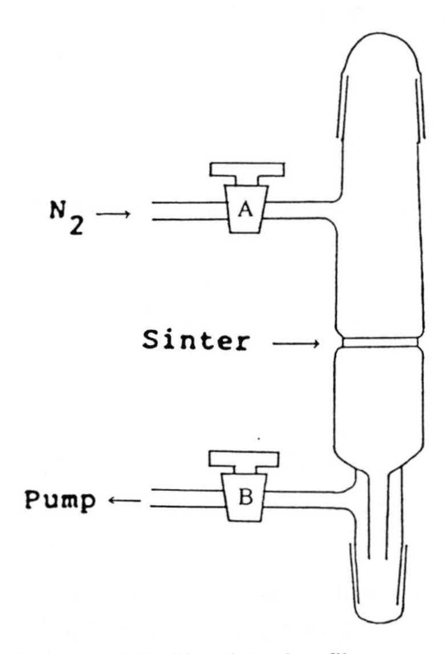

Figure 4.14-1. Closed suction filter.

end of the closed filter, invert the apparatus and open tap B. Wash the white solid on the filter twice with 10 cm^3 portions of CCl_4. (This should be done quickly with minimum exposure to the atmosphere. To release the vacuum in the filter, dry nitrogen may be admitted through tap A.) Suck the product dry on the filter. When most of the CCl_4 has been removed from the product, rapidly replace the lower flask (as above when washing) with an empty one and continue pumping. When the product is dry, close tap B on the filter, detach from the suction and remove the closed apparatus to the dry bag. Transfer the white solid to a dry sample tube and seal it thoroughly with parafilm.

While the complex is drying, record the infra-red spectrum of CH_3COCl as a neat liquid: place several drops of CH_3COCl on a KBr plate and cover with another plate to give a thin liquid film. Record the spectrum over the range 4000–1200 cm^{-1}. The infra-red spectrum of the complex should be obtained by preparing a sample in the dry bag as follows: remove a small quantity of the white solid from the sample tube and place it on the KBr plate. Then add a *small* amount of Nujol (liquid paraffin) and cover it with another plate. Rub the two plates together to obtain an even dispersion of the complex in the Nujol and quickly record the infra-red spectrum over the range 2500–1200 cm^{-1} (consult a demonstrator).

If Raman facilities are available, a low-frequency spectrum is also instructive. Place a small amount (about 1 cm depth) of the sample in a melting point capillary tube, seal the end with plasticine and remove from the dry bag. Record a Raman spectrum over the range 100–400 cm^{-1} (consult a demonstrator). Alternatively, a Raman spectrum may be supplied. On this basis, suggest a structure for the complex, assigning the major bands in the infra-red and Raman spectra.

References

D. Cassimatis, B. P. Susz, *Helv. Chim. Acta* **1961,** *44,* 943.
B. P. Susz, J. J. Wuhrmann, *Helv. Chim. Acta* **1957,** *40,* 722.
F. P. Boer, *J. Amer. Chem. Soc.* **1966,** *88,* 1572.
D. Cook, *Canad. J. Chem.* **1963,** *41,* 522.
P. N. Gates, D. Steele, *J. Mol. Structure* **1967,** *1,* 349.
K. Nakamoto, *Infrared Spectra of Inorganic and Coordination Compounds,* 4th ed., Wiley, New York, **1986.**

4.15 Arsenium Cations: Carbene Analogues

Neil Burford and Trenton M. Parks

Special Safety Precautions

Operations should be performed in a fume cupboard.

Carbenes and their analogues represent important synthetic building blocks in that they are small, simple molecules with coordinatively unsaturated sites and high reactivity. The phosphorus analogues (phosphenium cations) have been known since the 1970's, but the first arsenium cations were only recently identified. This project involves the preparation of a chloroarsolidine (**1**), which is the starting material for quantitative production of the arsolidinium (**2**) gallate. The reaction involves heterolytic cleavage of the covalent As – Cl bond using a Lewis acid to give the ionic salt. In addition, the novel arsolidine-arsolidinium complex (**3**) is prepared and all three compounds are characterised by their melting point and IR and NMR spectroscopic features.

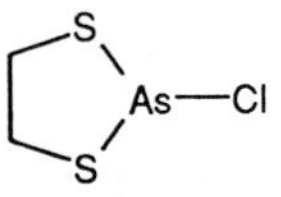

(**1**) 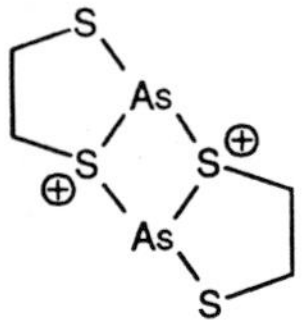 (**2**) 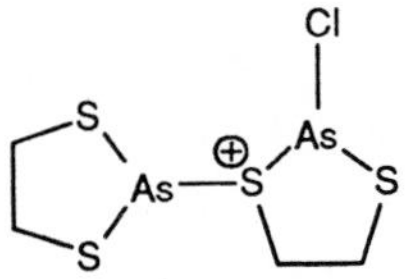 (**3**)

Experimental

*a) 2-Chloro-cyclo-1,3-dithia-2-arsapentane **1** **

Chemicals required:

1,2-Ethanedithiol	Due to the pungent odour of ethanedithiol, all procedures should be performed in an efficient fume cupboard. The apparatus should be rinsed with bleach following the procedure.
Arsenic trichloride	*Intensely poisonous!* Latex gloves are not effective protection against this compound. The most effective protection is to avoid contact. Arsenic trichloride is slowly hydrolysed in air.
Carbon tetrachloride	Poisonous and carcinogenic. Must be dried over P_2O_5.

* A modification of the procedure has been mentioned by W. H. C. Rueggeberg, A. Ginsburg, W. A. Cook, *J. Amer. Chem. Soc.* **1946,** *68,* 1860.

Apparatus required:

50 mL two-neck round-bottomed flask with stoppers, 50 mL pressure-equalising dropping funnel, stir motor and stir bar, source of dry nitrogen.

Under an atmosphere of dry nitrogen, 1.4 g of 1,2-ethanedithiol in 5 ml of CCl_4 is added dropwise over a period of 5 minutes to a stirred solution of 2.7 g of $AsCl_3$ in 10 ml of CCl_4. The reaction self-cooling and is accompanied by the liberation of HCl gas. The solution is stirred for 30 minutes, then stoppered and placed in the freezer (−20 °C) to promote crystallisation of the product. The cold supernatant liquid is decanted from the crystals and the remainder of the solvent is removed *in vacuo.* Typical yields: 80–90%. M.p.: 38.5–39.0 °C. The FT-IR spectrum is shown in Figure **4.15-1a**. NMR (CD_2Cl_2): ^1H, 3.74 ppm, multiplet; ^{13}C, 44.6 ppm.

b) Bis(1,3-dithia-2-arsolidinium tetrachlorogallate), [2][GaCl$_4$]$_2$

Chemicals required:

Gallium trichloride	Extremely moisture sensitive. Purified by vacuum sublimation onto a water cooled finger.
2-Chloro-*cyclo*-1,3-dithia-2-arsapentane	From part a), exposed to dynamic vacuum (10^{-3} torr) for at least 30 minutes. Slowly hydrolyses in air.
Methylene chloride	Dried over CaH_2, P_2O_5, and CaH_2; degassed by freeze-pump-thaw.

Apparatus required:

Dual compartment (100 ml each) reactor with two teflon stopcocks (A and B) and FETFE O-rings, featuring demountable compartments separated by stopcock B (Fig. **4.15-2**), a modification of the standard H-tube reaction vessel.
Further: Teflon coated magnetic stir bar, stirring motor, evacuated glass solvent bulb containing dry and degassed CH_2Cl_2, dry box (N_2), vacuum line.

The product and $GaCl_3$ are extremely moisture sensitive and must be handled, stored and manipulated under vacuum or in an inert atmosphere (dry box). The reaction is performed in a dual compartment reaction vessel, which is evacuated and flame dried (bunsen) under dynamic vacuum before use. The starting materials, 0.50 g of 2-chloro-*cyclo*-1,3-dithia-2-arsapentane and 0.45 g of $GaCl_3$, are introduced into separate compartments in the dry box, and a stir bar is included with the $GaCl_3$ (compartment 2). The sealed unit is removed from the dry box and evacuated (do not evacuate for long periods, as $GaCl_3$ is volatile). Methylene chloride solvent is static vacuum distilled onto each of the reactants (≈15 ml each) by cooling each compartment, in turn, with liquid nitrogen ($GaCl_3$ first). At room temperature, the solution of 2-chloro-*cyclo*-1,3-dithia-2-arsapentane is poured slowly (15 min) into the stirred solution of $GaCl_3$ giving an instantaneous bright yellow reaction mixture. After stirring for 15 minutes, ≈75% of the solvent is distilled from the solution by cooling empty compartment 1. Solid or oil may be observed at this time. Stopcock B is closed and the solid or oily material is redissolved by warming the bulb under the hot water tap (≈65 °C), accompanied by vigorous shaking. On slow cooling to room temperature, a yellow crystalline solid is formed. The solution is decanted from the crystals, and they are washed by cold spot (liquid

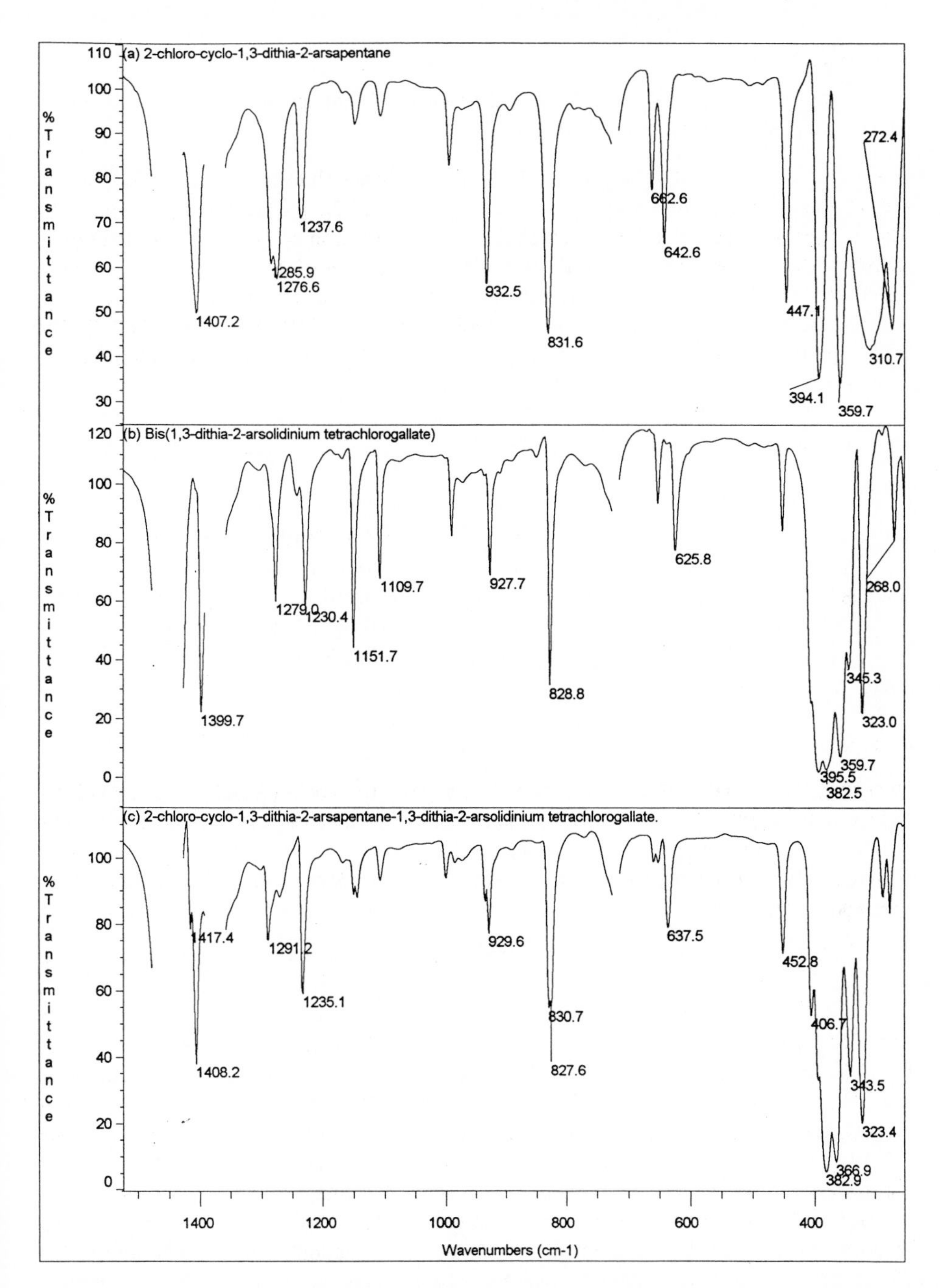

Figure 4.15-1. FT-IR spectra obtained as Nujol mulls between CsI plates on a Nicolet 510P spectrometer, Nujol regions have been blanked. a) 2-chloro-*cyclo*-1,3-dithia-2-arsapentane (**1**). b) Bis(1,3-dithia-2-arsolidinium tetrachlorogallate), [**2**][$GaCl_4$]$_2$. c) 2-chloro-*cyclo*-1,3-dithia-2-arsapentane-1,3-dithia-2-arsolidinium tetrachlorogallate, [**3**][$GaCl_4$].

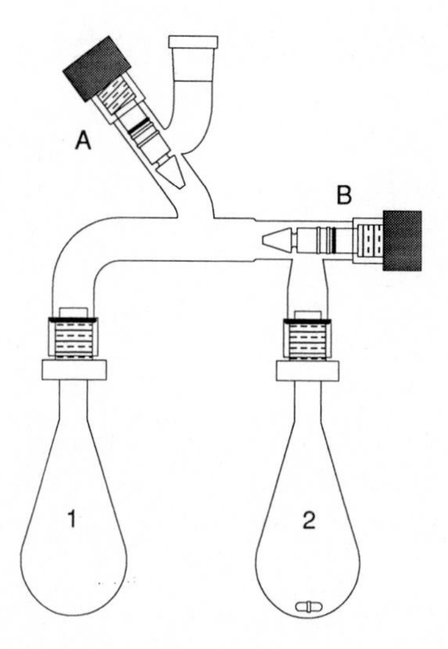

Figure 4.15-2. Dual compartment vacuum reactor.

nitrogen cotton swabs) back-distillation. The solvent is entirely removed from the vessel *in vacuo* and the crystalline material is isolated and manipulated in the dry box.
Typical yields: 80–90%. M.p.: 94.0–95.5 °C. The FT-IR spectrum is shown in Figure **4.15-1 b**, NMR (CD_2Cl_2): ^{1}H, 4.06 ppm; ^{13}C, 45.8 ppm.

c) 2-Chloro-cyclo-1,3-dithia-2-arsapentane-1,3-dithia-2-arsolidinium tetrachlorogallate, [3][GaCl$_4$]

Using the same materials and procedures described in part b), slowly add a solution of 0.45 g of $GaCl_3$ in 15 ml of CH_2Cl_2 to a stirred solution of 1.0 g of 2-chloro-*cyclo*-1,3-dithia-2-arsapentane in 15 ml of CH_2Cl_2. A yellow solution is initially formed and produces a pale yellow precipitate during the addition. The majority of the solvent (>75%) is removed *in vacuo* depositing more of the product, then the remainder of the supernatant solution is decanted from the precipitate, which is washed by cold spot (liquid nitrogen cotton swabs) back-distillation. The prodcut can be recrystallised from hot methylene chloride as described in part b). Typical yields: 80–90%. M.p.: 116–117 °C. The FT-IR spectrum is shown in Figure **4.15-1 c**. NMR (CD_2Cl_2): ^{1}H, 3.90 ppm; ^{13}C, 45.0 ppm.

References

M. Sanchez, M. R. Mazières, L. Lamandé, R. Wolf in *Multiple Bonding and Low Coordination in Phosphorus Chemistry,* (Eds.: M. Regitz, O. J. Scherer), Georg Thieme Verlag, Stuttgart, **1990**, p. 129.

N. Burford, T. M. Parks, B. W. Royan, B. Borecka, T. S. Cameron, J. F. Richardson, E. J. Gabe, R. Hynes, *J. Amer. Chem. Soc.* **1992,** *114,* 8147.

N. Burford, T. M. Parks, B. W. Royan, P. S. White, J. F. Richardson, *Can. J. Chem.* **1992,** *70,* 703.

N. Burford, J. Müller, T. M. Parks, *J. Chem. Educ.* **1995**, (in press).

A. L. Wayda, J. L. Dye, *J. Chem. Educ.* **1985,** *62,* 356.

4.16 Preparation and Some Reactions of the Folded $P_2N_4S_2$ Ring

Tristram Chivers, Daniel D. Doxsee and Robert W. Hilts

Special Safety Precautions

1. Chlorodiphenylphosphine (Ph_2PCl) is a noxious, acrid liquid which reacts readily with moist air. It should be stored in a Schlenk vessel under a nitrogen atmosphere and transferred in a fume cupboard. Spillages should be destroyed with an aqueous solution of sodium hypochlorite.
2. Liquid bromine and thionyl chloride ($SOCl_2$) are highly toxic and extremely corrosive liquids. These reagents must be handled with protective gloves inside a fume cupboard.
3. Organolithium reagents are likely to combust spontaneously on contact with air. These reagents must be transferred from 'sure-seal' bottles by using a calibrated syringe after a needle has been inserted through the septum of the container. Spillages should be destroyed with *n*-butanol.
4. Trimethylsilyl azide (Me_3SiN_3) is a highly flammable and poisonous liquid. It also must be handled with gloves in the fume cupboard.

The eight-membered ring in 1,5-$R_4P_2N_4S_2$ (R = aryl, alkyl) has a folded structure (**1**) with a cross-ring S–S bond of *ca.* 2.5 Å. This inorganic heterocycle is a hybrid of S_4N_4 and the well known cyclophosphazenes. It undergoes a wide range of reactions based on the sulfur or nitrogen centres. For example, oxidative addition with halogens gives S,S′-dihalogeno derivatives, while the interactions with Lewis or Brønsted acids produces N-bonded adducts with retention of the S–S bond. The $P_2N_4S_2$ ring in **1** exhibits a versatile coordination chemistry and the following bonding modes have been established: η^1-N, η^2-S,S′, η^2-N,S-μ-S′. The reaction of **1** (R = Ph) with organolithium reagents produces the adducts Li[$Ph_4P_2N_4S_2R$] (R = alkyl, aryl), which serve as a source of $Ph_4P_2N_4S_2R^-$ anions in the formation of η^1-S bonded complexes with late transition metals. The folded rings (**1**) exhibit ^{31}P NMR chemical shifts at anomalously low fields (110–140 ppm) compared to cyclophosphazenes as a result of the transannular S–S interaction. Consequently, ^{31}P NMR spectroscopy provides a diagnostic probe for the loss or retention of the S–S bond in reactions of **1**.

In this experiment, you will prepare **1** (R=Ph) from Li[$Ph_2P(NSiMe_3)_2$] (**2**) and investigate the reactions of this inorganic heterocycle with bromine and with a Brønsted acid. The phosphorus-nitrogen-silicon reagent (**2**) has a monomeric structure and may be used for the preparation of P–N heterocycles containing lanthanide or actinide metals.

1 (cyclic structure: S—S, N, N, N, N, PR_2, PR_2)

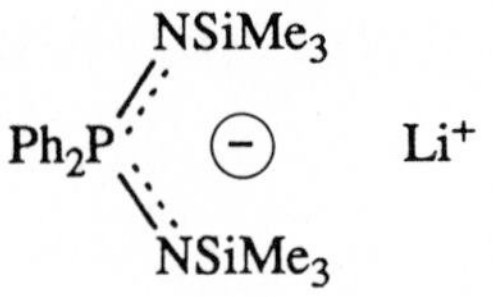

2 $Ph_2P(NSiMe_3)_2^-\ Li^+$

Note: If necessary, in view of time limitations, this experiment can be subdivided into parts a) and b) only or, if $Li[Ph_2P(NSiMe_3)_2]$ is provided, parts b)–d) only. The second option avoids delays imposed by the lengthy reflux necessary for the Staudinger reaction Eqn. 1.

Experimental

a) $Li[Ph_2P(NSiMe_3)_2]$ (2)

The preparation of the lithium reagent **(2)** involves the two steps in Eqns. (1) and (2).

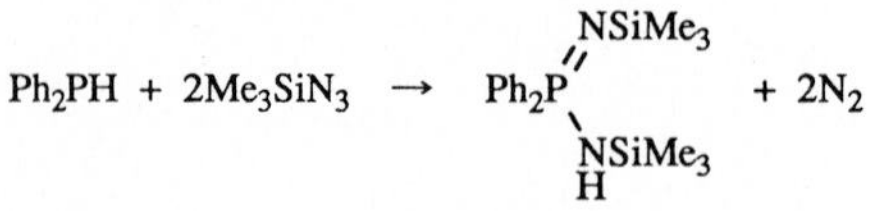

$$Ph_2PH + 2Me_3SiN_3 \rightarrow Ph_2P(=NSiMe_3)(NHSiMe_3) + 2N_2 \quad (1)$$

$$Ph_2P(=NSiMe_3)(NHSiMe_3) + nBuLi \rightarrow [Ph_2P(NSiMe_3)_2]^- Li^+ + n\text{-}C_4H_{10} \quad (2)$$

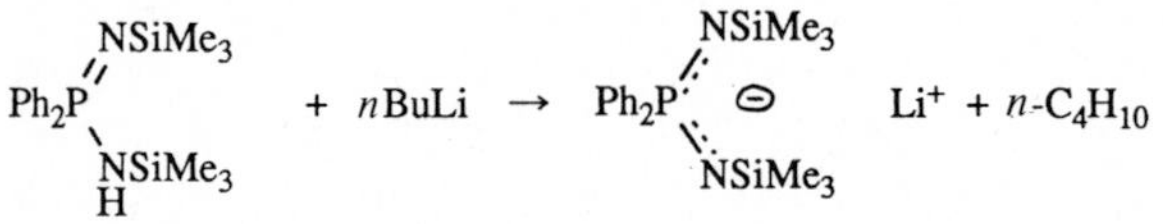

The procedure for Eqn. (1) is a modification of that described by Paciorek and Kratzer (see References).

Under an atmosphere of nitrogen, heat a mixture of Me_3SiN_3 (3.63 g, 31.5 mmol) and Ph_2PH (2.75 g, 15 mmol) in a 100 ml side-arm (Schlenk) flask fitted with a reflux condenser at 140 °C for 24 h. While still at 140 °C, apply a dynamic vacuum for at least 2 h to remove any residual Me_3SiN_3. Cool the flask to room temperature and record the yield of your product. This procedure will give 3–4 g of $[Ph_2P(NHSiMe_3)(NSiMe_3)]$ as a colourless liquid. Obtain the infra-red spectrum of the product on KBr plates. Look for a strong N–H band at *ca.* 2900 cm^{-1}. Record and interpret the ^1H NMR spectrum in $CDCl_3$. ^{31}P NMR ($CDCl_3$): singlet at 0.2 ppm (with reference to 85% H_3PO_4).

$Ph_2P(NSiMe_3)[N(H)SiMe_3]$ (6.07 g, 16.9 mmol) is transferred via cannula to a 250 ml side-arm (Schlenk) flask fitted with a septum and equipped with a magnetic stir bar. The flask should be previously evacuated and back-flushed with N_2 gas several times. Approximately 75 ml of dry hexanes are added to the flask, which is then placed in an ice-water bath. After about 10 minutes, an equimolar amount of *n*BuLi in hexanes (calculate the required volume based on the known concentration of this solution) is added dropwise via syringe (20 min) with stirring. The ice-water bath is then removed and the mixture is stirred until it reaches room temperature. The septum is then replaced (be sure N_2 gas is flowing into the vessel through the side-arm) with a reflux condenser equipped with a tap adapter which has N_2 gas flowing through it. With the tap adapter open and the side-arm tap closed, the mixture is heated at reflux for one hour by using a heating mantle. The mixture is then allowed to cool to room temperature, the condenser is replaced with a septum and the flask is placed in a freezer (*ca.* −20 °C) for 2-3 hours. This results in the formation of white crystals. The residual supernatant liquid is decanted with a cannula and discarded. The white crystals of $Li[Ph_2P(NSiMe_3)_2]$ (4.80 g, 13.1 mmol) (78% yield) are dried under vacuum. The product can be handled in air for short periods, but it should be stored under an atmosphere of dry nitrogen. Record and interpret the ^1H NMR spectrum of your product in $CDCl_3$. ^{31}P NMR (in CH_2Cl_2): singlet at 16.5 ppm (referenced to 85% H_3PO_4).

*b) 1,5-$Ph_4P_2N_4S_2$ (**1**, R=Ph)*

A colourless solution of freshly distilled $SOCl_2$ (0.97 g, 8.16 mmol) in dry CH_2Cl_2 (100 ml) cooled in an ice bath at 0 °C is added dropwise (*ca.* 15 min), by cannula, to a rapidly stirred solution of $Li[Ph_2P(NSiMe_3)_2]$ (3.0 g, 8.16 mmol) in dry CH_2Cl_2 (200 ml) at 0 °C in a 500 ml side-arm (Schlenk) flask. During the addition, the solution should change from colourless to yellow. The ice bath is removed and the mixture is stirred under N_2 for 18 hours. The solvent is removed under vacuum and then acetonitrile (30 ml) is added to the pale yellow residue. The resulting yellow suspension is stored overnight in the freezer at *ca.* −20 °C. The yellow supernatant solution is discarded and the cream coloured solid is dissolved in 25 ml of CH_2Cl_2 to give a cloudy, colourless solution. The precipitated LiCl is removed by using a filter cannula and the filtrate is taken to dryness under vacuum. The residue is washed with hexane (30 ml) and then dried under dynamic vacuum for 2 hours to give 1,5-$[Ph_4P_2N_4S_2]$ (1.20 g, 2.45 mmol) as a white crystalline solid. Yield: 60%. The product can be handled in air for short periods, but it is slowly hydrolysed upon exposure to moist air for several weeks. Obtain a ^{31}P {^1H} NMR spectrum for your product in toluene. If your sample is contaminated with phosphorus containing by-products, it can be purified by recrystallisation from CH_2Cl_2/hexane at 0 °C. Record the infra-red spectrum of your product as a Nujol mull and compare it with the literature values. Obtain the mass spectrum of your product and identify the parent ion. Provide an explanation of the unusually long S−S bond in **1**.

c) 1,5-$Ph_4P_2N_4S_2Br_2$ (3)

Note: In view of the moisture sensitivity of 1,5-$Ph_4P_2N_4S_2Br_2$, all samples of **3** must be handled in an inert atmosphere.

Liquid bromine (1.0 g, 0.6 mmol) is added dropwise by syringe to a stirred solution of 1,5-$[Ph_4P_2N_4S_2]$ (0.30 g, 0.6 mmol) in dry dichloromethane (15 ml) in a 100 ml Schlenk flask. (The syringe should be washed with hexanes immediately after use.) The resulting yellow solution is stirred for 30 minutes and then cooled to −20 °C to produce yellow moisture-sensitive crystals of 1,5-$Ph_4P_2N_4S_2Br_2$ (0.33 g, 0.5 mmol). Yield: 83%. Record and interpret the ^{31}P {^1H} NMR spectrum of your product in $CDCl_3$. Is your product contaminated with unreacted 1,5-$Ph_4P_2N_4S_2$? Record the infra-red spectrum of your product as a Nujol mull and compare it with the literature values. Where would you expect the S−Br stretching vibration to occur?

d) Reaction of 1,5-$Ph_4P_2N_4S_2$ with $HBF_4 \cdot Et_2O$

A colourless solution of $HBF_4 \cdot Et_2O$ (0.19 g, 1.2 mmol) in 8 ml of dry dichloromethane is added dropwise, over 30 minutes, by cannula, to a stirred solution of 1,5-[$Ph_4P_2N_4S_2$] (0.51 g, 1.0 mmol) in 25 ml of dichloromethane in a 100 ml Schlenk flask. Stir the mixture for 16 hours at 23 °C, and then filter the slightly cloudy solution and remove the solvent under vacuum. Dry hexane (50 ml) is added to the pale yellow residue and the mixture is rapidly stirred. Discard the supernatant solution and dry the colourless precipitate of 1,5-[$Ph_4P_2N_4S_2H$]BF_4 (0.43 g, 0.74 mmol) under dynamic vacuum. Record the $^{31}P\{^1H\}$ NMR spectrum of your product in $CDCl_3$ and propose a structure for the protonated derivative on the basis of the data that you obtain.

References

T. Chivers, D. D. Doxsee, M. Edwards, R. W. Hilts in *The Chemistry of Inorganic Ring Systems,* (Ed.: R. Steudel), Elsevier, **1992,** p. 271–294.

N. Burford, T. Chivers, M. N. S. Rao, J. F. Richardson, *Inorg. Chem.* **1984,** *23,* 1947.

T. Chivers, G. Y. Dénès, S. W. Liblong, J. F. Richardson, *Inorg. Chem.* **1989,** *28,* 3683.

T. Chivers, R. W. Hilts, *Inorg. Chem.* **1992,** *31,* 5272.

T. Chivers, M. Edwards, A. Meetsma, J. C. van de Grampel, A. van der Lee, *Inorg. Chem.* **1992,** *31,* 2156.

T. Chivers, M. Edwards, R. W. Hilts, A. Meetsma, J. C. van de Grampel, *J. Chem. Soc., Dalton Trans.* **1992,** 3053.

A. Recknagel, A. Steiner, M. Noltemeyer, S. Brooker, D. Stalke, F. T. Edelmann, *J. Organomet. Chem.* **1991,** *414,* 327.

A. Recknagel, M. Witt, F. T. Edelmann, *J. Organomet. Chem.* **1989,** *371,* C40.

H. Schmidbaur, K. Schwirten, H. Pickel, *Chem. Ber.* **1969,** *102,* 564; W. Wolfsberger, W. Hager, *Z. Anorg. Allg. Chem.* **1977,** *433,* 247.

K. L. Paciorek, R. H. Kratzer, *J. Org. Chem.* **1966,** *31,* 2426.

N. Burford, T. Chivers, J. F. Richardson, *Inorg. Chem.* **1983,** *22,* 1482.

4.17 Sulphur-Rich Homocycles and Heterocycles: Titanocene Pentasulphide, *Cyclo*-Heptasulphur and 1,2,3,4,5,6,7-Heptathionane

Jörg Albertsen and Ralf Steudel

Special Safety Precautions

1. Chloroform ($CHCl_3$): observe the general precautions for the handling of chlorinated hydrocarbons, wear gloves, do not inhale, longer exposition may result in health problems.
2. Carbon disulphide (CS_2): very toxic on inhalation and on skin contact, very flammable; always wear gloves; make sure that all glass joints are secured by springs.
3. Dichlorodisulfane (S_2Cl_2) and sulphuryl chloride (SO_2Cl_2): very cauterising, reacts with water to produce acid. Residues may be carefully added (dropwise) to ice-cooled aqueous sodium hydroxide (10%).
4. Bismercaptoethane: flammable, toxic, residues may be oxidised by stirring in aqueous sodium hypochlorite (NaClO) for 12 h.

Elemental sulphur occurs naturally as a mineral but is commercially produced in huge amounts from hydrogen sulphide (H_2S), which is a constituent of natural gas ("sour gas") and a product of the desulphurization of crude oil performed in refineries.

Natural and commercial sulphurs consist mainly of S_8 molecules but, in addition, they always contain traces of S_7 as a solid solution in the S_8 matrix. In this experiment, you will prepare pure, crystalline S_7 using the titanocene complex $(C_5H_5)_2TiS_5$ as a sulphur transfer reagent. This reagent is useful also to prepare sulphur-rich organic polysulphanes (R_2S_n, $n \geq 5$) like the cyclic $C_2H_4S_7$ which forms a nine-membered ring. Organic polysulphanes are intermediates in the industrial vulcanization of caoutchouc by elemental sulphur to procduce rubber on a huge scale.

Experimental

a) Titanocene pentasulfide

$$NH_3 + NH_4(HS) + 1/2\ S_8 \rightarrow (NH_4)_2S_5$$

$$(C_5H_5)_2TiCl_2 + (NH_4)_2S_5 \rightarrow (C_5H_5)_2TiS_5 + 2NH_4Cl$$

Sublimed sulphur (4.7 g) is placed in a 500 ml round-bottomed vessel and 36.8 ml of an aqueous ammonium sulphide solution (1 M) are added. The mixture is stirred until all the sulphur has dissolved (*ca.* 20 min). Then 300 ml $CHCl_3$ and 9.2 g $(C_5H_5)_2TiCl_2$ are added, resulting in two liquid phases which are stirred for at least 3 h. After transfer of the mixture into a separatory funnel (volume 500 ml), the lower $CHCl_3$ phase is separated and then washed with several portions of distilled water (total volume 1000 ml). The combined aqueous phases are extracted with *ca.* 800 ml $CHCl_3$ until the red colour of the aqueous phase has faded. The combined $CHCl_3$ phases are evaporated to dryness in a vacuum.

The black-violet residue is the desired product, but still contains *ca.* 1% S_8 impurity, which can be removed by recrystallisation from CS_2 using a Soxhlet apparatus. The purification process may be checked by reversed-phase HPLC analysis (C18 stationary phase, methanol as eluent), the retention time of S_8 is larger that that of $(C_5H_5)_2TiS_5$ (detection by UV absorption).

Titanocene pentasulphide melts at 201 °C with decomposition, the ^1H NMR spectrum in CS_2 shows two singlets at $\delta = 6.10$ and 6.42 (30 °C). Calculate the yield of your product and explain why two NMR signals are observed. The structure of $(C_5H_5)_2TiS_5$ is given by Epstein, *et al.* (see References).

b) Cyclo-heptasulphur

$$(C_5H_5)_2TiS_5 + S_2Cl_2 \rightarrow S_7 + (C_5H_5)_2TiCl_2$$

Titanocene pentasulphide (2.7 g) and carbon disulphide (90 ml) are stirred in a 100 ml round-bottomed vessel at 20 °C for 10 min. Then S_2Cl_2 (0.64 ml) is added from a pipette and the stirring is continued for 30 min. The now cherry red reaction mixture is filtered (folded filter) to remove the precipitated $(C_5H_5)_2TiCl_2$ and the volume is reduced to 45 ml in a vacuum evaporator (water bath temperature 70 °C). If necessary, the solution is filtered again, 70 ml of cold diethyl ether (−78 °C) are added and the mixture is rapidly cooled to −78 °C.

After 20 h, the precipitated crude S_7 is isolated by suction of the cold mixture through a paper filter. The crystals are rapidly transferred to a beaker and immediately dissolved in 80 ml of toluene. The golden yellow solution is dried with anhydrous $MgSO_4$ (*ca.* 5 g), filtered by suction and cooled to −78 °C. After 20 h, the needle-like crystals of S_7 are filtered off at 20 °C, quickly washed with 5 ml toluene and 10 ml *n*-pentane followed by pumping off of the solvents in a vacuum for 15 min. Since solid S_7 fairly rapidly polymerises at 20 °C and is light sensitive, it is stored at −78 °C in the dark.

The HPLC retention time (see above) of S_7 is intermediate between those of $(C_5H_5)_2TiS_5$ and S_8. S_7 melts reversibly at 39 °C but prolonged heating to temperatures above the melting point results in exothermic polymerisation. Record the infra-red spectrum of S_7 dissolved in CS_2 and compare it to the published spectrum. How can the structure of S_7 be understood compared with the well-known crown structure of S_8?

c) 1,2,3,4,5,6,7-Heptathionane

$$1,2\text{-}C_2H_4(SH)_2 + SO_2Cl_2 \rightarrow C_2H_4(SCl)_2 + SO_2$$

$$1,2\text{-}C_2H_4(SCl)_2 + (C_5H_5)_2TiS_5 \rightarrow C_2H_4S_7 + (C_5H_5)_2TiCl_2$$

1,2-Bismercaptoethane (2.4 g) is placed in a well-dried Schlenk tube (50 ml) which is then filled with a protecting atmosphere of dry N_2. After addition of 10 ml of dry CH_2Cl_2 and cooling to 5 °C in an ice bath, the sulphuryl chloride (4.9 ml; dissolved in 10 ml CH_2Cl_2) is added dropwise within 90 min. Eventually, the solvent and dissolved SO_2 are removed by evaporating to dryness in a rotary evaporator. The orange solid residue of $C_2H_4(SCl)_2$ "melts away" when exposed to moist air due to hydrolysis.

1,2-Ethane-bis-sulphenylchloride, $C_2H_4(SCl)_2$, (0.6 g), dissolved in 10 ml of dry methylene chloride, is placed in a 50 ml round-bottomed flask and 1.3 g of titanocene pentasulphide, dissolved in 10 ml CH_2Cl_2, are added at once. After stirring for 30 min, the precipitated titanocene dichloride is filtered off (paper filter) and the solution evaporated to dryness on a rotary evaporator. The residue is purified by extraction with *n*-pentane (100 ml) in a Soxhlet apparatus. The volume of the extract is reduced by 50% and the solution cooled to −78 °C. After 20 h, the crystals of $C_2H_4S_7$ are isolated by rapid suction through a paper filter followed by washing with a little *n*-pentane. The product forms yellow crystals which are stable in air and having a m.p. of 62.5 °C. The HPLC retention time of $C_2H_4S_7$, using a C18 stationary phase and methanol as an eluent, is intermediate between those of $(C_5H_5)_2TiS_5$ and S_7.

Try to draw the molecules of $C_2H_4S_7$ three-dimensionally starting from S_8 and compare your result to the actual molecular structure as determined by X-ray crystallography.

References

R. Steudel, B. Holz, *Z. Naturforsch., Part B* **1988,** *43 b,* 581.

M. Schmidt, B. Block, H. D. Block, H. Köpf, E. Wilhelm, *Angew. Chem. Int. Ed. Engl.* **1968,** *7,* 632.

H. Köpf, B. Block, M. Schmidt, *Chem. Ber.* **1968,** *101,* 272; R. Steudel, R. Strauss, *J. Chem. Soc. Dalton Trans.* **1984,** 1775.

R. Steudel, S. Förster, J. Albertsen, *Chem. Ber.* **1991,** *124,* 2357.

E. F. Epstein, I. Bernal, H. Köpf, *J. Organomet. Chem.* **1971,** *26,* 229.

R. Steudel, S. Paßlack-Stephan, G. Holdt, *Z. Anorg. Allg. Chem.* **1984,** *517,* 7.

R. Steudel, F. Schuster, *J. Mol. Struct.* **1978,** *44,* 143.

R. Steudel, R. Reinhardt, F. Schuster, *Angew. Chem. Int. Ed. Engl.* **1977,** *16,* 715.

R. Steudel, J. Steidel, J. Pickardt, F. Schuster, R. Reinhardt, *Z. Naturforsch. Part B* **1981,** *35 b,* 1378.

W. H. Mueller, M. Dines, *J. Heterocycl. Chem.* **1969,** *6,* 627.

M. Kustos, J. Pickardt, J. Albertsen, R. Steudel, *Z. Naturforsch.,* Part B **1993,** *48 b,* 928.

4.18 Reaction of Alkali Metal Polyselenides and Polytellurides with Group 6 Metal Carbonyls

Joseph W. Kolis

Special Safety Precautions

1. Alkali metals are very strong reducing agents and can burn in the presence of protic solvents such as water or alcohols.
2. Ammonia is a gas at room temperature with a vapor pressure of approximetely 10 atmospheres. Thus, is closed systems, it should be handled with extreme care. Any glass vessel containing liquid ammonia should always be kept below the boiling point of ammonia (−33 °C)
3. Ammonia solutions containing dissolved alkali metals are unstable and slowly decompose to form the alkali metal amide and H_2. Thus, care should always be taken to adequately vent any closed system from time to time to prevent H_2 buildup. The decomposition reaction is catalysed by trace amounts of water or other basic impurities. If clean, dry glassware is used initially, the H_2 buildup should be quite slow.
4. Transition-metal carbonyls are quite toxic and should always be handled in a well vented fume cupboard. Also, CO will be produced in the experiment so flasks and vacuum pumps should also be in well vented areas or in a fume cupboard.
5. Any time vacuum and Schlenk lines are used, there is a finite possibility of breakage so eye protection should be worn at all times.

The polyatomic dianions of the group 16 elements selenium and tellurium have been known for a very long time. These highly coloured compounds can be made easily by the reduction of the elements with alkali metals in liquid ammonia, or by direct reaction of the elements at high temperatures. Once prepared, these salts are quite soluble and stable in a number of polar solvents including water, alcohols, acetonitrile and dimethylformamide (DMF). However, they are extremely sensitive to oxidation and should be protected from air at all times, especially in solution. In solution, a number of extremely complex and poorly understood equilibria dominate the chemistry of these species. In general, any solution of these anions is best thought of as a mixture of the various possible chain lengths (E_n^{2-}) where E is S, Se, Te and $n = 2–8$) like

$$^{-}Se-Se-Se-Se-Se^{-}$$

These anions can be readily crystallised as pure salts by the addition of large organic counterions such as $[Ph_4P]^+$ and $[Et_4N]^+$. The chainlength of the anions in the salts are much more dependent upon the counterion and solvents than the nominal identity of the alkali metal salt used as starting material.[1]

Despite the ancient lineage and ease of preparation of these unusual polyatomic anions, their chemistry has been very poorly developed until recently. However, the last eight years has seen an explosive growth of the complexation chemistry of these anions with transition metals in particular. In fact their chemistry is often substantially different than that of the better known polysulphides. One specific class of transition-metal complexes which have a very rich and diverse coordination chemistry with the polychalcogenides is the simple metal carbonyls. In this experiment, the preparation of a salt of each of the polychalcogenides will be carried out, and their reactions with $Mo(CO)_6$ and $W(CO)_6$ investigated. The resultant products will be characterized by IR and visible spectroscopy. It will be seen that polyselenides readily oxidise the metal center to its highest oxidation state, leading to complete loss of all CO ligands and formation of the well known tetrathio and tetraseleno-metalates.

$$M(CO)_6 + Se_4^{2-} \longrightarrow [MSe_4]^{2-} + 6\,CO$$

In contrast, the polytellurides do not oxidise the metal center, but merely replace two CO ligands forming a chelating complex.

$$M(CO)_6 + Te_4^{2-} \longrightarrow [(CO)_4M(Te_4)]^{2-}$$

Experimental

a) Na_2E_4 (E = Se, Te)

All the alkali metal polychalcogenides can be prepared in the same manner and the preparation of Na_2Se_4 is detailed here. Two clean, dry flasks are attached to the vacuum line, one containing a small pellet of Na and the other containing 0.145 g (6.3 mmol) Na, 1.0 g (12.6 mmol) selenium powder and a magnetic stir bar. Both flasks are evacuated and the second flask sealed off temporarily. Approximately 15 ml of ammonia from a storage cylinder is distilled into the first flask using an acetone slush bath. The storage cylinder is closed off. The ammonia solution is allowed to warm slightly to form the inky blue solution characteristic of solvated electrons. The second flask is now opened to the line and the ammonia is transferred into it again using an acetone slush bath. WARNING: This is a closed system and the pressure should constantly be monitored using a Hg manometer or other pressure sensing device. If the pressure begins to approach one atmosphere, both flasks should be cooled and the vacuum line manifold evacuated before more ammonia is distilled. Once all the ammonia is transferred, the second flask is evacuated one more time, sealed and placed on a magnetic stirrer in a low Dewar flask with an isopropanol or acetone slush. The reaction is stirred for 45 minutes, with care being taken to replenish the slush in the cooling bath. The reaction should probably be done behind a blast shield or in a fume cupboard, although there is little danger of breakage if the flask is kept cool by the bath. After a bit, the solution should begin to assume a characteristic color, reddish orange for polysulphide, dark greenish red for polyselenide and deep purple for polytelluride. After 45 min to 1 hr, the flask is placed back on the vacuum line, the ammonia evacuated into the first flask which is removed to a fume cupboard where it is vented to the air and the ammonia allowed to evaporate. The second flask is subjected to dynamic vacuum for 10 minutes to remove the last traces of ammonia. Remaining in the flask should be a dry powder of the appropriate sodium polychalogenide (yellow-orange for Na_2S_4 and dark gray for Na_2Se_4 and Na_2Te_4). If an adequate glove box is available, the solids can be transferred to clean dry storage ampoules and stored under nitrogen for use at a later time. If this is not convenient, the powder can be stored in the closed flask under N_2 until it is used for further reactions.

b) $[PPh_4]_2[MSe_4]$ *(M = Mo, W)*

The student can choose between any combination of $Mo(CO)_6$ or $W(CO)_6$ and polyselenide. If desired, the same chemistry takes place with polysulphides and these could be used in a similar fashion. The reactions are performed in exactly the same way in each case. A detailed procedure is given for the preparation of a $MoSe_4^{2-}$ salt.

In a glove bag, a 100 ml Schlenk flask is charged with 200 mg (0.76 mmol) $Mo(CO)_6$, 274 mg (0.76 mmol) Na_2Se_4 and stir bar. The flask is attached to a Schlenk line and evacuated and purged with N_2 several times. Using a cannula or gas tight syringe, 12–15 ml DMF is added to the flask. The DMF should have previously been dried and de-airated by storage for 24 h over activated sieves and bubbled for 30 min with N_2. After the DMF is added, the flask should be evacuated (very important!) and sealed off. The flask is transferred to an oil bath or heating mantle and heated at 90 °C for 1 h. Although you are heating a closed system, the boiling point of DMF is 153 °C so if the flask is evacuated properly, little pressure is generated within. Also, the small amount of CO produced will not be sufficient to strain the glass flask. However, to be safe, the flask can be reattached to the Schlenk line once or twice during heating and carefully evacuated before returning it to the heat. If desired, the reaction can also be monitored during this time by withdrawing a small aliquot and obtaining an IR spectrum to watch the disappearance of bands in the CO stretching region (2100–1700 cm^{-1}).

After some heating, the reaction mixture will begin to undergo a noticable colour change. A combination of $Mo(CO)_6$ and polysulphide will change from greenish blue to orange, while the corresponding reaction with polyselenide changes from brown-green to blue. The tungsten analogous are yellow and cherry red for sulphide and selenide respectively. Once the reaction is complete, the flask is removed from the heat and allowed to cool. The flask is repurged with N_2 and, against the flow of N_2, 650 mg (1.50 mmol) PPh_4Br is added to the mixture. The solution is stirred for 5 min to allow the counterion salt to dissolve, then filtered into a clean Schlenk flask under N_2 using a fritted Schlenk filter. De-airated THF (10 ml) is added to the mixture to cause slow precipitation of the metal chalcogenide complex. To prevent the product from forming an oil, the THF is added via syringe and allowed to roll very *slowly* down the inside of the flask so as to form a second layer of THF on the DMF solution. The flask is placed in a refrigerator overnight, taking care not to disturb the two layers. The layers will slowly mix and the THF will cause precipitation of well formed crystals of the product. Occasionally, white crystals of NaBr will precipitate first and contaminate some of the product. These can be removed by Schlenk filtration and discarded. A fresh 10 ml layer of THF and storage overnight will lead to formation of a new crop of the desired product. The crystals are isolated by Schlenk filtration of the solution, which can then be discarded. The product is washed with a 3 ml aliquot of THF and dried by pumping on the Schlenk line. The compounds $[PPh_4]_2[ME_4]$ can be handled in the air for short periods of time but any prolonged manipulation should be done under a protective nitrogen atmosphere. The salts can be characterised by IR and visible spectroscopy, and the results compared to the values in Table **4.18-1**. The compounds are quite stable as their salts if stored under a protective atmosphere of N_2.

Table 4.18-1. Spectroscopic data.

	IR (cm^{-1})	UV-Vis (nm)
MoS_4^{2-}	458	467, 317
$MoSe_4^{2-}$	255	555, 359
WS_4^{2-}	480	392, 277
WSe_4^{2-}	281	463, 316
$[Cr(CO)_4Te_4]^{2-}$	1951, 1850, 1821, 1778	
$[Mo(CO)_4Te_4]^{2-}$	1971, 1950, 1821, 1780	
$[W(CO)_4Te_4]^{2-}$	1971, 1950, 1821, 1780	

c) $[PPh_4]_2[M(CO)_4Te_4]$ (M = Cr, Mo, W)

These chelate complexes can be prepared using a procedure exactly analogous to that for the formation of the ME_4^{2-} salts prepared above. However, the polytelluride does not induce oxidative decarbonylation of the metal center, but merely substitutes two *cis* CO ligands. In this case, $Cr(CO)_6$ can also be chosen as a starting material. ($Mo(CO)_6$ can be used but the Mo containing product often does not crystallise well.) The synthesis and workup occurs exactly as described above except that the solution colour changes from dark purple to dark brown. Also, two 10 ml layerings of THF are usually needed to induce crystallisation. The product can be isolated as described above, but it is more air sensitive than the ME_4^{2-} salts, so it should be handled under N_2 at all times. The IR spectrum in the CO stretching reagion is quite distinctive and contains the four bands predicted for a molecule with C_{2v} symmetry.

References

M. A. Ansari, J. A. Ibers, *Coord. Chem. Revs.* **1990,** *100,* 223.
L. C. Roof, J. W. Kolis, *Chem. Rev.* **1993,** *93,* 1037.
M. A. Draganjac, T. B. Rauchfuss, *Angew. Chem. Int. Ed. Engl.* **1985,** *24,* 742.
A. Müller, E. Diemann, *Adv. Inorg. Chem.* **1987,** *31,* 89.
J. W. Kolis, *Coord. Chem. Revs.* **1990,** *105,* 195.
S. C. O'Neal, J. W. Kolis, *J. Am. Chem. Soc.* **1988,** *110,* 1971.
A. Müller, E. Diemann, R. Jostes, H. Bögge, *Angew. Chem. Int. Ed. Engl.* **1981,** *20,* 934.
W. A. Flomer, S. C. O'Neal, D. Jeter, A. W. Cordes, J. W. Kolis, *Inorg. chem.* **1988,** *27,* 969.

4.19 Sulphur-Nitrogen Heterocycles and $[SNBr_{0.4}]_x$ Polymer

J. Derek Woollins

Special Safety Precautions

1. Sulphur monochloride (S_2Cl_2) is corrosive and very toxic by inhalation. *Always* use in a fume cupboard. Spillages should be treated with solid sodium bicarbonate followed by copious amounts of water.
2. Sulphur dichloride (SCl_2), which is produced and collected in the liquid nitrogen cold trap during the preparation of $(NSCl)_3$, is corrosive. After removal from the vacuum line, clamp the trap at the back of the fume cupboard and allow to warm to room temperature. Add *dropwise* aqueous sodium bicarbonate. The *treated* material can be washed down the drain.
3. Chlorine gas is toxic. The whole cylinder should be securely clamped inside a fume cupboard. If in any doubt whatsoever about safe operation of the cylinder, consult a demonstrator.
4. Carbon tetrachloride is toxic by inhalation or contact.
5. All of the sulphur-nitrogen compounds prepared in this experiment, except $[S_4N_3]Cl$, should be regarded as air and moisture sensitive and thus likely to hydrolyse to HCl and SO_2 if exposed to the atmosphere.
6. The residue from the preparation of $[S_3N_2Cl]Cl$ must be destroyed. You should slowly tip the contents onto solid sodium bicarbonate in a fume cupboard. Very slowly add water and allow to stand.
7. Make sure that you carefully grease all joints with silicone.

As a result of investigations begun over a hundred years ago, there is an extensive chemistry of inorganic sulphur-nitrogen compounds. Many of these compounds have remarkable chemical properties as well as unusual structures and bonding. Most recently, polymeric sulphur nitride, $(SN)_x$, has been found to be a one-dimensional conductor and a superconductor at low temperature. In this experiment, you will prepare three examples of SN heterocycles: $[S_3N_2Cl]Cl$, $(NSCl)_3$ and $[S_4N_3]Cl$, as well as the doped polymer $[SNBr_{0.4}]_x$. The first two are useful intermediates in the formation other SN compounds whilst the third is an example of a 10π aromatic system.

Experimental

a) $[S_3N_2Cl]Cl$

In a fume cupboard
Ammonium chloride (8.0 g), sulphur (4.0 g) and sulphur monochloride (24 cm^3; *care*) are placed in a dry 50 cm^3 B24 neck round-bottomed flask. The flask is fitted with a B24 straight walled condenser and a calcium chloride drying tube. The joints should be throughly greased with silicone grease. The mixture is heated to *gentle* reflux (isomantle). Deep red crystals of the product should sublime out of the reaction into the lower end of the condenser within 15–20 min. You should try to keep the heating to a minimum and it is often helpful to lag the top of the flask with glass wool insulation. Gentle refluxing is maintained for 6–8 hours, or overnight if necessary – do not attempt to restart the reaction with product in the condenser since the product is often washed back into the reaction by the refluxing S_2Cl_2. The reaction is allowed to cool and the $[S_3N_2Cl]Cl$ removed as follows.

A thoroughly dried, preweighed B24 Schlenk tube is prepared and securely clamped with N_2 flowing through it. The condenser containing $[S_3N_2Cl]Cl$ is rapidly transferred over (less than 5 seconds!). With continuous and steady N_2 flow, the $CaCl_2$ tube is removed and the product gently scraped into the Schlenk tube using a long handled spatula. The condenser is then returned to the reaction vessel and the Schlenk tube is stoppered (carefully grease the stopper with silicone grease). It may be necessary to pump on the product for a few minutes to remove traces of S_2Cl_2.

A second batch of $[S_3N_2]Cl$ (to be used in the synthesis of $[S_4N_3]Cl$) is prepared by addition of a further 10 ml of S_2Cl_2 to the reaction which is then refluxed as before.

Calculate the yield of your first crop of product and measure its IR spectrum (glove bag, Nujol mull, NaCl plates). Seal up a small sample in an ampoule.

b) $(NSCl)_3$

Using the first crop of $[S_3N_2Cl]Cl$, transfer the condenser to the top of a B24 Schlenk tube as before. With N_2 flow, remove the drying tube and connect the condenser to a scrubber unit. Turn off the N_2 and connect the apparatus to a Cl_2 cylinder (ask a demonstrator before using the chlorine cylinder). Pass Cl_2 gently through the system. The $[S_3N_2Cl]Cl$ will rapidly react and the product should be washed down into the Schlenk tube by the SCl_2 that is formed. Usually 5 minutes of slow Cl_2 flow is sufficient. Turn off the Cl_2 cylinder and pump away the SCl_2. If the product is still orange coloured, repeat the chlorination. Measure the IR spectrum (glove bag, NaCl and polythene plates, 1100–200 cm^{-1}) and seal some of your sample in a glass ampoule.

c) $[S_4N_3]Cl$

Use the $[S_3N_2Cl]Cl$ from your second crop. You may have to rescale the reaction.

$[S_3N_2Cl]Cl$ (0.4 g) is refluxed in dry CCl_4 (20 cm^3) and S_2Cl_2 (12 cm^3) under N_2 for 5 hours or until all of the dark solid has been converted to a bright yellow precipitate. Cool the reaction, filter off the product (in air on a sintered funnel) wash it with 3 × 5 cm^3 of CCl_4. Discard your CCl_4/S_2Cl_2 waste into the container provided (see point 1 of the Special Safety Precautions).

Measure the IR spectrum (Nujol mull, KBr plates, 1200–200 cm^{-1}). Measure the UV spectrum (400–200 nm) in *conc.* HCl (*Care:* if you spill any acid in the spectrometer, report it the demonstrator immediately). You should accurately weigh your sample so that you can report extinction coefficients. As a guide to the concentration required, the strongest band has $\varepsilon = 10000–15000\ dm^3\ mol^{-1}\ cm^{-1}$. There is no need to seal your product up as it is not particularly air sensitive. The ^{15}N NMR spectrum of 100% ^{15}N labelled $[S_4N_3]Cl$ is given in Figure **4.19-1.**

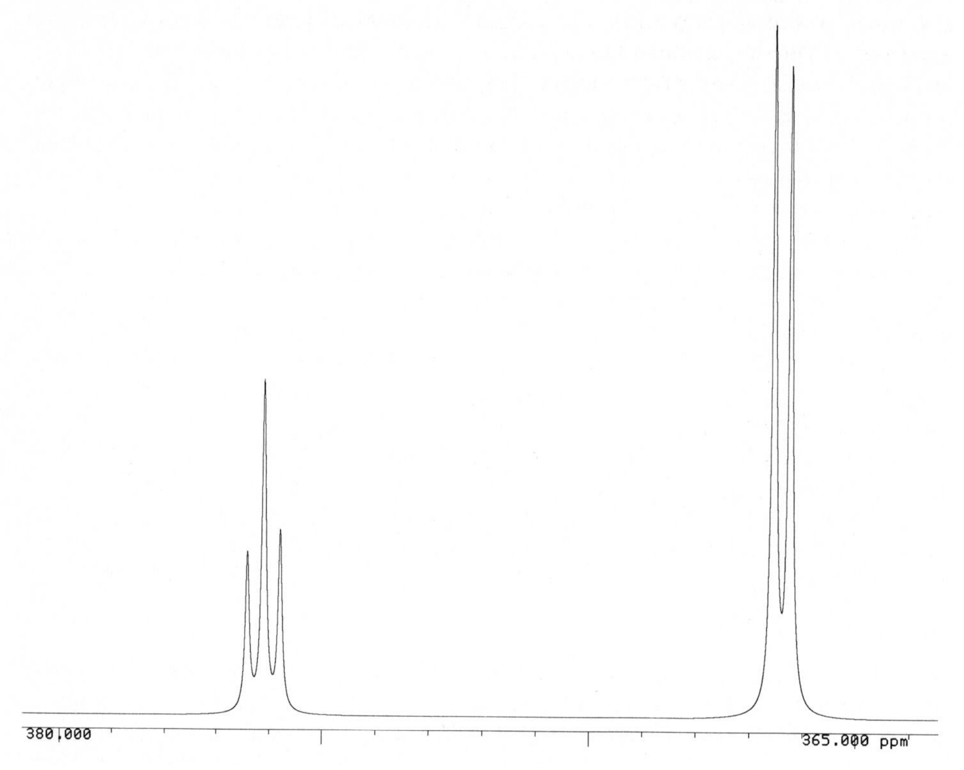

Figure 4.19-1. ^{15}N NMR of $[S_4N_3]Cl$ in conc. HCl at 27.36 MHz.

d) $[SNBr_{0.4}]_x$

$(NSCl)_3$ (1.34 g) is dissolved in 35 ml of dry CH_2Cl_2 under N_2. The reaction is cooled to −60 °C and Me_3SiBr (2.2 ml) is *slowly* added. After stirring for 15 min, the reaction is allowed to warm to room temperature and the resulting precipitate filtered, washed with CH_2Cl_2 (2 × 10 ml) and dried *in vacuo.*

Record the IR spectrum of your product and compare it with that in the literature. Prepare a pressed disc using the IR press and measure the conductivity of your solid sample. The properties of $(SN)_x$ and its halogenated derivatives have been reviewed by Labes, *et al.* (see References).

References

H. G. Heal, *The Inorganic Heterocyclic Chemistry of Phosphorus, Sulphur and Nitrogen,* Academic Press, London **1980.**

W. L. Jolly, K. D. Maguire, *Inorg. Synth.* **1967,** *9,* 102.

J. W. Waluk, J. Michl, *Inorg. Chem.* **1982,** *21,* 556.

A. J. Banister, J. A. Durant, I. B. Gorrell, R. S. Roberts, *J. Chem Soc., Faraday Trans. 2* **1985,** *81,* 1771.

A. J. Banister, *Nature Phys. Sci.* **1972,** *237,* 92.

I. R. Nevitt, H. S. Rzepa, J. D. Woollins, *Spectrochim. Acta* **1989,** *45A,* 367.

V. Demant, K. Dehnicke, *Z. Nat. B* **1986,** *41,* 929.

M. Labes, P. Love, L. F. Nicholls, *Chem. Rev.* **1979,** *79,* 1.

4.20 Chlorothionitrene and *Cyclo*thiazeno Complexes of Tungsten

Kurt Dehnicke

Special Safety Precautions

1. Tungsten hexachloride (WCl_6) and tungsten oxotetrachloride ($WOCl_4$) should be regarded as air and moisture sensitive and thus likely to hydrolyse to HCl (which is toxic by inhalation) and $WO_3 \cdot nH_2O$.
2. Chlorothionitrene tungsten tetrachloride ($[Cl_4W(NSCl)]_2$) and cyclothiazeno tungsten trichloride $[Cl_3W(N_3S_2)]_2$ are also moisture sensitive and thus likely to hydrolyse to NH_4Cl, $WO_3 \cdot nH_2O$ and sulphur chlorides. The latter are toxic by inhalation or contact.
3. Dichloromethane is toxic by inhalation or contact, just as the thionyl chloride formed by the reaction of $WOCl_4$ with $(NSCl)_3$. The solution of $SOCl_2$ in CH_2Cl_2, which remains after filtration of the $WCl_3(N_3S_2)$ sample, must be destroyed by adding an aqueous suspension of sodium bicarbonate in a fume cupboard. After separating the aqueous phase by means of a dropping funnel, the dichloromethane is regained by distillation.
4. Make sure that you carefully grease all joints with silicone. All procedures must be carried out under dry nitrogen.

Chlorothionitrene complexes are compounds with the $\left[\underline{\overline{N}} = S\diagup^{Cl}\right]^{2-}$ ligand, which is unknown as a free ion. Examples are prepared from molybdenum, tungsten, rhenium, ruthenium and osmium. They can easily be prepared by reaction of metal chlorides like $MoCl_5$ or WCl_6 with $(NSCl)_3$, which is in equilibrium with its monomer in solution. The reactions with metal chlorides are thus accompanied by the redox reaction $NSCl + 2e^- \rightarrow [NSCl]^{2-}$. Chlorothionitrene complexes are useful precursors for the syntheses of nitrido complexes, thionitrosyl complexes as well as of cyclothiazeno complexes.

Metallacyclothiazeno complexes are heterocycles of the type M(=N̄–S̄⊕–N|–S̄⊖=N̲) with planar six-membered rings, delocalised N–S bonds and metal-nitrogen double bonds. Examples are known from vanadium, molybdenum and tungsten, whilst with rhenium the heterocycle Re(=N–S̄|)(=N=S̲|) is formed.

Experimental

a) $[Cl_4W(NSCl)]_2$

Tungsten hexachloride (10.7 g) and dichloromethane (50 ml) are placed in a dry 100 ml Schlenk tube. To this suspension, a solution of $(NSCl)_3$ (2.20 g) in CH_2Cl_2 (20 ml) is slowly added dropwise with stirring (magnetic stirrer). The preparation is stirred for altogether 48 h at room temparature; the reddish brown precipitate is filtered under inert gas through a sintered glass frit, then washed with 10 ml CH_2Cl_2 and dried *in vacuo.* Yield: 77%. Single crystals can be prepared by cooling a saturated solution of the sample in CH_2Cl_2 to −18 °C, or by subliming *in vacuo* at 145 °C.

Calculate the yield of the dry precipitate, measure its IR spectrum (glove bag, Nujol mull, KBr plates) and compare it with the literature data. Seal up a small sample in an ampoule.

b) $[Cl_3W(N_3S_2)]_2$

This complex can be synthesised by reaction of the chlorothionitreno derivative described above with excess $(NSCl)_3$ according to Eqn. (1). In this reaction, by-products are also formed, which reduce the yield.

$$Cl_4W(NSCl) + 2/3\ (NSCl)_3 \rightarrow Cl_3W(N_3S_2) + SCl_2 + Cl_2 \qquad (1)$$

The complete conversion to the cyclothiazeno complex can be induced when tungsten oxotetrachloride (Eqn. 2) is the starting reactant.

$$WOCl_4 + (NSCl)_3 \rightarrow Cl_3W(N_3S_2) + SOCl_2 + Cl_2 \qquad (2)$$

For this route, $WOCl_4$ (6.24 g) is suspended in dichloromethane (50 ml) and mixed with $(NSCl)_3$ (4.60 g) with stirring at 20 °C; $(NSCl)_3$ must be dissolved in dichloromethane (40 ml). After only a few minutes, a change of colour can be observed from red-orange to brown. To complete the reaction, the preparation is stirred for a further 48 h. Thereafter, the dark brown precipitate is filtered under dry nitrogen, washed with 20 ml CH_2Cl_2 and dried *in vacuo.* 4.56 g of the product can be isolated (yield 63%). The yield can be increased by reducing the solvent of the filtrate.

Calculate the yield of the dry precipitate, measure its IR spectrum (glove box, Nujol mull, KBr plates) and compare it with the literature data. Seal up a small sample in an ampoule.

$[Cl_3W(N_3S_2)]_2$ reacts with PPh_4Cl in CH_2Cl_2, forming a red solution, from which $PPh_4[Cl_4W(N_3S_2)]$ can be precipitated by adding CCl_4. The cyclothiazeno complex also reacts with Lewis bases like THF or pyridine, forming the donor-acceptor complexes $[(THF)Cl_3W(N_3S_2)]$ and $[(Py)Cl_3W(N_3S_2)]$ respectively.

c) $WOCl_4$

$WOCl_4$, which is needed for the synthesis of $[Cl_3W(N_3S_2)]_2$, can easily be prepared by reaction of tungsten hexachloride with hexamethyldisiloxane or with trichloronitromethane.

For the synthesis from WCl_6 and $(Me_3Si)_2O$, add a dichloromethane (15 ml) solution of $(Me_3Si)_2O$ (2.05 g) dropwise to a suspension of WCl_6 (5.0 g) in CH_2Cl_2 (20 ml) at room temperature over a period of 75 min. The red solid is collected by removing the supernatant liquor by filtration, washed with petroleum ether (b.p.: 40–60 °C, 2 × 20 ml) and dried *in vacuo.* The yield is nearly complete. The reaction follows Eqn. (3).

$$WCl_6 + (Me_3Si)_2O \rightarrow WOCl_4 + 2\,ClSiMe_3 \quad (3)$$

Cl_3SiMe_3, which is contained in the filtrate, can be transformed into hexamethyldisiloxane again by its reaction with the calculated amount of water in the presence of such bases as triethylamine (Eqn. 4). After filtration of triethyl ammonium chloride, the solution can be used anew for the synthesis of $WOCl_4$.

$$2\,ClSiMe_3 + H_2O + 2\,Et_3N \rightarrow (Me_3Si)_2O + 2[Et_3NH]Cl \quad (4)$$

The synthesis of $WOCl_4$ from trichloronitromethane proceeds according to Eqn. (5). The emerging nitrosyl chloride is very toxic and must be destroyed by leading it into a NaOH solution.

$$WCl_6 + CCl_3NO_2 \rightarrow WOCl_4 + NOCl + CCl_4 \quad (5)$$

References

K. Dehnicke, U. Müller, *Comments Inorg. Chem.* **1985,** *4,* 213.

H. W. Roesky, *Rings, Clusters and Polymers of Main Group and Transition Elements,* Elsevier, Amsterdam, **1989.**

W. Hiller, J. Mohyla, J. Strähle, H. G. Hauck, K. Dehnicke, *Z. Anorg. Allg. Chem.* **1984,** *514,* 72.

U. Kynast, U. Müller, K. Dehnicke, *Z. Anorg. Allg. Chem.* **1984,** *508,* 26.

U. Kynast, E. Conradi, U. Müller, K. Dehnicke, *Z. Naturforsch.* **1984,** *39b,* 1680.

V. C. Gibson, T. P. Kee, A. Shaw, *Polyhedron* **1987,** *7,* 579.

H. Prinz, K. Dehnicke, U. Müller, *Z. Anorg. Allg. Chem.* **1982,** *488,* 49.

4.21 Selenium-Nitrogen and Tellurium-Nitrogen Compounds

Herbert W. Roesky and Judith Gindl

Special Safety Precautions

1. Most of the selenium and tellurium compounds are toxic. $Se[N(SiMe_3)_2]_2$, $Se[NCMe_3(SiMe_3)]_2$, $Te[N(SiMe_3)_2]_2$ and $Te[NCMe_3(SiMe_3)]_2$ are relatively volatile. All preparations should be carried out in a well ventilated fume cupboard.
2. All compounds are air and moisture sensitive. Therefore, experimental manipulations should only be performed under an atmosphere of dry nitrogen gas in Schlenk apparatus or in a dry box.
3. Butyllithium reacts violently with H_2O, the solutions should only be handled under dry nitrogen gas.
4. $(ClTeNSN)_3N$ may decompose violently when heated over 207 °C. $(FTeNSN)_3N$ may explode at room temperature when exposed to mechanical strain.

In the last ten decades, a large number of sulphur-nitrogen compounds could be obtained, but only a few selenium-nitrogen and tellurium-nitrogen compounds are known. Most are unstable and some of them decompose with explosion. Another problem is the lack of suitable stable precursors for the selenium- and tellurium-nitrogen chemistry. In this experiment you will find that some stable selenium-nitrogen and tellurium-nitrogen compounds ($Se[N(SiMe_3)_2]_2$, $Te[N(SiMe_3)_2]_2$, $Se[(NCMe_3(SiMe_3)]_2$ and $Te[NCMe_3(SiMe_3)]_2$) are useful precursors for preparing new compounds with Se – N and Te – N bonds. $Se(NCMe_3)_2SnCl_4$ is a stable selenodiimide. $(ClTeNSN)_3N$ and $(FTeNSN)_3N$ are tellurium nitrides that are stable at room temperature. Most of the known tellurium nitrides are very explosive. All compounds described have been characterised by X-ray structural analysis.

Experimental

a) $Se[N(SiMe_3)_2]_2$

$LiN(SiMe_3)_2$ (3.08 g) is dissolved in 40 ml of dry *n*-hexane. The solution is cooled to −78 °C and Se_2Cl_2 (2.11 g) is slowly added. The resulting yellow solution is stirred for 1 h and then allowed to warm up to room temperature. It is further stirred for 12 h, the solution is filtered and the volatiles are removed *in vacuo* by using a liquid nitrogen trap system. The orange residue is sublimed at 30–40 °C under dynamic vacuum (<0.1 torr), which yields 2.75 g (75%) of the yellow crystalline product having a melting point of 64–65 °C.

Measure the ^1H NMR spectrum (dry box, CCl_4 as solvent, $SiMe_4$ as reference). ^1H NMR: δ 0.27 (s). Mass spectrum (EI) [*m/z* (^{80}Se peaks, %)]: 400 (M^+, 80), 275 ($M^+ - SeMe_2 - Me$, 45), 239 ($M^+ - HN(SiMe_3)_2$, 40).

b) *Te[N(SiMe$_3$)$_2$]$_2$*

A solution of $LiN(SiMe_3)_2$ (1.10 g) in dry *n*-hexane is cooled to −78 °C, then $TeCl_4$ (0.51 g), dissolved in toluene, is added slowly. The reaction mixture is stirred for 1 h at −78 °C and after being allowed to warm up to room temperature, it is further stirred for 12 h. The solution is filtered and the volatiles are removed *in vacuo* using a liquid nitrogen trap system. Sublimation of the residue at 30–40 °C under dynamic vacuum (<0.1 torr) gives 0.43 g of the product. In order to obtain analytically pure product, the orange crystals are sublimed once again, now melting at 69–71 °C.

Measure the ^1H NMR spectrum (dry box, CCl_4 as solvent, $SiMe_4$ as reference). ^1H NMR: δ 0.24 (s). Mass spectrum (EI) [*m/z* (^{130}Te peaks, %)]: 450 (M^+, 50), 289 ($M^+ - HN(SiMe_3)_2$, 20), 275 ($M^+ - TeMe_2 - Me$, 100).

c) *Se[NCMe$_3$(SiMe$_3$)]$_2$*

To a solution of $HNCMe_3(SiMe_3)$ (3.08 g) in 40 ml dry *n*-hexane are added 16 ml of a 1.6 M *n*BuLi/hexane solution at 0 °C. The solution is stirred for a few hours at room temperature, then cooled to −78 °C and Se_2Cl_2 (3.0 g) is slowly added. After allowing to warm up to room temperature, the solution is stirred further for 12 h then filtered through Celite. The volatiles are removed *in vacuo* using a liquid nitrogen trap system. The orange residue is sublimed at 40–50 °C under vacuum (<0.1 torr). Sublimation gives 3.2 g (65%) yellow crystals, melting at 72–73 °C. Further sublimation gives colorless crystals.

Measure the ^1H NMR spectrum (dry box, CCl_4 as solvent, $SiMe_4$ as reference) and the ^{77}Se NMR spectrum ($CDCl_3$ as solvent, $SeMe_2$ as reference). ^1H NMR: δ 0.35 (s, $SiMe_3$), 1.45 (s, CMe_3). ^{77}Se NMR: δ 1071 (s). Mass spectrum (EI) [*m/z* (^{80}Se peaks, %)]: 368 (M^+, 100), 297 ($M^+ - NCMe_3$, 30), 224 ($M^+ - NCMe_3(SiMe_3)$, 60).

d) *Te[NCMe$_3$(SiMe$_3$)]$_2$*

To a solution of $HNCMe_3(SiMe_3)$ (3.18 g) in 40 ml dry *n*-hexane are slowly added 16 ml of a 15% *sec*BuLi/cyclohexane solution at 0 °C. The solution is stirred for a few hours, then cooled to −78 °C and $TeCl_4$ (1.47 g), dissolved in toluene, is added slowly. After allowing to warm up to room temperature, the solution is stirred for 12 h and filtered through Celite. The volatiles are removed *in vacuo* using a liquid nitrogen trap system. The crude product is purified by sublimation at 40–50 °C under dynamic vacuum (<0.1 torr). The bright yellow product (0.81 g; 35%) melts at 67–70 °C.

Measure the ^1H NMR spectrum (dry box, CCl_4 as solvent, $SiMe_4$ as reference) and the ^{125}Te NMR spectrum ($CDCl_3$ as solvent, $TeMe_2$ as reference). ^1H NMR: δ 0.35 (s, $SiMe_3$), 1.47 (s, CMe_3). ^{125}Te NMR: δ 1742 (s). Mass spectrum (EI) [*m/z* (^{130}Te peaks, %)]: 418 (M^+, 60), 347 ($M^+ - NCMe_3$, 20), 274 ($M^+ - NCMe_3(SiMe_3)$, 90).

e) *Se(NCMe$_3$)$_2$SnCl$_4$*

To a solution of $Se[NCMe_3(SiMe_3)]_2$ (2.16 g) from part c) in 20 ml dry CH_2Cl_2 is added $SnCl_4$ (3.03 g) in 10 ml CH_2Cl_2, the reaction mixture is stirred for 3 d at room temperature.

The green precipitate is separated by filtration and the residue redissolved in 500 ml dry THF. Insoluble by-products are separated by filtration, the solvent is removed *in vacuo*. A yellow product remains (0.87 g, 31%), decomposing at 142 °C.

Alternatively, the following procedure can be employed. To a suspension of $SeCl_4$ (0.55 g) in 20 ml dry diethyl ether is slowly added H_2NCMe_3 (1.1 g). The solution is filtered, the solvent removed *in vacuo* and the residue dissolved in 8 ml dry CH_2Cl_2. To this solution is added dropwise $SnCl_4$ (0.65 g) in 10 ml CH_2Cl_2 at room temperature. The immediately formed yellow solid is filtered off, dissolved in 200 ml dry THF and red by-products removed by filtration. After removing the solvent *in vacuo*, 0.35 g (29%) of the product remain, decomposing at 141 °C.

Measure the ^1H NMR spectrum (dry box, $(CD_3)_2SO$ as solvent, $SiMe_4$ as reference) and the ^{77}Se NMR spectrum ($(CH_3)_2SO$ as solvent, $SeMe_2$ as reference). The product is not stable in this solvent, the measurements must be performed immediately. ^1H NMR: δ 1.57 (s). ^{77}Se NMR: δ 1392.0 (s).

f) $(ClTeNSN)_3N$

3.50 g $TeCl_4$ are dissolved in 120 ml of dry toluene. A solution of 3.10 g $S(NSiMe_3)_2$ in 40 ml dry toluene is slowly added at 0 °C. The yellow precipitate is separated by filtration, washed with dry $CHCl_3$ and dried *in vacuo* to get 2.2 g (74%) of the product, decomposing at 207 °C.

Measure the IR spectrum (dry box, Nujol mull, KBr plates).
IR [cm^{-1}]: 1120 (vs sh), 1090 (s), 1060 (vs sh), 680 (s), 570 (s), 520 (s).

g) $(FTeNSN)_3N$

A solution of $S(NSiMe_3)_2$ (1.04 g) in 30 ml ether is slowly added to a suspension of TeF_4 (0.94 g) in 30 ml dry ether and the reaction mixture is stirred for 8 h. The red solid is filtered off, redissolved in dry pyridine/toluene (3 : 1) and allowed to crystallise at −10 °C, giving 0.50 g (51%) of the product after drying *in vacuo*. It decomposes at 112 °C.

Measure the ^{19}F NMR spectrum (pyridine/C_6D_6 as solvent, $CFCl_3$ as reference) and the ^{125}Te NMR spectrum (pyridine/C_6D_6 as solvent, $TeMe_2$ as reference). ^{19}F NMR: δ −42.6 (s). ^{125}Te NMR: δ 1157.

References

M. Björgvinsson, H. W. Roesky, *Polyhedron* **1991,** *10,* 2353.

M. Björgvinsson, H. W. Roesky, F. Pauer, D. Stalke, G. M. Sheldrick, *Inorg. Chem.* **1990,** *29,* 5140.

M. Björgvinsson, H. W. Roesky, F. Pauer, D. Stalke, G. M. Sheldrick, *Eur. J. Solid State Inorg. Chem.* **1992,** *29,* 759.

J. Gindl, M. Björgvinsson, H. W. Roesky, C. Freire-Erdbrügger, G. M. Sheldrick, *J. Chem. Soc. Dalton Trans.* **1993,** 811.

H. W. Roesky, J. Münzenberg, M. Noltemeyer, *Angew. Chem.* **1990,** *102,* 73; *Angew. Chem. Int. Ed. Engl.* **1990,** *29,* 61.

J. Münzenberg, H. W. Roesky, S. Besser, R. Herbst-Irmer, *Inorg. Chem.* **1992,** *31,* 2986.

4.22 Low Temperature Synthesis of Barium Amides and Phenoxides

Simon R. Drake, Philip Hall and David J. Otway

Special Safety Precautions

1. Ammonia and tetrahydrofuran are toxic and should be handled in a well ventilated fume cupboard, using disposable gloves and standard Schlenk techniques.
2. Tetrahydrofuran and toluene are initially dried over 4A molecular sieves, then purified by distillation under an atmosphere of dry, oxygen free nitrogen from calcium hydride or sodium benzophenone ketyl. The hexamethyldisilamide is dried over 4A molecular sieves, then distilled prior to use. These solvents are toxic and should be handled with care, using disposable gloves in a fume cupboard.
3. The ammonia should be "puratronic" grade and used as received with a pressure regulator, a section of plastic tubing and ~30 cm needle attached. Adequate flushing of the cylinder should be catered for by placing the cylinder (adequately clamped) in the fume cupboard. Ammonia gas is toxic.
4. Barium metal granules (2M5) are kept in either a glove box or glove bag during use, and should be handled with disposable gloves. All wastes must be destroyed by hydrolysis in clearly labelled containers.
5. The NMR samples were prepared in 5 mm NMR tubes inside either the glove bag or glove box, using predried deutrated solvents (dried over 4A mol. sieves).
6. The materials described here are all air and moisture sensitive, and they should be routinely prepared under dry nitrogen.
7. All joints should be carefully greased to ensure that moisture/oxygen do not attack the complexes.

The organometallic chemistry of the heavier alkaline earth metals calcium, strontium and barium is currently undergoing a renaissance. Reasons for the renewed interest in this previously dormant area are threefold. Firstly, the growing interest in either sol-gel technology or organometallic chemical vapor deposition (OMCVD) techniques for preparing thin films of speciality materials, *e. g.* piezoelectrics ($BaTiO_3$ and $BaMG_{0.33}$ $Ta_{0.66}O_3$), semiconductors and superconductors (*e. g.* $K_{0.6}Ba_{0.4}BiO_3$, $La_{2-x}Sr_xCuO_4$ and $YBa_2Cu_3O_{7-x}$), requires the synthesis of monomeric, volatile single source molecular precursors. Secondly, many of the "simplest" alkoxides, alkyls, aryls, silyls, etc. remain either poorly characterised or are presently unknown and, as such, they stand out as attractive synthetic targets. Finally, a limited range of barium organometallics (*e. g.* $Ba(CPH_3)_2(THF)_2$) have found application in catalytic processes, notably anionic polymerisation and as selective reagents for organic synthesis.

A series of metal trimethylsilylamides, $[M\{N(SiMe_3)_2\}_2(THF)_x]_n$ (where $x = 2$ and $n = 1$; $x = 1$ and $n = 2$; $x = 0$ and $n = 2$), have been synthesised by the use of a low temperature ammonia ether solvent route at −40 °C. These materials are all extremely reactive owing to the presence of the labile amide $(Me_3Si)_2N$ ligands. The amido complexes may be easily and routinely prepared from the reaction of bulk metal granules with the appropriate metal amide in tetrahydrofuran (or other etheral solvents) in the presence of liquified ammonia at −40 °C. The reaction is found to proceed very cleanly under these relatively mild conditions via a multistep pathway, and is found to be a high yield synthetic route to a wide range of barium (or indeed calcium or strontium) organometallics. These materials have found immediately application in reactions with aryloxides at ambient temperature.

This mixed solvent route is thus also a useful synthetic pathway to halide-free products, in contrast to previous routes using metathesis exchange of an alkali metal organometallic with a group 2 dihalide. This reaction is also substantially faster than the recently reported transmetallation reactions with tin or mercury organometallics, and has the added advantage that the only side product is ammonia. Other alternative routes have utilised finely divided "Rieke" metal powders, prepared by potassium reduction of the corresponding metal iodides which were subsequently reacted *in situ* with the ligand; and also the well known metal vapour synthesis (MVS) technique.

These metal amides have been found to be important starting materials for reactions with phenols, β-diketonates, sterically hindered alcohols, crown ethers, cyclic aromatics, and for the preparation of acetylide and siloxide complexes.

Experimental

a) $[Ba(N\{SiMe3\}_2)_2(THF)_2]$ (1)

Into a thoroughly flame dried Schlenk tube (3 vacuum-gas cycles) is placed 1.37 g (10 mmol) of barium metal granules inside a nitrogen atmosphere glove bag. The Schlenk tube is then stoppered, and the vessel attached to a vacuum line in a well ventilated fume cupboard. The Schlenk is then evacuated/flushed 3 times with nitrogen.

Tetrahydrofuran (25 cm^3), toluene (5 cm^3) and hexamethyldisilamide (9.88 cm^3; 40.0 mmol, a 2-fold excess) are added to the Schlenk tube via a degassed syringe. The top of the Schlenk is then removed under a positive gas flow and quickly replaced with a rubber "suba-seal". At this point, the *rapidly* stirred suspension is cooled by means of an external dry-ice/acetone bath to *ca.* −40 °C. The fume cupboard sash is pulled down as far as possible for the rest of the experiment. Ammonia gas is then slowly bubbled into the solution for 10–12 min, during which time the solution turns a grey colour. After this period of time, the mixture is slowly allowed to warm up to room temperature. Initially, bronze pieces of $[Ba(NH_3)_6]$ are formed as the metal reacts, but these readily dissolve over a period of 1–2 h. No further flush with NH_3 is normally necessary. The rate of stirring of the solution is then reduced to a minimum and the reaction mixture allowed to stir for *ca.* 8–12 h at ambient temperature. After this period of time, the solution may have a pale grey colour. At this point, the stirrer is turned off and the reaction mixture allowed to stand so a small amount (*ca.* 50 mg) of finely divided metal can settle out. The solution is then filtered through a celite pad and all the solvent removed *in vacuo.* The resulting residue is dissolved in a minimum of pentane and left to stand in the fridge. A large crop of colourless crystals will deposit overnight. The mixture is filtered, using a filter stick and the resulting crystalline solid is isolated in the glovebag and stored in a screw top vial. The remaining pentane solution is reduced to dryness, yielding a small quantity of extra product as a white solid. Calculate your yield.

Bis(trimethylsilylamido)barium bistetrahydrofuran is obtained as a white crystalline solid, which should be stored at ambient temperature in a sealed container under a nitrogen atmosphere. This complex is soluble in a range of aliphatic and aromatic hydrocarbon solvents (*e.g.* *n*-pentane, hexane, benzene and toluene) and polar solvents (*e.g.* THF, DME, HMPA, Et_2O, DMSO and CH_3CN). This material is very air and moisture sensitive, both in the solid state and in solution. Record the yield, m.p., IR, ^1H NMR and ^{13}C NMR spectra.

The mass spectra, using either electron impact or chemical ionisation, show substantial decomposition. A single crystal X-ray structure of $Ba\{N(SiMe_3)_2\}_2(THF)_2$ has been undertaken.

b) $[Ba(N\{SiMe_3\}_2)_2(THF)_2$ *(2)*

Into a thoroughly flame dried Schlenk tube (3 vacuum-gas cycles) is placed 1.00 g (1.65 mmol) of $[Ba(N\{SiMe_3\}_2)_2(THF)_2]$ inside a nitrogen atmosphere glove bag. The Schlenk tube is then stoppered, and the vessel attached to a vacuum line in a well ventilated fume cupboard. The Schlenk is then evacuated/flushed 3 times with nitrogen.

Toluene (25 cm^3) is added to the Schlenk tube via a degassed syringe. (The solvent used here must have a boiling point higher than that of THF). The solution is then left to stir for *ca.* 30 min at ambient temperature and then stripped to dryness using and external solvent trap to produce a colourless oily material. At this stage, the product may be crystallised from pentane and left to stand in the fridge overnight. Complete removal of the pentane gives a >95% yield of a free-flowing, white crystalline solid which is isolated in the glovebag and stored in a screw top vial. Yield is essentially quantitative.

Bis(trimethylsilylamido)barium tetrahydrofuran is obtained as a white crystalline solid, which should be stored at ambient temperature in a sealed container under a nitrogen atmosphere. This complex is soluble in a range of aliphatic and aromatic hydrocarbon solvents (*e.g.* *n*-pentane, hexane, benzene and toluene) and polar solvents (*e.g.* THF, DME, HMPA, Et_2O, DMSO and CH_3CN). This material is also found to be very air and moisture sensitive, both in the solid state and in solution. Record the IR, ^1H NMR and ^{13}C NMR spectra. A single crystal structure has recently been determined for this complex.

c) $[Ba\{N(SiMe_3)_2\}_2]_2$ *(3)*

Into a thoroughly flame dried sublimation vessel (3 vacuum-gas cycles) is placed 0.5 g (0.937 mmol) of $[Ba(N\{SiMe_3\}_2)_2(THF)]_2$ inside a nitrogen atmosphere glove bag. The apparatus is then stoppered, and the vessel attached to a vacuum line in a well ventilated fume cupboard. The sublimation flask is placed into a thermostated oil bath and heated slowly to 150 °C, under a vacuum of 10^{-2} torr. Small, colourless crystals and a fine white powder slowly form on the base of the cold finger. The resulting crystalline solid is isolated in the glovebag and stored in a screw top vial. Record the yield, m.p., IR, ^1H NMR and ^{13}C NMR spectra.

The complex is stable at ambient temperature in a sealed container under a nitrogen atmosphere. This complex is soluble in a range of aliphatic and aromatic hydrocarbon solvents (*e.g.* *n*-pentane, hexane, benzene and toluene) and polar solvents (*e.g.* THF, DME, HMPA, Et_2O, DMSO and CH_3CN). This material is exceedingly air and moisture sensitive, both in the solid state and in solution. A single crystal structure of $[Ba(N\{SiMe_3\}_2)_2]_2$ and its related Ca and Sr derivatives have recently been determined.

d) $[Ba(OC_6H_2tBu_3)_2(THF)_3 \cdot THF$ (4)

Into a thoroughly flame dried Schlenk tube (3 vacuum-gas cycles) is placed 0.5 g (0.86 mmol) of $[Ba(N\{SiMe_3\}_2)_2(THF)_2]$ inside a nitrogen atmosphere glove bag. The Schlenk tube, is then stoppered, and vessel attached to a vacuum line in a well ventilated fume cupboard. The Schlenk tube apparatus is then evacuated/flushed 3 times with nitrogen.

Toluene (40 cm^3) is added to the Schlenk tube via a degassed syringe, and then 2,4,6-tri-*tert*-butylphenol is added rapidly (0.45 g, 1.72 mmol). The mixture is then held at reflux temperature for 10 h and subsequently left to stir at room temperature for 4 h. The solvent is then removed using an external solvent trap to produce a pale green solid. This is washed several times with pentane (to remove any unreacted starting material). The solid is then redissolved in hot tetrahydrofuran (20 ml) and stirred for 15 min, the solvent then removed using an external trap, warm toluene (8 ml) added and the flask placed at $-25\,°C$ overnight. This results in a large crop of pale green crystals. The solution is removed with a cannula to yield the crystalline product which is isolated in the glovebag and stored in a screw top vial. Calculate the yield.

Bis(phenoxo)barium tetrakis-tetrahydrofuran is obtained as a pale green crystalline solid, which should be stored at ambient temperature in a sealed container under a nitrogen atmosphere. This complex is soluble in range of aromatic hydrocarbon solvents (*e.g.* benzene and toluene) and polar solvents (*e.g.* THF, HMPA, DMSO, and CH_3CN). This material is very air and moisture sensitive, both in the solid state and in solution. Record the yield, IR, ^1H NMR and ^{13}C NMR spectra. Single crystal structures of related calcium and strontium derivatives have recently been determined.

e) $[Ba(OCEt_3)_2]_n$ (5)

Into a thoroughly flame dried Schlenk tube (3 vacuum-gas cycles) is placed 0.5 g (0.086 mmol) of $[Ba(N\{SiMe_3\}_2)_2(THF)_2]$ inside a nitrogen atmosphere glove bag. The Schlenk tube is then stoppered, and the vessel attached to a vacuum line in a well ventilated fume cupboard. The Schlenk tube apparatus is then evacuated/flushed 3 times with nitrogen.

Toluene (40 cm^3) is added to the Schlenk tube via a degassed syringe and then triethylmethanol added rapidly (0.2 g, 0.176 mmol). The mixture is held at reflux temperature for 9 h and subsequently left to stir at room temperature for 4 h. The solvent is then removed using a external solvent trap to produce a colourless solid. This is washed several times with pentane (to remove any unreacted starting material). The product is isolated in the glovebag and stored in a screw top vial.

Bis(triethylmethoxo)barium is obtained as a white solid, which should be stored at ambient temperature in a sealed container under a nitrogen atmosphere. This complex is soluble in a range of a polar solvents (*e.g.* THF, Me_2CO, DME, HMPA and DMSO). This material is air and moisture sensitive, both in the solid state and in solution. Record the yield, IR, ^1H NMR and ^{13}C NMR spectra.

References

L. G. Hubert-Pfalzfraf, *New. J. Chem.* **1987,** *11,* 663.

C. N. R. Rao, B. Raveau, *Acc. Chem. Res.* **1989,** *22,* 106.

W. E. Lindsell, F. C. Robertson, I. Soutar, D. H. Richards, *Eur. Polym.* **1983,** *19,* 115 and references therein.

S. R. Drake, D. J. Otway, *J. Chem. Soc. Chem. Commun.* **1991,** 517.

T. P. Hanusa, Polyhedron, **1990,** *9,* 1345.

J. M. Boncella, C. J. Coston, J. K. Cammack, *Polyhedron* **1991,** *10,* 769.

M. Westerhausen, *Inorg. Chem.* **1991,** *30,* 96.

D. C. Bradley, M. B. Hursthouse, A. A. Ibrahim, K. M. A. Malik, M. Motevalli, R. Moseler, H. Powell, J. D. Runnacles, A. C. Sullivan, *Polyhedron* **1991,** *9,* 2959.

M. J. McCormick, K. B. Moon, S. R. Jones, T. P. Hanusa, *J. Chem. Soc. Chem. Commun.* **1990,** 779.

F. G. N. Cloke, P. B. Hitchcock, M. F. Lappert, G. A. Lawless, B. Royo, *J. Chem. Soc., Chem. Commun.* **1991,** 724.

S. R. Drake, J. A. Darr, P. Hall, unpublished results.

S. R. Drake, D. J. Otway, *Polyhedron* **1992,** *11,* 745.

D. F. Shriver, *The Manipulation of Air-Sensitive Compounds* McGraw Hill, New York, **1969.**

B. A. Vaartstra, J. C. Huffman, W. E. Sreib, K. G. Caulton, *Inorg. Chem.* **1991,** *30,* 121.

M. Westerhausen, W. Schwarz, *Z. Anorg. Allg. Chem.* **1992,** *609,* 39.

S. R. Drake, M. B. Hursthouse, K. M. A. Malik, D. J. Otway, *Polyhedron* **1992,** *11,* 1995.

Ba + 4 $HN\{SiMe_3\}_2$ —(THF /Toluene /NH_3, - 40 °C)→ (1) —(Toluene /Δ)→ (2) —(Sublimation)→ (3)

(1) —(2 Ar`OH, Toluene)→ (4)

(1) —(2 Et_3COH, Toluene)→ $[Ba(OCEt_3)_2]_n$ (5)

4.23 Structural, Electrical and Magnetic Properties of Perovskite Ceramics

Colin Greaves

Special Safety Precautions

Barium salts are very toxic. Due to the involatile nature of the barium compounds studied in this experiment, the use of a fume cupboard is unnecessary, but care should be taken to avoid ingestion during all handling operations.

The perovskite structure (Fig. **4.23-1**), which is adopted by many oxides with formula ABO_3, is very versatile, and many perovskites have useful technological applications (*e. g.* as ferroelectrics, catalysts, sensors and superconductors). In this structure, the A and O ions together form a cubic close-packed array, and the B ions occupy 1/4 of the octahedral holes. This experiment involves the synthesis of four compounds which are structurally closely related to perovskite, but have very different physical properties. Control of the types of cations in the large sites (A) allows some variation of the oxidation state of the smaller octahedral cations (B), and this confers the variable physical characteristics.

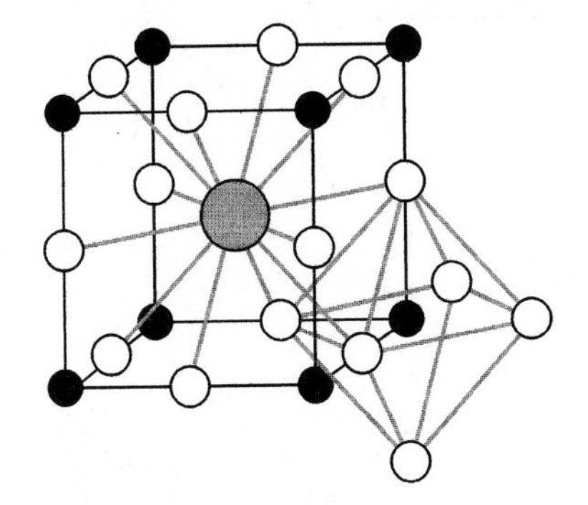

Large cation (A)

Octahedral cation (B)

Oxygen

Figure 4.23-1. The perovskite structure adopted by many ABO_3 compounds.

Transition-metal ions with unpaired electrons are paramagnetic, provided interactions between neighbouring ions are weak; this "magnetically dilute" situation occurs in solutions and many solids. The perovskite structure allows quite strong interactions to occure *via* covalence in the M – O – M bonds, and this may result in ordering of the magnetic moments to give "ferromagnetic" or "antiferromagnetic" materials. At elevated temperatures, both classes are paramagnetic (random arrangement of magnetic moments), but below a critical temperature, the moments order in a parallel (ferromagnetic) or anti-parallel (antiferromagnetic) fashion. This temperature is known as the Curie temperature (T_c) for ferromagnets and the Néel temperature (T_N) for antiferromagnets.

The synthesis of solid state ceramic samples may be achieved by a variety of techniques. The simplest exploits the finite ionic or atomic diffusion processes which occur in mechanically mixed reactants at elevated temperatures and, in general, this allows precise control of the product's stoichiometry. An alternative approach involves the precipitation of a precursor, which can easily be converted to the final product by heating. This method assures intimate cation mixing prior to heating such that lower temperature my often be used for the final heating stage. In this experiment, both methods are used.

$CaMnO_3$, $La_{0.85}Sr_{0.15}MnO_3$ and $La_{0.7}Sr_{0.3}MnO_3$ are prepared from precipitated precursors, whereas $YBa_2Cu_3O_7$ is prepared from a mechanical mixture of Y_2O_3, $BaCO_3$ and CuO. For $YBa_2Cu_3O_7$, it is not easy to ensure the correct cation ration by simple precipitation methods, and a ceramic grinding and sintering technique is preferred. $CaMnO_3$ is a paramagnetic insulator at room temperature, whereas $La_{0.85}Sr_{0.15}MnO_3$ is paramagnetic but electrically conducting, and $La_{0.7}Sr_{0.3}MnO_3$ is both electrically conducting and ferromagnetic. $YBa_2Cu_3O_7$ is metallic at room temperature, but becomes a superconductor with zero resistance to d. c. currents below 93K. Superconductors are perfectly diamagnetic, and it is this property which is examined in this experiment.

Experimental

a) $YBa_2Cu_3O_7$

If is preferable to use reagents (Y_2O_3, CuO, $BaCO_3$) which have been dried (*e. g.* 2 hours at 400 °C in a muffle furnace). Accurately weigh out about 0.5 g of $BaCO_3$ and the corresponding amounts of Y_2O_3 an CuO to give a Y : Ba : Cu ratio of 1 : 2 : 3 (0.1430 g and 0.3023 g respectively for 0.5000 g of $BaCO_3$). Grind the materials together in a clean pestle and mortar until no white steaks are observed on grinding (*ca.* 10 min). Press 1 or 2 pellets (1–2 mm thick, 13 mm diameter) of the mixture at *ca.* 5000 kg and place the pellets in an alumina boat. Using a furnace with a programmable controller, subject the pellects to the following thermal program in air:

1) Heat to 930 °C and hold for 12 h
2) Cool to 500 °C and hold for 1 h
3) Cool to 400 °C at 50 °C h^{-1}
4) Cool to room temperature

When the furnace temperature is below 400 °C, the samples may be removed using tongs and placed on an insulating board until cold.

b) $CaMnO_3$

Dissolve 2.36 g of $Ca(NO_3)_2 \cdot 4H_2O$ and 2.87 g of $Mn(NO_3)_2 \cdot 6H_2O$ in 50 cm^3 of distilled water. Whilst stirring the solution (magnetic stirrer), slowly add 100 cm^3 of 1 M KOH using a separating funnel (about 3 min). After standing for 15 minutes, the brown precipitate should be filtered using a large Büchner funnel and washed thoroughly with distilled water (to remove excess KOH and KNO_3). Pump the sample as dry as possible, transfer to a watch glass and dry at 200 °C in a drying oven (*ca.* 1 h). Grind the sample and press two pellets 1–2 mm thick. The pellets should be placed in a porcelain or alumina boat and heated at 1000 °C for about 12 h. When the furnace has cooled to below 400 °C, the boat may be removed using tongs and placed on an insulating board until cold.

c) $La_{0.85}Sr_{0.15}MnO_3$ and $La_{0.7}Sr_{0.3}MnO_3$

Repeat the procedure described for $CaMnO_3$, but use the following reagents:
$La_{0.85}Sr_{0.15}MnO_3$: 3.68 g of $La(NO_3)_3 \cdot 6H_2O$; 0.32 g of $Sr(NO_3)_2$; 2.87 g of $Mn(NO_3)_2 \cdot 6H_2O$
$La_{0.7}Sr_{0.3}MnO_3$: 3.03 g of $La(NO_3)_3 \cdot 6H_2O$; 0.63 g of $Sr(NO_3)_2$; 2.87 g of $Mn(NO_3)_2 \cdot 6H_2O$.

d) Physical Properties

Place one of the pellets of each oxide in turn on a piece of paper and note its behaviour when a bar magnet is placed under the paper. Cool the pellet of $La_{0.85}Sr_{0.15}MnO_3$ in liquid nitrogen using nylon forceps and re-examine its response to the magnet. If necessary, the pellets may be ground in order to examine the magnetic properties. Grind one of the pellets of $CaMnO_3$ using a clean pestle and mortar. Determine the magnetic susceptibility and effective magnetic moment of Mn^{4+} using any suitable method (*e.g.* a Johnson Matthey magnetic balance).

Cool one of the pellets of $YBa_2Cu_3O_7$ in liquid nitrogen, and quickly place a small magnet (Nd-Fe-B or Sm-Co) above it. The diamagnetic properties of a superconductor should allow you to float the magnet above the sample.

For each sample, measure the electrical resistance of one of the pellets using a suitable method (qualitatively, differences should be observable using a simple DVM in resistance mode, but quantitative measurements will require the use of a conventional 4-probe dc method, if available).

$CaMnO_3$ gives an X-ray powder diffraction trace with the first six reflections at 2θ values of 23.85°, 34.00°, 41.93°, 48.83°, 55.06° and 60.82° (wavelength 1.542Å). If X-ray diffraction facilities are available, the pattern can be recorded for the $CaMnO_3$ sample prepared. Confirm the primitive cubic structure of perovskite and determine the unit cell size.

Exercises

The structures of all four compounds are related to perovskite (Fig. **4.23-1**). If available, a model of the structure should be examined. There are two independent cation arrays: the first consists of octahedrally coordinated ions, which are Ti^{4+} in the parent $CaTiO_3$ and Mn/Cu ions in the compounds synthesised; the second array has 12-coordinate ions which are Ca^{2+} in $CaTiO_3$, La/Sr in $La_{0.85}Sr_{0.15}MnO_3$, etc. In fact, the Mn perovskites prepared all show minor deviations from the ideal cubic structure, due to size and electronic effects.

For an ideal, undistorted perovskite ABO_3 in which each cation (ionic radii r_A and r_B) contacts the coordinating O^{2-} ions (radius r_O), show that

$$r_A + r_B = \sqrt{2}\,(r_B + r_O)$$

Generally, some tolerance is allowed such that

$$r_A + r_B = t\sqrt{2}\,(r_B + r_O)$$

where the tolerance factor t is 0.8–1.0. For undistorted perovskites, t is high, *e.g.* 0.99 for $SrTiO_3$. From a table of ionic radii, determine t for $CaMnO_3$.

Mn^{3+} has the electron configuration $t_{2g}^3e_g^1$ and is therefore likely to show a substantial Jahn-Teller distortion. In fact, the distortion is cooperative as shown in Figure **4.23-2.** In the layer shown, for example, each Mn has 2 short Mn – O bonds and 2 long bonds. If the O ions above and below the Mn ions have short bonds, all the Mn ions can achieve a similar distorted stereochemistry. Explain why the distortion shown in Figure **4.23-2** should stabilise the structure.

The electrical and magnetic properties of Mn perovskites are critically dependent on the Mn oxidation state. What is the formal oxidation state of Mn in $CaMnO_3$, $La_{0.85}Sr_{0.15}MnO_3$ and $La_{0.7}Sr_{0.3}MnO_3$? The conductivity of many transition-metal oxides may be related to a simple mechanism involving the hopping of electrons between two transition-metal ions. Using such a mechanism and considering the energy involved when an electron hops from one Mn ion to a neighbouring Mn ion, explain the difference in conductivity between $CaMnO_3$, $La_{0.85}Sr_{0.15}MnO_3$ and $La_{0.7}Sr_{0.3}MnO_3$. In the determination of the effective magnetic moment, μ_{eff}, for Mn^{4+}, the Curie Law is used:

$$\chi_a = \mu^2/8T$$

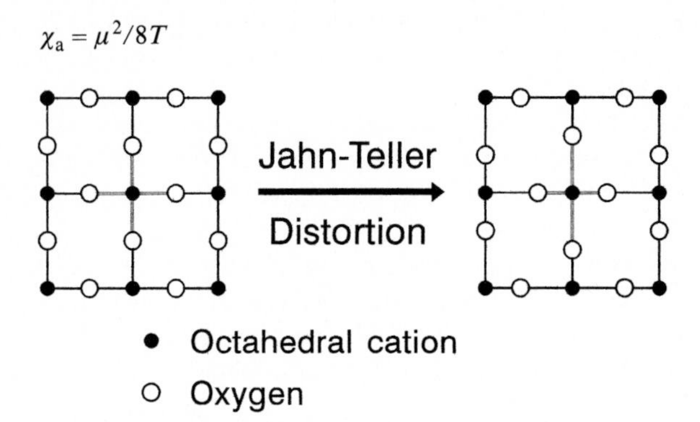

Figure 4.23-2. Cooperative Jahn-Teller distortions in a layer of the perovskite structure.

Compare your value μ_{eff} with μ_{so}, the spin-only moment, which is the magnetic moment expected if only electron spins contribute to μ_{eff}. The main reason for the disagreement is that $CaMnO_3$ is antiferromagnetic at low temperatures (T_N *ca.* 120 K). Interactions between magnetic moments are still apparent at higher temperatures and result in an apparent reduction in μ_{eff} due to a deviation from the Curie Law:

$$\chi_a = \mu_{so}^2/8(T + \theta)$$

Using your value of χ_a and the formula above, determine a value for θ.

Notice that whereas $CaMnO_3$ is antiferromagnetic, $La_{0.85}Sr_{0.15}MnO_3$ and $La_{0.7}Sr_{0.3}MnO_3$ are ferromagnetic. What do your measurements on $La_{0.85}Sr_{0.15}MnO_3$ and $La_{0.7}Sr_{0.3}MnO_3$ tell you about the change in Curie temperature (associated with the onset of ferromagnetic behaviour) with Mn oxidation state in this system?

Superconductors are materials which lose all electrical resistivity below a certain temperature, the critical temperature, T_c; above T_c, they are generally metallic in nature. Until 1986, when "high temperature superconductors" were discovered, the highest T_c was 23 K for Nb_3Ge. $YBa_2Cu_3O_7$ was the first material discovered with T_c above the temperature of liquid nitrogen, 77 K. When pure, and having its maximum possible oxygen content, this material becomes superconducting at 93 K. An important property of superconductors is that below T_c, magnetic fields are expelled from within the material – it becomes a perfect diamagnet. This is achieved by setting up currents on the surface of the bulk material to oppose the applied magnetic field. In this way, it is possible to float a superconductor above strong magnets, using the induced opposing field for levitation.

The structure of $YBa_2Cu_3O_7$ (Fig. **4.23-3** and model if available) comprises three perovskite-like unit cells in a row; the Y and Ba ions occupy the large cation positions but in this material, not all the oxygen sites are occupied, which reduces the coordination numbers for all the cations. Layers of 5-coordinate (square pyramidal) and chains of 4-coordinate (coplanar) Cu ions are formed. What is the ratio of 5-coordinate Cu to 4-coordinate Cu in the structure (remember that certain sites in the unit cell are shared with other unit cells)? What is the ratio of Cu^{2+} to Cu^{3+} ions in $YBa_2Cu_3O_7$? A square pyramidal crystal field influences the *d*-orbital energies of a transition-metal ion in the same way as an elongated octahedral (tetragonal) field, which is common for Jahn-Teller distorted ions. On the basis of Crystal Field Theory, discuss the preferred distribution of the Cu^{2+} and Cu^{3+} ions between the available sites in $YBa_2Cu_3O_7$.

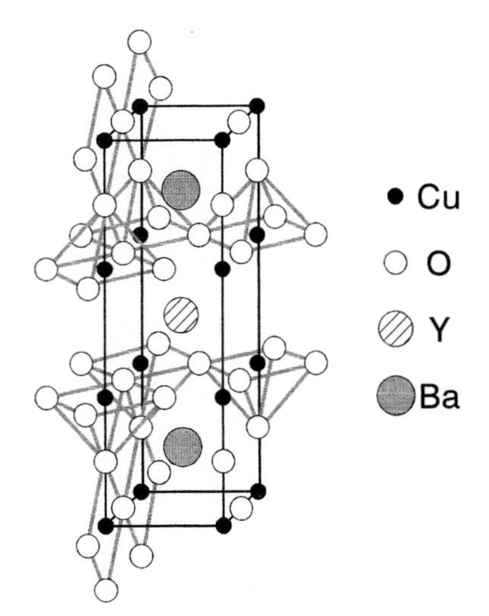

Figure 4.23-3. The unit cell of $YBa_2Cu_3O_7$ highlighting the Cu stereochemistry.

References

J. H. Van Santen, G. H. Jonker, *Physica* **1950,** *16,* 599.
G. H. Jonker, J. H. Van Santen, *Physica* **1950,** *16,* 337.
E. A. Wollan, W. C. Koehler, *Phys. Rev.* **1955,** *100,* 545.
V. A. Bokov, N. A. Grigoryan, M. F. Bryzhina, *Phys. Stat. Sol.* **1967,** *20,* 745.
J. G. Bednorz, K. A. Müller, *Z. Phys. B* **1986,** *64,* 189.

4.24 Metal-containing Liquid Crystals. The Synthesis of *trans*-bis(4-Alkyl-4′-cyano-1,1′-biphenyl)dichloroplatinum(II) and Related Species – The Use of Melt Syntheses

Duncan W. Bruce

Special Safety Precautions

Cyanobiphenyls	Are effectively non-toxic and non-irritants.
Platinum(II) Chloride	Is an eye and skin irritant, but should prevent no hazard in the quantities to be used. Gloves should be worn as a precaution.
Chloroform	Handle in a ventilated area and avoid exposure to skin.
Diethylether	Flammable. Use in ventilated area and avoid skin contact.
Celite	Is a silica dust and so inhalation should be avoided.

Almost everyone will now own some device or other containing a liquid crystal display (LCD), be it a wristwatch, a calculator, a lap-top computer or even the display on the washing machine. The wholesale commercialisation of LCD's began in the early 1970's following the synthesis of the alkylcyanobiphenyls, alkoxycyanobiphenyls and alkycyanoterphenyls (shown below) by Gray and coworkers at the University of Hull.

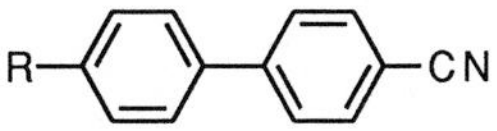

R—C₆H₄—C₆H₄—CN   RO—C₆H₄—C₆H₄—CN

Alkylcyanobiphenyls 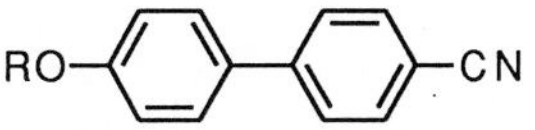   Alkoxycyanobiphenyls

R—C₆H₄—C₆H₄—C₆H₄—CN

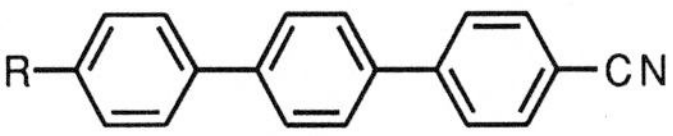

Alkylcyanoterphenyls

These materials were important as they had the correct combination of physical properties and chemical inertness for exploitation. Research on liquid crystals has boomed since that time and groups are internationally active in all areas of the subject from biology (cell membranes are liquid crystals) to device engineering, from the applied to the pure and from the experimental to the theoretical.

One of the more recent areas to come to prominence involves the design and synthesis of metal complexes which are liquid crystalline and these are discussed in four review articles. The area is of considerable fundamental interest, but one of the driving forces is the fact that certain properties of metals can be incorporated into liquid crystalline media in this way, for example paramagnetism, polarisability or even colour.

The complexes described in this synthesis are platinum(II) complexes of the original cyano-biphenyls, being readily prepared from $PtCl_2$ and the ligand directly.

The liquid crystalline state is a true state of matter which exists between the liquid and the solid state and as such, has properties reminiscent of each. Thus, in common with a liquid it is fluid, while in common with a solid, it possess order – a useful way to think about these systems is as ordered liquids. There are several different classifications of liquid crystals of which the broadest are *Thermotropic,* in which the solid-liquid crystal-liquid transitions are thermally induced, and *Lyotropic,* where they are solvent induced. We will concentrate solely on thermotropic systems.

Thermotropic liquid crystals can be further classified as polymeric (which will not concern us), discotic (*i. e.* disc-shaped, which will also not concern us) and calamitic (or rod-shaped) which are the type with which you will be dealing. The point about rod-like molecules is that they are structurally anisotropic and hence there are anisotropic dispersion forces existing between molecules. It is these rather weak forces which stabilise liquid crystal phases. This is not the place to discuss the various structural criteria which will promote the formation of liquid crystalline behaviour, but for our purposes, it is sufficient to say that rods built up of a rigid core (usually aromatic with at least two rings either linked directly or *via* groups such as esters, imines or vinyls) and terminated at one end by a flexible alkyl (or alkoxy) chain and at the other either by the same or by a small polar group (*e. g.* $-CN$, $-NO_2$, $-OMe$) are often liquid crystalline.

Such molecules generally form two general types of liquid crystal phase, *Nematic* and *Smectic.* If our rods were in a fluid state, with one-dimensional orientational order and no positional order, then they would describe the nematic phase (N) (Fig. **4.24-1**). The nematic phase is the most disordered type of mesophase and is the one used in most display applications.

Figure 4.24-1. Schematic picture of a nematic phase (Reproduced from *Inorganic Materials* by permission of John Wiley & Sons)

If we then introduce partial positional order in addition to the orientational order, a family of smetic phases is generated, which are characterised by having some layering of the molecules. In the smectic A (S_A) phase (Fig. **4.24-2 a**), the molecules are loosely associated within layers and point on average in a direction perpendicular to the layers. In the smectic C phase (S_C), the situation is similar (Fig. **4.24-2 b**), except that the molecules now make some angle, θ, to the layers. There is, however, no positional correlation between molecules within the S_A or S_C layers and there is considerable fluidity within, and easy diffussion between, the layers. Other types of smectic phase exist but will not be described here.

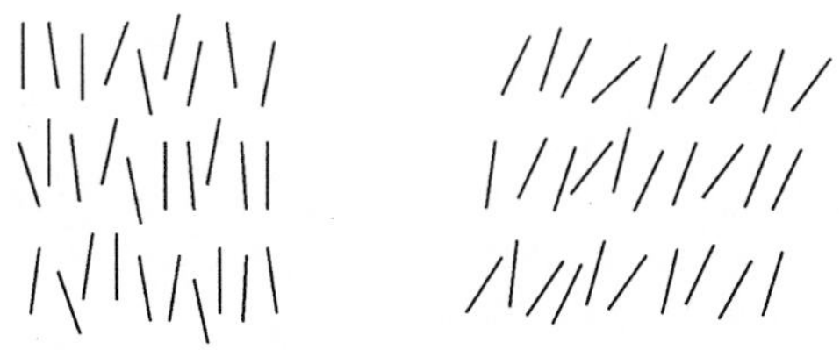

Figure 4.24-2. Schematic representation of a) the S_A and b) S_C phases (Reproduced from *Inorganic Materials* by permission of John Wiley & Sons)

Experimental

This is a very general procedure which is applicable to a whole host of cyanobiphenyl-type liquid crystals. It is described in detail here for 4-octyl-4′-cyanobiphenyl.

An oil bath is heated to 140 °C. On reaching the temperature, a round-bottomed flask (capacity 25 or 50 cm^3 and preferably with a B24 socket) equipped with a small magnetic stir bar and containing 4-octyl-4′-cyanobiphenyl (873 mg, 3 equivalents; abbreviated 8CB) is placed in the oil. Stirring is commenced and $PtCl_2$ (269 mg, 1 equivalent) is added. The temperature is maintained for about 30 minutes during which time the mixture solidifies, turning a dirty yellow. After cooling to room temperature, the solid mixture is dissolved in chloroform ($\approx$15–20 cm^3) and filtered twice through celite to remove unreacted $PtCl_2$. A large excess of diethyl ether is added to precipitate the complex. Ideally, the precipitated solid is recovered by centrifugation (as the precipitate tends to be quite fine), but it can be obtained by decanting off the mother liquor after standing (don't let all the ether evaporate off!). A small amount of residual solvent is not too important at this stage. The precipitate is then crystallised from hot chloroform/diethyl ether to give yellow crystals which are recovered by filtration, washed with diethyl ether and air dried.

Obtain the ^1H NMR spectrum in CD_2Cl_2 at 250 MHz and the infra-red spectrum (4000–200 cm^{-1}) as a Nujol mull between CsI plates. The 250 MHz ^1H NMR spectrum of *trans*-[$PtCl_2(8CB)_2$] is shown in Figure **4.24-3** for reference. Analyse the spectrum.

In the ^{13}C NMR spectrum, the cyanide carbon is seen at δ 117, but coupling to ^{195}Pt ($^2J_{Pt-C}$ = 289 Hz) is seen at 80 MHz with a little difficulty. There is no advantage in going to higher field. Why?

The infra-red spectrum can be used to confirm the geometry of the complex and to say something about its purity. Consult the literature, identify ν_{CN} for the bound ligand in your spectrum and determine whether your sample contains any free ligand. Similarly, confirm the *trans* geometry of the complex by locating ν_{Pt-Cl} and identifying the number of stretching vibrations. If you are familiar with the use of group theory, then identify the points groups for *cis*-and *trans*-[$PtCl_2(8CB)_2$] and determine the predicted number of Pt–Cl stretching vibrations which are active in the infra-red and Raman spectra. Compare these predictions with the spectrum you obtained.

Ideally, liquid crystal phase characterisation is carried out by a combination of techniques, namely polarising optical microscopy, differential scanning calorimetry and X-ray scattering. The most immediately useful of these is microscopy which is described below.

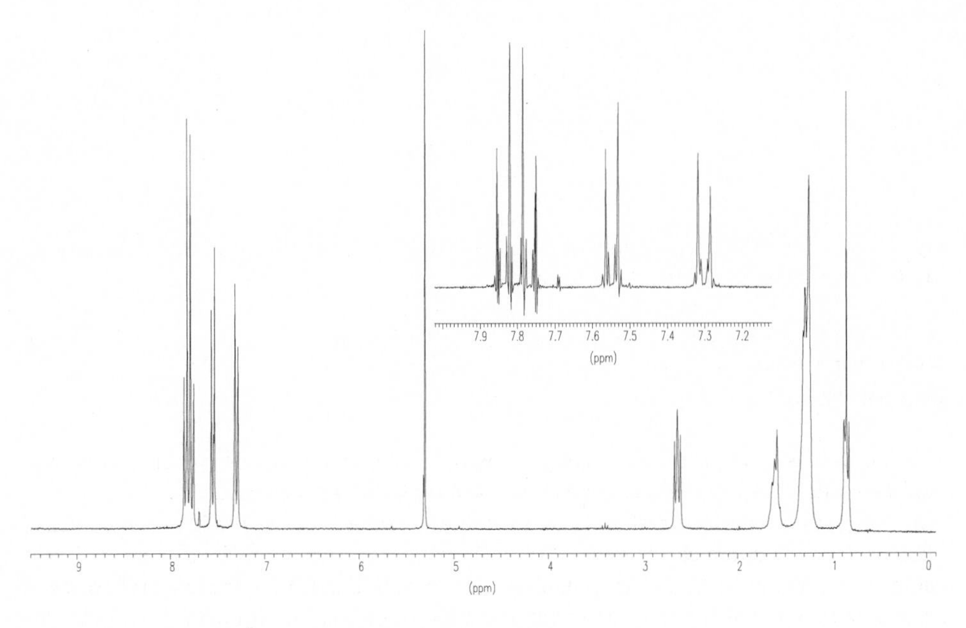

Figure 4.24-3. 250 MHz ^{1}H NMR spectrum of $[PtCl_2(8CB)_2]$

A sample is placed between two glass slides which are then placed on a hot stage, mounted on a microscope stage. The light which falls on the sample is plane polarised. Between the sample and the objective lens, there is another piece of polaroid whose polarisation direction is at right angles to that of the first polariser (Fig. **4.24-4**). In the absence of any sample, or in the presence of a 'normal' liquid such as water, the observer would see nothing as no light would pass through. However, when a material is in a liquid crystal phase, its physical parameters become anisotropic (*i. e.* different in different directions). One of these anisotropic properties is refracitive index and liquid crystalline phases have two of these as shown in Figure **4.24-5**.

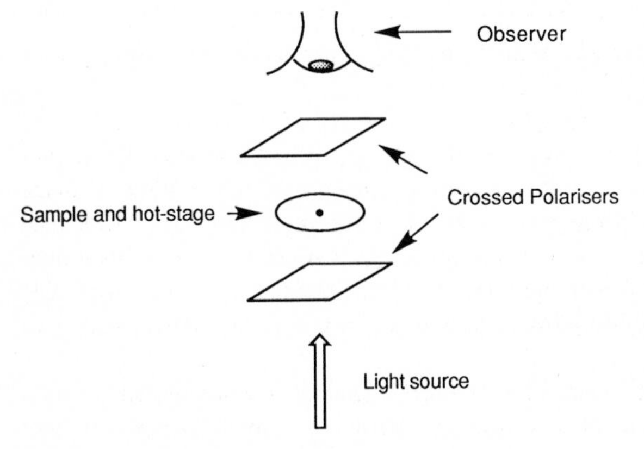

Figure 4.24-4. Schematic diagram of a polarising microscope (Reproduced from *Inorganic Materials* by permission of John Wiley & Sons)

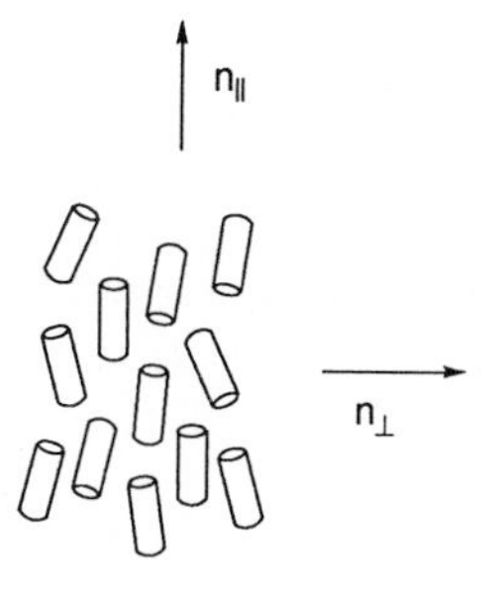

Figure 4.24-5. Representation of the molecular organisation in a nematic phase showing the origin of the two refractive indices.

Thus, it is convenient to think that two refracted rays are produced when light is incident on the sample in its liquid crystal phase. These two rays can then interfere with one another to produce an interference pattern which is now not absorbed by the polariser. These patterns, known as *textures,* are characteristic of individual liquid crystal phases, although in certain circumstances a given liquid crystal phase can exhibit more than one type of texture. A characteristic texture of the nematic phase is shown in Figure **4.24-6.**

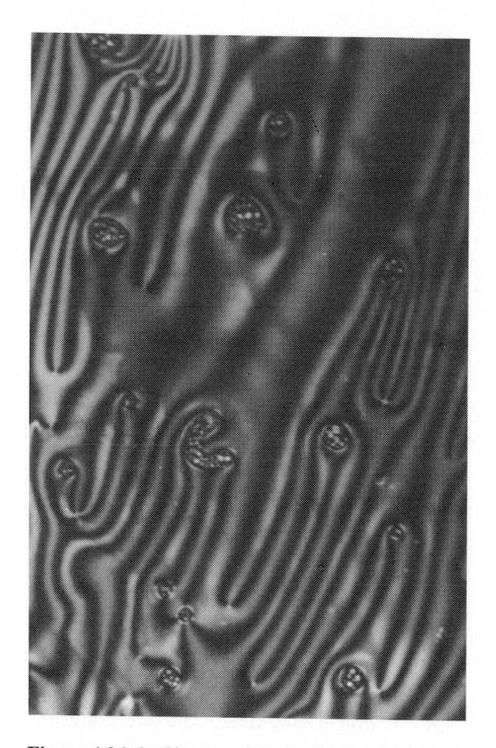

Figure 4.24-6. Characteristic optical texture of the nematic phase (reproduced from J. Chem. Soc., Chem. Commun., 1994, 729, by permission of the Royal Society of Chemistry)

The complex you have just made has the following phase behaviour:

$$\mathrm{K} \xrightarrow{166} \mathrm{N} \xrightarrow{176} \mathrm{I}$$

This means that the complex melts from the crystal phase (K – from the German, Kristall) into the nematic phase at 166 °C. The nematic phase then persists until 176 °C, when the fluid becomes isotropic (I – *i. e.* a 'normal' liquid). If held in this latter state for any length of time, the material will decompose.

Although you may not have access to the correct equipment to see this properly, a standard melting point experiment should show the crystals melt to a turbid fluid at 166 °C.

The synthetic method described above has been used for compounds with the structures shown below, with *n* varying typically between 1-12 when there is no cyclohexyl ring present, and taking the values 2, 3, 5 and 7 when there is a cyclohexyl ring. Not all of these results are published as yet, but interested parties are invited to contact the author for advice if they wish to look at other derivatives. All of these materials are commercially available. The method also works with $PtBr_2$, although the yields are smaller unless the conditions are modified slightly. PtI_2 does not work with these ligands. Further, this method works with other ligands and readers are directed to the literature for further details.

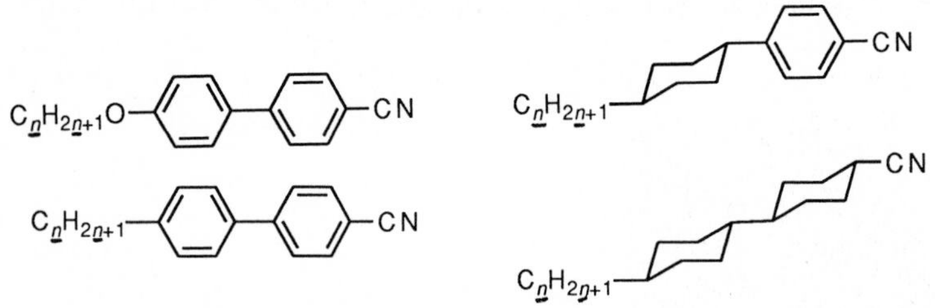

It is also possible to make palladium equivalents of these platinum systems, although there are drawbacks. They are synthesised by stirring two equivalents of the desired ligand for three hours at room temperature with $[PdCl_2(PhCN)_2]$ in acetone, removing the solvent on a rotary evaporator, chilling the residue in the freezer for two hours, and then triturating with ethanol to leave the pure, solid product after drying; yields are in the range 50–90%. This route is superior and generally more reliable than that published by Adams, *et. al.* Unfortunately, these complexes are labile in solution and so cannot be readily crystallised, which is why the 'melt' route is inappropriate. However, the melt route can be used with palladium if more inert complexes are to be synthesised.

The palladium complexes are also liquid crystalline and at lower temperatures than the platinum homologues. However, for complexes with alkylcyanobiphenyls, the nematic phase is monotropic, meaning that it is less stable that the crystal phase and so is found on supercooling the complex from its isotropic state (*i. e.* crystals melt straight to 'normal' liquid and the liquid crystal phase is found on supercooling). Therefore, it is suggested that alkoxycyanobiphenyls be used with palladium for convenience; all complexes with ligand chains up to and including octyloxy show only a nematic phase.

It would be possible to construct two general types of phase diagrams from the materials obtained above. In one scenario, the experiment could be made into a class experiment with different groups using ligands of different chain lengths. Thus, for one particular series, a phase diagram with chain length plotted on the x axis and temperature plotted on the y axis could be constructed, which would show the limits of liquid crystal phase stability.

In another scenario, it would be possible to construct binary phase diagrams at different percentage compositions. The clearing point of a mixture consisting of complexes with similar shape and type is simply expressed as:

$$T_c = \sum_i a_i T_{ci}$$

where T_c = clearing point of the mixture; a_i is the percentage of component i in the mixture and T_{ci} is its clearing point, *i.e.* it is a linear function of composition. A typical phase diagram for a mixture of two nematic compounds is shown below:

This shows the linear behaviour of the clearing point and the eutectic behaviour of the melting point. Such phase diagrams can typically be constructed at compositional increments of 10%. Further, they can be used to effect if one or both of the components has a monotropic phase as the drop in melting point can lead to an enantiotropic phase (*i.e.* observed on heating *and* cooling) for the mixture within a particular composition range. An example of this is the binary phase diagram constructed for the Pd complexes of propyl- and pentyl-bicyclohexylcarbonitrile (abbreviated CCH3 and CCH5 respectively), although in this case the lability of the palladium complexes meant that all three possible complexes (*i.e.* $[PdCl_2(CCH3)_2]$, $[PdCl_2(CCH5)_2]$ and $[PdCl_2(CCH3)(CCH5)]$ were present in statistical proportions. In our experience, mixtures for binary phase diagram studies are best made by mixing the two components in the correct proportion in solution (*e.g.* $CHCl_3$) and then removing the solvent.

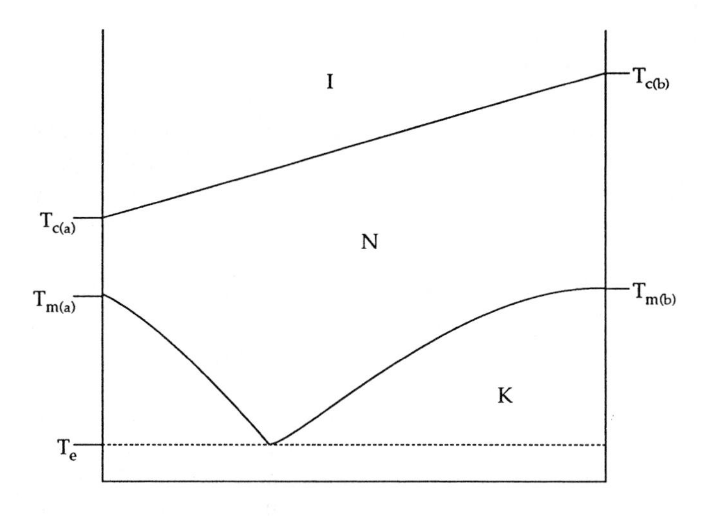

Figure 4.24-7.

Note

The idea for this synthesis came from examination of the synthesis of $[PtCl_2(PhCN)_2]$. This complex can be made in its *cis* form by reaction of aqueous K_2PtCl_4 with PhCN at room temperature. Mixtures of *cis*- and *trans*-$[PtCl_2(PhCN)_2]$ can be made by direct reaction of $PtCl_2$ with PhCN and the *trans/cis* ratio increases with the increased temperature of the reaction, although in this method, pure *trans* isomer is never recovered and the mixture must be separated by chromatography. See also Fanizzi, *et al.* for a discussion of the formation of *cis*- and *trans*-$[PtCl_2(MeCN)_2]$. This melt synthesis produces exclusively the *trans* isomer in good yield, although with different ligand types (*e.g.* phosphines), exclusively *cis* complexes can be similarly obtained.

References

D. W. Bruce in *Inorganic Materials,* (Eds.: D. W. Bruce, D. O'Hare), Wiley, Chichester, **1992.**

D. W. Bruce, *J. Chem. Soc. Dalton Trans.* **1993,** 2983.

P. Espinet, M. A. Esteruelas, L. A. Oro, J. L. Serrano, E. Sola, *Coord. Chem. Rev.* **1992,** *117,* 215.

A.-M. Giroud-Godquin, P. M. Maitlis, *Angew. Chem. Int. Ed. Engl.* **1991,** *30,* 402.

D. W. Bruce, B. Donnio, A. A. Maggs, J. R. Marsden, *Inorg. Chim. Acta.* **1991,** *188,* 41.

T. Uchiyama, Y. Toshiyasu, Y. Nakamura, T. Miwa, S. Kawaguchi, *Bull. Chem. Soc. Jpn.* **1981,** *54,* 181.

H. Adams, N. A. Bailey, D. W. Bruce, D. A. Dunmur, E. Lalinde, M. Marcos, C. Ridgway, A. J. Smith, P. Styring, P. M. Maitlis, *Liq. Cryst.* **1987,** *2,* 381.

J. R. Doyle, P. E. Slade, H. B. Jonassen, *Inorg. Synth.* **1960,** *6,* 218.

F. P. Fanizzi, F. P. Intini, L. Maresca, G. Natile, *J. Chem. Soc. Dalton Trans.* **1990,** 199.

4.25 Magnetochemistry Experiments: Exchange Coupling in Copper(II)-Nickel(II) Bimetallic Complexes and Spin Transition in Iron Complexes

E. Coronado and J. J. Borrás-Almenar

The aim of this experiment is to illustrate two phenomena of current interest in molecular magnetism, namely magnetic exchange interaction in bimetallic complexes and spin transition*, both of which can easily be detected by means of magnetic susceptibility measurements at variable temperatures. In the former case, two bimetallic coordination isomers of the *trans*-cyclohexane-1,2-diamine-*N,N,N',N'*-tetra-acetate (CDTA) containing Cu(II)-Ni(II) pairs are prepared and magnetically characterised. These compounds belong to an extensive series of bimetallic materials, the so-called EDTA family, in which the ability of the EDTA-like ligands to selectively coordinate to metal ion while still providing carboxylate bridges to link a second metal provides the possibility of preparing a wide variety of ferrimagnetic systems of variable dimensionality. Furthermore, the inertness of these CDTA complexes in solution makes these systems of particular interst to illustrate the possibility of kinetic control of their syntheses.

In the second case, the compound di-*iso*thiocyanatobis(1,10-phenantroline) iron(II), $Fe(phen)_2(NCS)_2$, is particularly suitable since it shows an abrupt $S=0 \Leftrightarrow S=2$ spin transition at a critical temperature close to 175 K, which may be detected by magnetic measurements.

Experimental

a) The bimetallic complexes $NiCu(cdta) \cdot 6H_2O$ *(1) and* $CuNi(cdta) \cdot 7H_2O$ *(2)*

Both compounds are prepared following the same general procedure. To 2 mmol of H_4CDTA acid (Titriplex IV), a concentrated solution containing 8 mmol of NaOH is added dropwise with stirring until complete deprotonation of the acid is effected (final pH *ca.* 11.7). To this solution, an equimolar solution of the relevant metal nitrate (Cu or Ni respectively) is added. After allowing for a short period (*ca.* 15 min) to ensure complexation of this metal ion, an equimolar aqueous solution of the second metal nitrate is added. Then, acetone is added dropwise with stirring until incipient turbidity. The resultant solutions are filtered and stored at *ca.* 5 °C. Blue crystalline solids of **1** and **2** appear after a few hours.

Caution: In the preparation of **2** small modifications of the experimental conditions (for example, heating the solutions) and allowing long periods of time for crystallization (slow

* Equivalent terms found in the literature are spin crossover, magnetic crossover or spin equilibrium. For a review see P. Gütlich, *Struct. Bonding (Berlin)* **1981,** *44,* 83; see also E. König, G. Ritter, S. K. Kalshreshtha, *Chem. Rev.* **1985,** *85,* 219.

evaporation of the solution at room temperature with no addition of acetone) can lead to a different compound formulated as $Cu_3Ni_2(cdta)_2(NO_3)_2 \cdot 15H_2O$. This compound can easily be identified by the presence of nitrates in its IR spectrum (sharp absorption at *ca.* 1390 cm^{-1}). Record the IR spectra of the obtained compounds as KBr pellets in order to verify this point.

The structures of **1** and **2** (see Fig. **4.25-1**) show the presence of bimetallic dimers with two different coordination sites in which the two metal atoms are linked through a carboxylate bridge. In **1**, the CDTA hexacoordinates the Cu^{II} ion while the Ni^{II} ion is in an octahedral site formed by five water molecules and an oxygen atom from the carboxylate bridge. In **2**, the CDTA site is occupied by Ni^{II}, while Cu^{II} is now occupying a square pyramidal site with four water molecules and an oxygen atom.

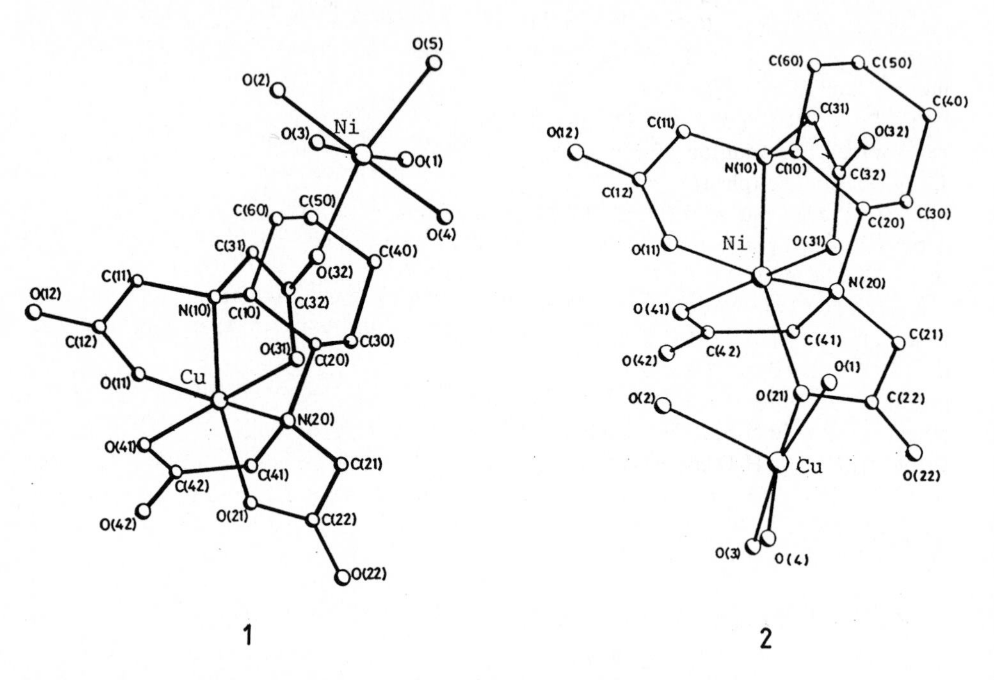

Figure 4.25-1. Structures of **1** and **2**.

Determine the magnetic susceptibilities of the two products as a function of the temperature with a Faraday balance equipped with a helium or nitrogen cryostat. Correct the experimental data for the diamagnetic contributions of the constituent atoms using the Pascal tables and from T.I.P. contributions of the metal ions. Plot the product magnetic susceptibility times the temperature (proportional to μ_{eff}^2), vs. the temperature.

The magnetic behaviour of **1** follows a Curie law with a constant value of the product χT ($\approx$1.6 emu K mol^{-1}) over the temperature range 4–300 K. Conversely, as the temperature is lowered, **2** exhibits a continuous decrease of χT down to 15 K. Below this temperature a constant value of χT ($\approx$0.4 emu K mol^{-1}) is observed.

Derive the susceptibility expression for an exchange-coupled pair with spins 1/2–1 in terms of the exchange coupling, *J*, and the Landé factor, *g*. Use this expression to fit the magnetic data of **2**. Explain the different behaviours of the two compounds on the basis of the presence or absence of magnetic exchange interactions between the $Cu^{II}-Ni^{II}$ pairs. Correlate this result with the structural differences in the bridging network and in the two coordination sites of the two complexes. For a general discussion see Fuertes et al.

b) $Fe(phen)_2(NCS)_2$

This is one of the spin-crossover systems that have been most investigated through various techniques including magnetic and calorimetric measurements, Mossbauer, IR, UV-visible, NMR and XPS spectrometries, and X-ray diffraction and absorption. Depending on the method of synthesis, two different polymorphs can be prepared.

Method A. All operations are carried out under inert atmosphere. A suspension of potassium thiocyanate K(SCN) (8 mmol) and hydrated iron(II) sulphate, $Fe(SO_4) \cdot 6H_2O$, (4 mmol) in methanol is stirred until the reaction is complete. The colourless solution containing Fe^{2+} and SCN^- is separated from the white precipitate of potassium sulphate by filtration. The eventual colouring of the solution indicates the presence of traces of Fe^{3+}, and can be eliminated by addition of some crystals of ascorbic acid. A stoichiometric amount of 1,10-pheanthroline (phen) in methanol is added dropwise to the above stirred solution. The pink-violet precipitate of $Fe(phen)_2(NCS)_2$ is filtered off, washed several times with methanol and dried in an argon stream. The purity of the compound can be checked by IR (the presence of a broad band at 1100 cm^{-1} is indicative of sulphate impurities).

Method B. In this case $Fe(phen)_2(NCS)_2$ is prepared by extracting a phenanthroline group from $Fe(phen)_3(NCS)_2 \cdot H_2O$ in a Soxhlet apparatus using acetone. The extraction is carried out for a period of 3 weeks under argon atmosphere. $Fe(phen)_3(NCS)_2 \cdot H_2O$ can be prepared by adding the stoichiometric amount of phen to a solution of Fe^{2+} and SCN^- prepared as described in method A. A more simple method for preparing $Fe(phen)_3(NCS)_2 \cdot H_2O$ consists of adding a saturated aqueous solution of K(SCN) to a mixture of $Fe(SO_4) \cdot 6H_2O$ and phen (in stoichiometric amounts) dissolved in the minimum quantity of water. The precipitate is washed with acetone and ether. In such a case the use of an inert atmosphere is not necessary.

c) Magnetic properties of $Fe(phen)_2(NCS)_2$.

Determine the magnetic susceptibilities on the polycrystalline samples in the temperature range 77–300 K. Correct the molar susceptibilities for diamagnetism and plot the χT product vs. T. An abrupt increase associated to the singlet ⇔ quintet spin transition should be observed around $T_c \approx 176$ K; such a transition is much sharper in sample B than in sample A. Determine the relative amounts of high- and low-spin isomers above and below T_c in sample A. For a general discussion see Gallois *et al.* and Ganguli *et al.*

References

E. Coronado in *Magnetic Molecular Materials,* NATO ASI Series E-198 (Eds.: D. Gatteschi, O. Kahn, J. S. Miller, F. Palacio), Kluwer Academic Publishers, Dordrecht, **1991,** p. 267.

A. Fuertes, C. Miravitlles, E. Escrivá, E. Coronado, D. Beltrán, *J. Chem. Soc. Dalton Trans.* **1987,** 1847.

F. Sapiña, E. Escrivá, J. V. Folgado, A. Beltrán, D. Beltrán, A. Fuertes, M. Drillon, *Inorg. Chem.* **1992,** *31,* 3851.

See E. Sinn, *Coord. Chem. Rev.* **1970,** *5,* 313.
B. Gallois, J. A. Real, C. Hauw, J. Zarembowitch, *Inorg. Chem.* **1990,** *29,* 1152 and references therein.
E. König, A. Madeja, *Inorg. Chem.* **1967,** *6,* 48.
P. Ganguli, P. Gütlich, E. W. Muller, W. Irler, *J. Chem. Soc. Dalton Trans* **1981,** 441.

Index